VOLUME SEVEN HUNDRED AND THIRTY THREE

METHODS IN ENZYMOLOGY

Histone Deacetylases in Neurodegenerative Disease - Part A

METHODS IN ENZYMOLOGY

Editors-in-Chief

KAREN N. ALLEN, Ph.D.

Professor and Chair
Department of Chemistry
Professor of Material Science and Engineering
Boston University

DAVID CHRISTIANSON

Roy and Diana Vagelos Laboratories
Department of Chemistry
University of Pennsylvania
Philadelphia, PA

Founding Editors

SIDNEY P. COLOWICK and NATHAN O. KAPLAN

VOLUME SEVEN HUNDRED AND THIRTY THREE

METHODS IN ENZYMOLOGY

Histone Deacetylases in Neurodegenerative Disease - Part A

Edited by

SUBASHCHANDRABOSE CHINNATHAMBI

Additional Professor Department of Neurochemistry
National Institute of Mental Health and Neurosciences Hospital
Bangalore, Karnataka, India

ELSEVIER

ACADEMIC PRESS
An imprint of Elsevier

Academic Press is an imprint of Elsevier
50 Hampshire Street, 5th Floor, Cambridge, MA 02139, United States
125 London Wall, London, EC2Y 5AS, United Kingdom

First edition 2026

ISBN: 978-0-443-47332-6
ISSN: 0076-6879

For information on all Academic Press publications
visit our website at https://www.elsevier.com/books-and-journals

Publisher: Zoe Kruze
Editorial Project Manager: Saloni Vohra
Production Project Manager: James Selvam
Cover Designer: Muthukumaran Thangaraj

Contents

Contributors *xvii*

1. Experimental strategies for evaluating the impact of HDACs and HDAC inhibitors on the epigenome. ChiP-seq **1**

Martina Minisini, Alessio Bertozzo, Eros Di Giorgio, and Claudio Brancolini

1. Introduction 2
   - 1.1 HDACs and HDACs inhibitors a brief introduction 2
   - 1.2 The basic principle of the ChIP-seq methodology 3
2. Rationale 5
3. Experimental settings 5
   - 3.1 Note 5
   - 3.2 Equipment 5
   - 3.3 Reagents 6
   - 3.4 Required solutions 6
   - 3.5 Safety note 6
   - 3.6 Additional note 7
   - 3.7 Procedure 7
4. Normalization 11
5. Concluding remarks 13

Acknowledgments 13
References 14

2. Methods for the kinetic analysis of time-dependent HDAC6 inhibitors **17**

Edoardo Cellupica, Gianluca Fossati, Christian Steinkühler, and Gianluca Caprini

1. Introduction 18
2. HDAC6 Activity assay 19
   - 2.1 Reagents 20
   - 2.2 Buffers and solutions 21
   - 2.3 Equipment 21
3. Assay setup 22
   - 3.1 Procedure for progress-curve analysis 22
   - 3.2 Procedure for jump dilution assay 23

4. Data analysis 24
4.1 Progress-curve analysis 24
4.2 IC_{50} determination 25
4.3 Mechanism interpretation based on k_{obs} 25
4.4 Jump dilution assay analysis 25
5. Representative results 26
6. Notes 28
7. Summary 29
Appendix A. Control for compound-dependent quenching/enhancement (deacetylated product standard ± compound) 29
References 30

3. Synthesis and functional characterisation of small molecule activator of p300/CBP lysine acetyltransferase activity: Implications in spinal injury repair **33**

Amrish Rai, Prajjval Mishra, Akash Kumar Singh, Priyanka Pandey, and Tapas K. Kundu

1. Introduction 34
2. Synthesis and characterization of the KAT activator for sufficient HAT activation 37
2.1 Synthesis of TTK21 37
2.2 Synthesis of glucose-derived carbon nanosphere (CSP) 39
2.3 Conjugation of CSP to TTK21 41
2.4 Biological characterization of CSP-TTK21 43
3. KAT activator molecule: Implication in spinal injury repair 46
3.1 Creation of mid-thoracic dorsal hemisection model of spinal cord injury 46
3.2 Oral administration of KAT activator CSP-TTK21 after spinal cord injury 48
3.3 Assessment of functional recovery after spinal cord injury (locomotor test) 49
3.4 Studying the induction of acetylation after spinal cord injury 51
Acknowledgments 52
References 53

4. Chromatin immunoprecipitation assay for analyzing histone H3-mediated regulation of copper homeostasis **55**

Sakshi Singh, Rakesh Kumar Sahu, and Raghuvir Singh Tomar

1. Introduction 56
2. Before you begin 58

2.1 Yeast strains and growth conditions 58
2.2 Buffer recipes 58
3. Key resources table 59
4. Materials and equipment 61
5. Step-by-step method details 62
5.1 Assays of copper tolerance in yeast strains 62
5.2 Analysis of metallothionine gene induction 63
5.3 Chromatin immunoprecipitation assay to investigate the mechanisms of metallothionein gene induction 64
6. Quantification and statistical analysis 71
7. Expected outcomes 72
8. Advantages and limitations 73
9. Optimization and troubleshooting 73
10. Concluding remarks 74
Acknowledgements 75
Author contributions 75
References 75

5. Zinc-finger ubiquitin-binding domain of histone deacetylase 6 modulates Tau structure and function 81

Abhishek Ankur Balmik and Subashchandrabose Chinnathambi

1. Introduction 82
1.1 Pathological hallmarks of AD 83
1.2 Role of acetylation as a post-translational modification in AD 83
1.3 Histone deacetylases – classification and functions 84
1.4 HDACs in AD pathology 86
2. Rationale 89
3. Analysis of HDAC6 ZnF UBP function in modulation of Tau structure and function 92
3.1 Plasmid preparation and protein expression 92
3.2 Protein analysis 100
3.3 Tau – HDAC6 interaction analysis 101
3.4 Aggregation studies 104
3.5 Structural and morphological analysis of Tau 105
4. Concluding remarks 107
References 108

6. Sirtuins in mitochondrial physiology and energy homeostasis **111**

Arshdeep Kaur and Ullas Kolthur Seetharam

1. Introduction 112
2. Rationale and experimental challenges in studying mitochondrial sirtuins 114
   2.1 Relative abundance of mitochondrial sirtuins 114
   2.2 NAD^+ availability and mitochondrial compartmentalization 115
   2.3 Substrate specificity overlapping 116
   2.4 Non-canonical localization and the importance of tissue specificity 116
   2.5 Post-translational regulation of mitochondrial sirtuins 117
3. Experimental models and design considerations 118
   3.1 Pharmacological vs genetic approaches 118
   3.2 Metabolic state as a critical experimental variable 119
4. Measurement of sirtuins enzymatic activity 120
   4.1 Measuring catalytic activity 120
   4.2 Assessing NAD^+ dependence 126
   4.3 Monitoring acylation dynamics using chemical biology approaches 129
5. Assessment of mitochondrial protein acylation 129
   5.1 Mitochondrial respiration and bioenergetics 130
   5.2 Mitochondrial redox balance and reactive oxygen species 132
   5.3 Mitochondrial dynamics and quality control 134
6. Cellular energy state and whole-body homeostasis 135
7. Conclusion 136

Acknowledgements 137

References 137

7. DNA methylation and histone modifications in tumor hypoxia **143**

Shruti Ganesh Dhamdhere, Parik Kakani, and Sanjeev Shukla

1. Introduction 144
2. Rationale 147
3. Analyses of differential DNA methylation and histone modifications in hypoxic cancers as compared to their normoxic counterparts 148
4. Isolation and immunoprecipitation of genomic DNA to assess methylation patterns 149
   4.1 Equipments 149
   4.2 Reagents 149
   4.3 Procedure 150
   4.4 Notes 151

5. Chromatin immunoprecipitation (ChIP) to determine occupancy of histone marks at specific locus 151
5.1 Equipments 151
5.2 Reagents 151
5.3 Procedure 155
5.4 Notes 157
6. Effect of presence of epigenetic modifications on transcription factor binding and gene regulation 157
7. De novo maintenance of methylation via dCas9-DNMT3A CRISPR systems under hypoxic condition 158
7.1 Equipments 158
7.2 Reagents 158
7.3 Procedure 158
7.4 Notes 159
8. De novo DNA methylation under hypoxia abolishes recruitment of hypoxic transcription factors HIF1α and CTCF to binding sites 159
9. De novo histone dissymmetric methylation under hypoxia by dCas9-PRMT5 regulates splicing outcome 159
10. Challenges and limitations 160
11. Concluding remarks 160
References 161

8. Epigenetic modulators as therapeutics for neurodegenerative disorders/Alzheimer's disease 163

Debojyoti Halder, Denish Prajapati, Tonmoy Banerjee, Swati Biswas, and Balaram Ghosh

1. Introduction 164
2. Epigenetic regulation of the nervous system 167
2.1 DNA methylation in neural regulation 167
2.2 Histone modifications and transcriptional regulation 168
2.3 Chromatin remodeling in neuronal function 168
2.4 Non-coding RNA-mediated regulation 169
2.5 Integration of epigenetic mechanisms in synaptic plasticity 169
2.6 Epigenetic dysregulation in neurodegenerative disorders 171
3. Epigenetic dysfunction in AD 171
3.1 DNA methylation in AD 172
3.2 Histone modifications in AD 173
3.3 Non-coding RNAs in AD 174
3.4 Epigenetic regulation of peripheral immune responses in AD 174

3.5 Experimental assays and analytical methods used in epigenetic studies of AD 176
4. Epigenetic modulators for neurodegenerative disorders 196
4.1 Classification of epigenetic modulators 196
5. Prospects and emerging therapeutics 212
Acknowledgements 214
Declaration of generative AI and AI-assisted technologies in the writing process 214
Conflict of interest 214
References 214

9. Integrated SDH, COX histochemistry, and MHC fibre typing in skeletal muscles: Methods and interpretation 223

Aastha Munjal, Ridhi Singh, Ashwini Bhelave, Shwetha Sekar, Ninitha Asirvatham-Jeyaraj, and Nagalingam R. Sundaresan

1. Introduction 224
2. Sirtuins and skeletal muscle metabolism: Applications of SDH/COX and MHC fibre typing in metabolic research 225
2.1 Exercise training 225
2.2 Obesity and insulin resistance 227
2.3 Ageing and sarcopenia 227
2.4 Genetic manipulation of metabolic regulators 229
3. Biochemical principles underlying SDH/COX histochemistry and MHC fibre typing in skeletal muscles 229
3.1 Succinate dehydrogenase (SDH) activity assay 229
3.2 Cytochrome c oxidase (COX) histochemistry 230
3.3 Muscle fibre typing using MHC immunohistochemistry 230
4. Methodology 231
4.1 Tissue sectioning 231
4.2 SDH staining 233
4.3 COX staining 234
4.4 MHC fibre typing 235
5. Quantitative image analysis 238
5.1 Imaging acquisition standards 238
5.2 Statistical considerations 239
6. Interpretation of results 239
6.1 SDH/COX staining 239
6.2 MHC fibre typing 244

7. Strengths and limitations of SDH/COX and MHC fibre typing histochemical approaches 244
7.1 Strengths 244
7.2 Limitations 244
8. Troubleshooting 245
8.1 SDH Staining 245
8.2 COX staining 245
8.3 MHC immunostaining 245
9. Future directions 248
References 248

10. Analysis of copper toxicity by western blotting of secretory proteins translocated via the Sec61 channel **253**

Smriti Anand, Prateeksha Thakur, Vandana Anjana, Santoshi Acharjee, Rajshree Pal, Sugandh Sharma, and Raghuvir Singh Tomar

1. Introduction 254
2. Rational 259
3. Methodology 260
3.1 Transformation 260
3.2 Yeast growth condition 265
3.3 Whole cell extract preparation 266
3.4 Western blotting 268
4. Data analysis and interpretation 273
5. Conclusion 274
Acknowledgments 275
Author's contribution 275
Data availability statement 275
Conflict of interest 275
References 275

11. Histone deacetylase as a potential therapeutic target for memory dysfunction during aging and neurodegenerative disorders **279**

Padmanabh Singh and Mahendra Kumar Thakur

1. Introduction 280
1.1 Histone acetylation 281
1.2 Histone deacetylases 282
1.3 Class I HDACs 282

1.4 Histone deacetylase 2 282
1.5 Therapeutic potential of HDAC inhibitors 284
2. Expression of HDAC2 upregulated in the hippocampus of old mice during age associated memory decline 284
2.1 Animal 284
2.2 Equipment 285
2.3 Reagents 285
2.4 Procedure 285
3. Histone acetylation level declined in the hippocampus of old mice during age associated memory decline 287
3.1 Animal 287
3.2 Equipment 287
3.3 Reagents 288
3.4 Procedure 288
4. HDAC2 inhibition improved learning and memory in old mice 289
4.1 Animal 289
4.2 Equipment 289
4.3 Reagents 289
4.4 Procedure 289
5. HDAC2 inhibition increased histone acetylation level in the hippocampus of old mice 290
5.1 Animal 290
5.2 Equipment 290
5.3 Reagents 290
5.4 Procedure 291
6. HDAC2 inhibition increased histone acetylation level at the promoter of synaptic plasticity genes in the hippocampus of old mice 292
6.1 Animal 292
6.2 Equipment 292
6.3 Reagents 292
6.4 Procedure 292
7. Conclusion 293
Acknowledgment 294
References 294

12. Synergetic inhibition of deubiquitinases and acetyltransferases to regulate tau acetylation and aggregation in SH-SY5Y cell line model of Alzheimer's disease **297**

Basavaraju K C, Harshitha D K, and Poornima Priyadarshini

1. Introduction 298
   1.1 Deubiquitinases 299
   1.2 Acetyltransferases and tau acetylation 302
2. Rationale 305
3. Methods 306
   3.1 Chemicals 306
   3.2 Procedure 307
   3.3 Notes 308
   3.4 Protocol for microscopy 308
   3.5 Gene expression studies 312
   3.6 cDNA synthesis: (iScript™ cDNA synthesis Kit-Biorad) 312
4. Notes 313
5. Conclusions 313

Acknowledgement 314
Funding 314
Data availability 314
Ethical approval and consent to participate 314
CRediT authorship contribution statement 314
Consent for publication 314
References 315

13. Histone deacetylases in neurodegeneration and neuronal plasticity **321**

Bhupesh Mehta, Sinjitha Nambiar, Omkar Shirke, Satyarth Pandita, and Yogananda S. Markandeya

1. Introduction 322
   1.1 HDACs in the nervous system: An overview 322
   1.2 HDAC nucleocytoplasmic shuttling: How it works 323
   1.3 HDACs and neuronal plasticity 325
   1.4 Dysfunction of HDACs in neurodegenerative disease 326
   1.5 Scope and organization of this chapter 331
2. Core experimental methods for investigating HDAC function 332
   2.1 Enzymatic assays 332
   2.2 Cellular assays 334
   2.3 In vivo approaches 337
   2.4 Validation of HDAC subcellular localization 339

2.5 CRISPR-based manipulation of HDAC localization 342
3. Data analysis and interpretation 352
3.1 Quantification of HDAC localization 352
3.2 Statistical considerations 353
3.3 Integration of multi-modal data 354
4. Future methodological directions and outstanding challenges in HDAC research 355
4.1 Emerging experimental technologies 355
4.2 Translational and precision experimental strategies 356
4.3 Unresolved mechanistic questions and experimental strategies 357
4.4 Integrative perspective 358
5. Conclusion 358
References 359

14. ChIPmentation-based mapping of histone modifications in aging: Experimental design and analytical considerations **367**

Amalanandan Johna, Sreedarsanam Sreya, Sengupta Tiyas, Pavithra Elumalai, Avik Sengupta, Rahul Kumar, and Mayilaadumveettil Nishana

1. Introduction 368
2. Equipments 372
3. Consumables 373
4. Procedure 374
4.1 Crosslinking of cells 374
4.2 Cell lysis and chromatin preparation 375
4.3 Chromatin sonication 375
4.4 Immunoprecipitation 376
4.5 Washing of immunocomplexes 376
4.6 ChIPmentation 376
4.7 Elution, reverse crosslinking, and protein digestion 376
4.8 DNA purification 377
4.9 Library amplification 377
4.10 Library size selection 378
4.11 Quality control and quantification 378
4.12 Library pooling and sequencing 378
4.13 Computational analysis 378
5. Conclusion and future perspectives 379
Acknowledgments 380

Ethical declaration 380
Data availability 380
References 380

15. Histone Decatylase 6 and Tau-induced microglial chemotaxis by trans-well migration and wound-scratch assay **385**

Subashchandrabose Chinnathambi, and Smita Eknath Desale

1. Introduction 386
   1.1 Rationale 388
2. Materials 388
   2.1 Stock and working buffer solutions 388
   2.2 Cell culture 390
   2.3 Tau species preparation 390
   2.4 2-D migration by wound-scratch assay 390
   2.5 3-D migration by Trans-well assay 390
   2.6 Instruments 391
   2.7 Softwares 392
3. Methods 392
   3.1 Preparation of Tau protein 392
   3.2 Recombinant HDAC6 ZnF UBP preparation 394
   3.3 Mammalian cell culture and maintenance 397
   3.4 Preparation of Tau aggregates for migration assays 398
   3.5 Wound-scratch assay- 2D migration assay 398
   3.6 Transwell-migration assay- 3D migration assay 400
4. Notes 402
   4.1 2-D migration by wound-scratch assay 402
   4.2 3-D migration by Trans-well assay 403
5. Concluding remarks 404

Acknowledgements 404
References 404

16. Experimental techniques to evaluate IA-MSCs–mediated alleviation of vasogenic edema following ischemic stroke **407**

Aishika Datta, Nikita Malik, Gautam Karmarkar, Anirban Barik, Deepaneeta Sarmah, Vikas Khajekar, Akshada Dubey, Ushmita Mukherjee, Anita Kumari, Swagata Bhattacharya, Bijoyani Ghosh, and Pallab Bhattacharya

1. Introduction 408
2. Rationale 410

3. IA-MSCs in combination with SIRT-1 Activator and NFkB Inhibitor: Implication in post stroke edema 411
3.1 Induction of middle cerebral artery occlusion model 411
3.2 Administration of intra-arterial mesenchymal stem cells (IA-MSCs), SIRT 1 activator, inhibitor and NFkB Inhibitor after ischemic stroke 414
3.3 Histological assessment 416
3.4 Behavioural assessment 417
3.5 Cerebral infarct size and percentage edema measurement 419
3.6 Western blotting 419
3.7 RT-PCR 423
3.8 Immunofluorescence analysis 424
3.9 Oxidative stress measurement 425
3.10 Mitochondrial functional assessment in post-stroke vasogenic edema 426
4. Conclusion 429
Acknowledgement 429
References 429

Contributors

Santoshi Acharjee
Department of Biological Sciences, Indian Institute of Science Education and Research, Bhopal, Madhya Pradesh, India

Smriti Anand
Department of Biological Sciences, Indian Institute of Science Education and Research, Bhopal, Madhya Pradesh, India

Vandana Anjana
Department of Biological Sciences, Indian Institute of Science Education and Research, Bhopal, Madhya Pradesh, India

Ninitha Asirvatham-Jeyaraj
Department of Biotechnology, Bhupat and Jyoti Mehta School of Biosciences, Indian Institute of Technology Madras, Chennai, India

Abhishek Ankur Balmik
Department of Pharmacology and Chemical Biology, Emory University School of Medicine, Atlanta, GA, United States

Tonmoy Banerjee
Nanomedicine Research Laboratory, Department of Pharmacy, Birla Institute of Technology and Science Pilani Hyderabad Campus, Shamirpet, Hyderabad, India

Anirban Barik
Department of Pharmacology and Toxicology, National Institute of Pharmaceutical Education and Research (NIPER), Ahmedabad, Gandhinagar, Gujarat, India

Alessio Bertozzo
Laboratory of Epigenomics, Department of Medicine, Università degli Studi di Udine, Udine, Italy

Pallab Bhattacharya
Department of Pharmacology and Toxicology, National Institute of Pharmaceutical Education and Research (NIPER), Ahmedabad, Gandhinagar, Gujarat, India

Swagata Bhattacharya
Department of Pharmacology and Toxicology, National Institute of Pharmaceutical Education and Research (NIPER), Ahmedabad, Gandhinagar, Gujarat, India

Ashwini Bhelave
Department of Microbiology and Cell Biology, Indian Institute of Science, Bengaluru, India

Swati Biswas
Nanomedicine Research Laboratory, Department of Pharmacy, Birla Institute of Technology and Science-Pilani Hyderabad Campus, Shamirpet, Hyderabad, India

Claudio Brancolini
Laboratory of Epigenomics, Department of Medicine, Università degli Studi di Udine, Udine, Italy

Gianluca Caprini
Italfarmaco Group, New Drug Incubator Department, Cinisello Balsamo, Italy

Edoardo Cellupica
Italfarmaco Group, New Drug Incubator Department, Cinisello Balsamo, Italy

Subashchandrabose Chinnathambi
Department of Neurochemistry, National Institute of Mental Health and Neuro Sciences (NIMHANS), Institute of National Importance, Bangalore, Karnataka, India

Harshitha D K
Department of Molecular Nutrition, CSIR-Central Food Technological Research Institute (CSIR-CFTRI), Mysore, Karnataka; Academy of Scientific and Innovative Research (AcSIR), Ghaziabad, India

Aishika Datta
Department of Pharmacology and Toxicology, National Institute of Pharmaceutical Education and Research (NIPER), Ahmedabad, Gandhinagar, Gujarat, India

Smita Eknath Desale
Department of Neurochemistry, National Institute of Mental Health and Neuro Sciences (NIMHANS), Institute of National Importance, Bangalore, Karnataka, India

Shruti Ganesh Dhamdhere
Department of Biological Sciences, Indian Institute of Science Education and Research, Bhopal, Madhya Pradesh, India

Eros Di Giorgio
Laboratory of Biochemistry, Department of Medicine, Università degli Studi di Udine, Udine, Italy

Akshada Dubey
Department of Pharmacology and Toxicology, National Institute of Pharmaceutical Education and Research (NIPER), Ahmedabad, Gandhinagar, Gujarat, India

Pavithra Elumalai
School of Biology, Indian Institute of Science Education and Research Thiruvananthapuram, Thiruvananthapuram, India

Gianluca Fossati
Italfarmaco Group, New Drug Incubator Department, Cinisello Balsamo, Italy

Balaram Ghosh
Epigenetic Research Laboratory, Department of Pharmacy, Birla Institute of Technology and Science-Pilani, Hyderabad Campus, Shamirpet, Hyderabad, India

Bijoyani Ghosh
Department of Pharmacology and Toxicology, National Institute of Pharmaceutical Education and Research (NIPER), Ahmedabad, Gandhinagar, Gujarat, India

Debojyoti Halder
Epigenetic Research Laboratory, Department of Pharmacy, Birla Institute of Technology and Science-Pilani, Hyderabad Campus, Shamirpet, Hyderabad, India

Amalanandan Johna
School of Biology, Indian Institute of Science Education and Research Thiruvananthapuram, Thiruvananthapuram, India

Basavaraju K C
Department of Molecular Nutrition, CSIR-Central Food Technological Research Institute (CSIR-CFTRI), Mysore, Karnataka; Academy of Scientific and Innovative Research (AcSIR), Ghaziabad, India

Parik Kakani
Department of Biological Sciences, Indian Institute of Science Education and Research, Bhopal, Madhya Pradesh, India

Gautam Karmarkar
Department of Pharmacology and Toxicology, National Institute of Pharmaceutical Education and Research (NIPER), Ahmedabad, Gandhinagar, Gujarat, India

Arshdeep Kaur
Tata Institute of Fundamental Research Mumbai, Mumbai, Maharashtra, India

Vikas Khajekar
Department of Pharmacology and Toxicology, National Institute of Pharmaceutical Education and Research (NIPER), Ahmedabad, Gandhinagar, Gujarat, India

Rahul Kumar
Department of Biotechnology, Indian Institute of Technology Hyderabad, Kandi, Telangana, India

Anita Kumari
Department of Pharmacology and Toxicology, National Institute of Pharmaceutical Education and Research (NIPER), Ahmedabad, Gandhinagar, Gujarat, India

Tapas K. Kundu
Molecular Biology and Genetics Unit, Transcription and Disease Laboratory, Jawaharlal Nehru Centre for Advanced Scientific Research, Jakkur, Bengaluru; Neuroscience Unit, Jawaharlal Nehru Centre for Advanced Scientific Research, Bangalore, India

Nikita Malik
Department of Pharmacology and Toxicology, National Institute of Pharmaceutical Education and Research (NIPER), Ahmedabad, Gandhinagar, Gujarat, India

Yogananda S. Markandeya
Department of Biophysics, National Institute of Mental Health and Neurosciences (NIMHANS), Bengaluru, India

Bhupesh Mehta
Department of Biophysics, National Institute of Mental Health and Neurosciences (NIMHANS), Bengaluru, India

Martina Minisini
Laboratory of Epigenomics, Department of Medicine, Università degli Studi di Udine, Udine, Italy

Prajjval Mishra
Molecular Biology and Genetics Unit, Transcription and Disease Laboratory, Jawaharlal Nehru Centre for Advanced Scientific Research, Jakkur, Bengaluru, India

Ushmita Mukherjee
Department of Pharmacology and Toxicology, National Institute of Pharmaceutical Education and Research (NIPER), Ahmedabad, Gandhinagar, Gujarat, India

Aastha Munjal
Department of Microbiology and Cell Biology, Indian Institute of Science, Bengaluru, India

Sinjitha Nambiar
Department of Biophysics, National Institute of Mental Health and Neurosciences (NIMHANS), Bengaluru, India

Mayilaadumveettil Nishana
School of Biology, Indian Institute of Science Education and Research Thiruvananthapuram, Thiruvananthapuram, India

Rajshree Pal
Department of Biological Sciences, Indian Institute of Science Education and Research, Bhopal, Madhya Pradesh, India

Priyanka Pandey
Molecular Biology and Genetics Unit, Transcription and Disease Laboratory, Jawaharlal Nehru Centre for Advanced Scientific Research, Jakkur, Bengaluru, India

Satyarth Pandita
Department of Biophysics, National Institute of Mental Health and Neurosciences (NIMHANS), Bengaluru, India

Denish Prajapati
Epigenetic Research Laboratory, Department of Pharmacy, Birla Institute of Technology and Science-Pilani, Hyderabad Campus, Shamirpet, Hyderabad, India

Poornima Priyadarshini
Department of Molecular Nutrition, CSIR-Central Food Technological Research Institute (CSIR-CFTRI), Mysore, Karnataka; Academy of Scientific and Innovative Research (AcSIR), Ghaziabad, India

Amrish Rai
Division of Neuroscience and Ageing Biology, CSIR-Central Drug Research Institute, Lucknow; Academy of Scientific and Innovative Research (AcSIR), Ghaziabad, India

Rakesh Kumar Sahu
Laboratory of Biochemistry and Molecular Biology, National Cancer Institute, National Institutes of Health, Bethesda, MD, United States

Deepaneeta Sarmah
Department of Pharmacology and Toxicology, National Institute of Pharmaceutical Education and Research (NIPER), Ahmedabad, Gandhinagar, Gujarat, India

Ullas Kolthur Seetharam
Tata Institute of Fundamental Research Mumbai, Mumbai, Maharashtra; Tata Institute of Fundamental Research Hyderabad; Centre for DNA Fingerprinting and Diagnostics, Hyderabad, Telangana, India

Shwetha Sekar
Department of Biotechnology, Bhupat and Jyoti Mehta School of Biosciences, Indian Institute of Technology Madras, Chennai, India

Avik Sengupta
Department of Biotechnology, Indian Institute of Technology Hyderabad, Kandi, Telangana, India

Sugandh Sharma
Department of Biological Sciences, Indian Institute of Science Education and Research, Bhopal, Madhya Pradesh, India

Omkar Shirke
Department of Biophysics, National Institute of Mental Health and Neurosciences (NIMHANS), Bengaluru, India

Sanjeev Shukla
Department of Biological Sciences, Indian Institute of Science Education and Research, Bhopal, Madhya Pradesh, India

Akash Kumar Singh
Molecular Biology and Genetics Unit, Transcription and Disease Laboratory, Jawaharlal Nehru Centre for Advanced Scientific Research, Jakkur, Bengaluru; Neuroscience Unit, Jawaharlal Nehru Centre for Advanced Scientific Research, Bangalore, India

Padmanabh Singh
Biochemistry and Molecular Biology Laboratory, Department of Zoology, Institute of Science, Centre of Advanced Study, Banaras Hindu University, Varanasi, India

Ridhi Singh
Department of Microbiology and Cell Biology, Indian Institute of Science, Bengaluru, India

Sakshi Singh
Division of Molecular and Cellular Biology, Eunice Kennedy Shriver National Institute of Child Health and Human Development, National Institutes of Health, Bethesda, MD, United States

Sreedarsanam Sreya
School of Biology, Indian Institute of Science Education and Research Thiruvananthapuram, Thiruvananthapuram, India

Christian Steinkühler
Italfarmaco Group, New Drug Incubator Department, Cinisello Balsamo, Italy

Nagalingam R. Sundaresan
Department of Microbiology and Cell Biology, Indian Institute of Science, Bengaluru, India

Mahendra Kumar Thakur
Biochemistry and Molecular Biology Laboratory, Department of Zoology, Institute of Science, Centre of Advanced Study, Banaras Hindu University, Varanasi, India

Prateeksha Thakur
Department of Biological Sciences, Indian Institute of Science Education and Research, Bhopal, Madhya Pradesh, India

Sengupta Tiyas
School of Biology, Indian Institute of Science Education and Research Thiruvananthapuram, Thiruvananthapuram, India

Raghuvir Singh Tomar
Department of Biological Sciences, Indian Institute of Science Education and Research Bhopal, Madhya Pradesh, India

CHAPTER ONE

Experimental strategies for evaluating the impact of HDACs and HDAC inhibitors on the epigenome. ChiP-seq

Martina Minisini[a], Alessio Bertozzo[a], Eros Di Giorgio[b], and Claudio Brancolini[a,*]

[a]Laboratory of Epigenomics, Department of Medicine, Università degli Studi di Udine, Udine, Italy
[b]Laboratory of Biochemistry, Department of Medicine, Università degli Studi di Udine, Udine, Italy
*Corresponding author. e-mail address: claudio.brancolini@uniud.it

Contents

1. Introduction 2
   1.1 HDACs and HDACs inhibitors a brief introduction 2
   1.2 The basic principle of the ChIP-seq methodology 3
2. Rationale 5
3. Experimental settings 5
   3.1 Note 5
   3.2 Equipment 5
   3.3 Reagents 6
   3.4 Required solutions 6
   3.5 Safety note 6
   3.6 Additional note 7
   3.7 Procedure 7
4. Normalization 11
5. Concluding remarks 13
Acknowledgments 13
References 14

Abstract

HDACs (histone deacetylases) are components of multiprotein complexes that remove the acetyl group from lysine residues (and other acyl groups) to regulate protein function. This important post-translational modification can affect the local hydrophobic environment of a specific protein surface. When the protein substrates are histones, the regulation of lysine acetylation influences chromatin accessibility and gene expression. Defining the different genomic regions under the influence of specific HDACs is fundamental to understanding how these enzymes control the epigenetic status of cells. Furthermore, it is important to clarify the mechanisms of action of HDAC inhibitors from

Methods in Enzymology, Volume 733
ISSN 0076-6879, https://doi.org/10.1016/bs.mie.2026.05.031

a therapeutic perspective. ChIP-seq (Chromatin Immunoprecipitation followed by sequencing) is a powerful technique to identify where histone acetylations regulated by HDACs or influenced by HDAC inhibitors occur in the genome. Here, we describe a protocol to perform a ChIP-seq experiment to detect the genome-wide distribution of histone acetylation.

1. Introduction

1.1 HDACs and HDACs inhibitors a brief introduction

The human genome is organised in a complex and highly structured architecture, allowing the compaction of a vast amount of genetic information. Specifically, 220 base pairs of DNA define a nucleosome, of which 146 are wrapped around a protein core formed by a histone octamer (H2A, H2B, H3, H4). These are known as classical or standard histones and are accompanied by various isoforms that may differ by a few or several amino acids. Completing the scenario is histone H1 (and its different isoforms), which binds the linker DNA between two nucleosomes and is essential for higher-order chromatin compaction (Davey et al., 2002; Luo et al., 2026). Histones are a fundamental component of chromatin. This complex structure is both dynamic and highly regulated, thanks to epigenetic regulators: erasers, writers, and readers, which respectively remove post-translational modifications (PTMs), catalyse their addition, or recognise and impose modifications, thereby influencing chromatin state and transcription (Ausio, 2006; Helin & Dhanak, 2013; Klemm et al., 2019; Luger et al., 1997). The epigenetic modifications that most impact chromatin structure and function are DNA methylation and histone PTMs. Several histone PTMs have been described that can influence the assembly of multi-protein complexes onto DNA, the interaction between nucleosomes, and the interaction between nucleosomes and DNA (Cricchi et al., 2025; Joseph & Young, 2023; Villar-Garea & Imhof, 2006). Histone acetylation, governed by histone acetyltransferases (HATs) and histone deacetylases (HDACs), is associated with increased DNA accessibility, open chromatin, and often gene transcription. HDACs comprise 18 enzymes divided into five main classes: class I (HDAC1, HDAC2, HDAC3, HDAC8), class IIa (HDAC4, HDAC5, HDAC7, HDAC9), class IIb (HDAC6, HDAC10), class III or sirtuins (SIRT1–7), and class IV, which consists solely of HDAC11 (Brancolini et al., 2022; Brancolini, 2025; Clocchiatti et al., 2011; Haberland et al., 2009). HDACs are a heterogeneous group of proteins with several biological functions, including differentiation, cell proliferation, cancer progression, apoptosis, and DNA damage response, due to their role in controlling acetylation states. Given the number of HDAC enzymes and the variety of functions performed by each class and member, different HDAC inhibitors (HDACi) have been developed (Delcuve et al., 2012; Di Giorgio et al., 2015). Several HDACi are being tested in clinical and pre-clinical trials, and to date, four have been approved by the Food and Drug Administration: SAHA/Vorinostat, a pan-HDACi, for cutaneous T-cell lymphoma; Romidepsin, an HDAC1/2 inhibitor, for cutaneous and peripheral T-cell lymphoma; Belinostat, a pan-HDACi, for peripheral T-cell lymphoma; and Panobinostat, another pan-HDACi, for multiple myeloma (Eckschlager et al., 2017). Other HDACi have been developed and studied in different contexts. Although some are specific for a class of HDACs, such as TMP-195 and TMP-269, which selectively inhibit class IIa members, most are pan-HDACi and may also have several off-target effects, with the risk of broad and side effects (Brancolini et al., 2022; Di Giorgio et al., 2015). Among subclass-targeted HDACi developed to specifically target a class of HDACs, NKL-54 shows some specificity for class I enzymes: HDAC1, HDAC2, and HDAC3 (Jayathilaka et al., 2012; Minisini et al., 2022).

1.2 The basic principle of the ChIP-seq methodology

The first evidence of genome-wide mapping of chromatin binding sites dates to 2001, when Peter Brown's group reported the binding profiles of the transcription factors SBF and MBF in yeast (Iyer et al., 2001). This approach, later termed ChIP-on-chip, was based on the purification of chromatin–protein complexes followed by hybridisation of the associated genomic DNA onto custom-designed microarrays, enabling large-scale identification of protein–DNA interactions despite limitations in resolution and coverage. The subsequent integration of chromatin immunoprecipitation with next-generation sequencing (ChIP-seq), first reported in 2007, represented a major technological advance, allowing a more comprehensive, high-resolution, and unbiased analysis of chromatin landscapes (Barski et al., 2007). ChIP-seq is now widely used to profile DNA-binding proteins and histone modifications across the genome, providing a detailed view of protein–DNA interactions and chromatin states (Park, 2009). The ChIP-seq protocol is adapted to the specific aim of the study, and a key initial decision is the choice between native ChIP (N-ChIP) and cross-linking ChIP (X-ChIP). N-ChIP is typically used to investigate stable protein–DNA interactions, such as those involving histones, which are tightly associated with DNA. In this approach, chromatin is analysed under native conditions without chemical cross-linking, reducing the risk of cross-linking-induced artefacts and preserving the physiological state of chromatin (O'Neill, 2003). In contrast, X-ChIP requires cross-linking with formaldehyde, as the proteins of interest, such as transcription factors or epigenetic regulators, may not be tightly bound to chromatin or DNA, or the interaction may be highly dynamic. Formaldehyde cross-links proteins to DNA by forming reversible covalent bonds, primarily through reactions with amino and imino groups of proteins and nucleic acids, resulting in the generation of methylene bridges ($-CH_2-$) (Lee et al., 2006). These cross-links can further stabilise into more complex linkages, including amide-like bonds under certain conditions, thereby preserving protein–DNA interactions. For this reason, formaldehyde-based cross-linking is widely used even when studying histone modifications. Importantly, the duration and conditions of cross-linking must be carefully controlled, as excessive cross-linking can increase background noise and cause false-positive signals by capturing transient or non-specific protein–DNA interactions (Orlando et al., 2014).

The choice also depends on the method of chromatin shearing: N-ChIP is typically coupled with micrococcal nuclease digestion, which generates nucleosome-sized fragments, whereas X-ChIP relies on sonication, producing randomly sheared DNA fragments whose size depends on sonication parameters such as time and intensity. In practice, integrating histone modification profiles with transcription factor binding data has led to the widespread adoption of X-ChIP-seq, even when N-ChIP could be technically suitable, due to its greater versatility in capturing both stable and transient protein–DNA interactions within a unified experimental framework. More recently, alternative approaches such as CUT&RUN and CUT&Tag have emerged as powerful methods that do not rely on conventional chromatin fragmentation. Although not addressed in this study, these techniques are based on antibody-targeted cleavage or tagmentation of chromatin, using protein A/G–conjugated nucleases or transposases to selectively process genomic regions bound by specific chromatin-associated proteins, thereby increasing resolution and reducing background (Kaya-Okur et al., 2019; Skene & Henikoff, 2017).

In traditional X-ChIP-seq workflows, after chromatin shearing, the protocol includes immunoprecipitation with a specific antibody against the protein of interest, treatment with RNase A to remove contaminating RNA, reversal of cross-links by heat combined with proteinase K digestion, and subsequent purification of the recovered DNA fragments.

2. Rationale

Given the complexity of the chromatin and epigenetic landscape, including protein-DNA interactions, a specific method capable of detecting and revealing this intricate scenario is required, and ChIP-seq has become a key technique in the field. In the context of histone PTMs, ChIP-seq has been extensively used to study chromatin reorganisation under specific conditions. Moreover, multiple studies have employed ChIP-seq to map the genomic distribution of different histone PTMs, leading to a more precise functional definition of the genome based on specific PTMs. ChIP-seq is also important for understanding the effects of epigenetic drugs, such as HDACi, thereby providing greater insight into the impact of these molecules on the chromatin landscape and enhancing our knowledge of their influence on specific genomic loci, related genes, and cis-regulatory elements (Egan et al., 2016; Minisini et al., 2022; Wu et al., 2021).

3. Experimental settings

3.1 Note

The experimental procedure described here is based on an X-ChIP-seq performed to immunoprecipitate the histone mark H3K27ac in the high-grade leiomyosarcoma cell line SK-UT-1 treated with the HDAC inhibitor NKL54. H3K27ac is a well-known acetylation marker of active cis-regulatory elements, and is enriched at promoters, enhancers, and super-enhancers associated with active transcription (Creyghton et al., 2010; Zhang et al., 2020).

The materials and equipment suppliers listed are those used in our laboratory. However, reagents from other suppliers or similar equipment from other brands may also be suitable.

3.2 Equipment

Bioruptor Plus sonicator (Diagenode).
Lab centrifuge (Eppendorf).
Thermocycler (Eppendorf).
Rotator (Orma Scientific Instruments).
Qubit fluorometer (Thermo Fisher).
Magnetic holder DynaMag (Thermo Fisher).
4200 TapeStation System (Agilent Technologies).

3.3 Reagents

- Cell line of interest, culture medium, HDACi, DMSO, plastic plates
- PBS 1X
- Glycine (Sigma-Aldrich, G8790)
- UltraPure Formaldehyde 37 % (PanAeac. AppliChem, ITW Reagents 131328)
- Dynabeads Protein A beads (Thermo Fisher, 10001D)
- Anti-H3K27ac antibody (Abcam, ab4729)
- PureLink RNAse A (Invitrogen, 12091021)
- Proteinase K solution (Thermo Fisher, 26160)
- Chip clean and concentrator kit (Zymo Research, D5201)
- Qubit 1X dsDNA High Sensitivity Assay Kit (Thermo Fisher, Q33230)

3.4 Required solutions

- PBS glycine 125 mM
- SDS lysis buffer (1 % SDS, 10 mM EDTA, 50 mM Tris pH 8.0)
- ChIP Dilution Buffer (0.01 % SDS, 1.1 % Triton X-100, 1.2 mM EDTA, 16.7 mM Tris pH 8.0, 165 mM NaCl)
- Low salt immune complex wash buffer (0.1 % SDS, 1 % TRITON, 20 mM Tris-HCl pH 8.0, 150 mM NaCl)
- High salt wash buffer (0.1 % SDS, 1 % TRITON, 2 mM EDTA, 20 mM Tris-HCl pH 8.0, 500 mM NaCl)
- LiCl immune complex wash buffer (0.25 M LiCl, 1 % NP40, 0.2 % sodium deoxycholate, 1 mM EDTA, 10 mM Tris-HCl pH 8.0).
- Reverse Crosslinking Salt Mixture (250 mM TRIS pH 7.5, 62.5 mM EDTA pH 8.0, 1.25 M NaCl, 0,5 % SDS)
- TE pH 8.0 (Tris pH8, 1 mM EDTA pH 8.0)

3.5 Safety note

For hazard information and safety warnings, please refer to the safety data sheet for each reagent. Formaldehyde is highly toxic by skin contact, ingestion, and inhalation, and is classified as a known carcinogen. SDS causes eye damage, and respiratory and skin irritation; a liquid or diluted stock solution (e.g. 10 %) is generally safer than SDS powder, which is also considered a flammable solid. PMSF is toxic if inhaled, ingested, or absorbed through the skin. Wear a lab coat and gloves and work in a chemical hood.

During the sonication step, hearing protection must be worn, as the sonicator generates high-frequency noise that can damage hearing. Aerosol exposure must also be avoided. Even when using the Bioruptor Plus sonicator, which is a closed system that minimises these risks, safety precautions should be maintained, and pregnant women should avoid prolonged exposure.

3.6 Additional note

SDS precipitates at low temperatures, appearing as white crystals. Therefore, it is better to store it at room temperature and, if possible, to slightly heat solutions before adding SDS.

3.7 Procedure

3.7.1 Before starting

Cells of interest, in this case SK-UT-1 cells, are seeded in a 92 mm plate at the desired concentration, here 1.5 × 10^5 cells/ml, and left in the incubator at atmospheric O_2 (20 %) and CO_2 (5 %) until 80–90 % confluence is reached. For this experiment, cells are treated with the HDACi NKL-54 (5 μM) or dimethyl sulfoxide (DMSO) for 14 h.

3.7.2 Crosslinking

The first step of an X-ChIP-seq experiment is crosslinking, which involves treating cells with formaldehyde to allow covalent binding between proteins and DNA. A 37 % formaldehyde solution is directly added to the cell culture medium to achieve a final concentration of 1 %. An important consideration is the duration of treatment (longer for transcription factors and shorter for histone markers).

Note.

This duration should be determined experimentally in a ChIP endpoint assay. In our conditions for histone PTMs, the optimal treatment time is 10 min.

Next, crosslinking is stopped by adding 2.5 ml of PBS/glycine 125 mM for 5 min. The addition of glycine is necessary to quench the action of formaldehyde. Next, two PBS 1X washes are performed, followed by scraping and collection of cells with 1 ml of PBS 1X. After centrifugation at 1500 rcf for 5 min at 4 °C and removal of the supernatant, the cell pellet can be processed immediately or stored at −80 °C.

Pause.

At this step, it is possible to pause. Particularly, when samples are collected at different times (e.g. in a time-course experiment), it is better to store the samples at −80 °C and proceed when all samples have been collected.

3.7.3 Lysis

The correct choice of lysis buffer is essential for a successful experiment. In this protocol, samples are lysed using SDS lysis buffer (supplemented with PIC and PMSF) for 10–15 min at 4 °C. SDS lysis buffer contains the strong anionic detergent SDS, that efficiently lyses nuclei and releases chromatin.

Note:

Notably, SDS is a detergent with strong denaturing action, and its use at low concentrations (1 % or less) ensures effective nuclear lysis and chromatin solubilisation. Furthermore, when used at the appropriate concentration, SDS increases sonication efficiency. The use of SDS in combination with other detergents and EDTA also mitigates its damaging effects (Kidder et al., 2011).

3.7.4 Sonication

A fundamental step in a ChIP-Seq experiment is sonication. Sonication fragments DNA into pieces whose size is determined by the duration and power of sonication. To obtain 200–500 bp fragments, the desirable DNA fragment size for ChIP-Seq experiments, the appropriate sonication time and power must be experimentally defined. In this protocol for SK-UT-1 cells, chromatin sonication is performed for 30 min using a Bioruptor Plus, with cycles of 30 s on and 30 s off at high intensity. After sonication, samples are centrifuged at 10,000 rpm for 10 min at 4 °C, and the supernatant is diluted with dilution buffer (10x). The pellet, containing insoluble precipitates, is discarded.

Note:

As previously mentioned, sonication is a critical step in the ChIP-seq protocol due to its high variability. The first parameter that must be controlled is temperature, as sonication generates energy and heat; therefore, the water bath in which the samples are placed must be kept cold (4 °C). Another important variable is time, which should be optimised before starting the ChIP-seq protocol. Notably, the sample volume can also affect sonication; for the Bioruptor Plus, the maximum volume is 300 μl, and it is important that all samples have the same volume. Another fundamental factor to consider is the viscosity of the sample; samples that are too viscous must be diluted. Furthermore, sonicators equipped with a water bath not only keep the samples cold but also ensure uniform sonication across different samples. Achieving homogeneous sonication between samples reduces technical variation that could influence the results. Problems arising from incorrect timing, differing sample volumes, or non-uniform sonication may lead to over-shearing or under-shearing, resulting in misinterpretation of the results.

3.7.5 Preclearing

To remove non-specific binding to protein A that could affect the final result, the samples are incubated for 1 h on a rotating wheel at 4 °C after the addition of 5 μl of magnetic Dynabeads (initial concentration of 30 mg/ml).

3.7.6 Immunoprecipitation

The antibody is prepared according to the targets of interest and the number of initial cells. The supernatant is collected, and protein A from preclearing is discarded using a magnetic holder. Before immunoprecipitation, chromatin concentration is quantified using the Qubit fluorometer to ensure a defined and reproducible amount of input material. In our protocol, immunoprecipitation is performed with 25–35 μg of total chromatin. Sample volumes are adjusted so that equal amounts of chromatin are used across all experimental conditions. Importantly, INPUT (1/50 of the volume) is set aside before immunoprecipitation and stored at −20 °C until step 7. In our settings, 2 μg of anti-H3K27ac antibody are used to immunoprecipitate each sample. Specifically, a mixture containing 8 μl of magnetic Dynabeads, the antibody of interest (anti-H3K27ac), and ChIP dilution buffer is prepared and added to each sample. Immunoprecipitation is performed overnight on a rotating wheel at 4 °C.

3.7.7 Washes

Washes at increasing stringency are performed as follow: 500 μl of low salt immune complex wash buffer, 500 μl of high salt immune complex wash buffer, 500 μl of LiCl immune complex wash buffer and 1 ml of TE buffer. Each wash step is conducted for 5 min on a rotating wheel at 4 °C, followed by the isolation of Dynabeads containing the immune complex through a magnetic holder. It is suggested to spin-down the samples before the use of magnetic holder, in order to avoid sample residues on the cap.

3.7.8 RNase A treatment

Inputs are unfrozen and, together with beads/IP, are incubated with 1 μl of RNase A (starting concentration of 20 mg/ml) diluted in TE buffer for 30 min at 37 °C to degrade any RNA molecules.

3.7.9 Reversion of crosslinking

To revert crosslinking, 8 μl of proteinase K (starting concentration of 20 mg/ml) diluted in 100 μl of reverse crosslinking salt mixture buffer is added to each INPUT and beads/IP sample. Crosslink reversion is performed in a thermocycler (Eppendorf) at 37 °C for 5 h, 68 °C for 10 h, and 10 °C for 1 h. Cooling to 10 °C is preferable to 4 °C to avoid SDS precipitation. The combination of proteinase K and high temperature enables the reversion of formaldehyde covalent bonds, thereby releasing the DNA.

3.7.10 DNA purification

After crosslink reversion, DNA is free from protein and can be extracted or purified using the ChIP and Clean concentrator kit and quantified with a Qubit fluorimeter. The resulting products of ChIP are fragments of genomic DNA that were associated with the immunoprecipitated protein via the antibody (anti-H3K27ac). After purification, genomic DNA can be processed for NGS.

Note.

Sonication can be evaluated using various methods that allow visualisation of the size of the resulting fragments. The gold standard for assessing ChIP sonication efficacy is capillary electrophoresis (Bioanalyzer, TapeStation); alternatively, agarose gel electrophoresis can be used. Fig. 1 shows examples of poor and good sonication, represented as electropherogram peaks analysed using Agilent 4200 TapeStation technology.

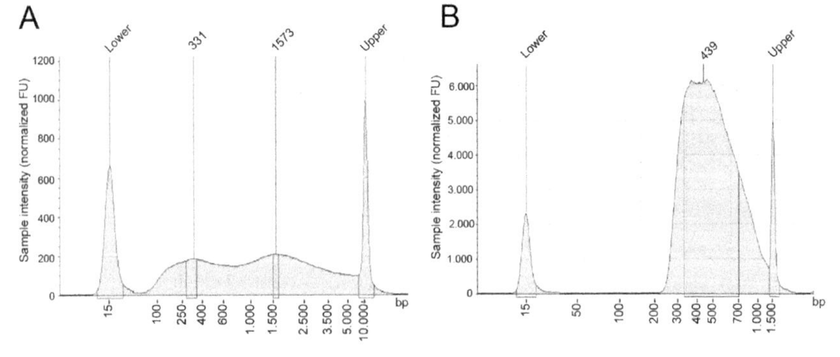

Fig. 1 Bioanalyzer electropherograms show an example of poor sonication (A) and a good sonication peak at 439 bp (B). The analysis was performed using Agilent 4200 TapeStation technology.

Note.

Before proceeding with library preparation and DNA sequencing, a small amount of ChIP sample can be used to evaluate the quality of the ChIP by qPCR on well-known targets that should be enriched in the experiments performed.

4. Normalization

To achieve consistent quantitative analysis, a fundamental and critical step in ChIP-seq is data normalization, which has been debated for years. An important internal control in ChIP-seq is the input sample, as it undergoes all the steps of the protocol except immunoprecipitation, making it a good reference for analysis. Additionally, a negative control for immunoprecipitation, such as an IgG antibody, is sometimes used to filter out negative signals from the analysis (Colando et al., 2025).

Normalization is essential for reliably comparing different samples and experimental conditions, and for overcoming common and unavoidable biases, particularly when several variables are involved and more than two groups are compared (Meyer & Liu, 2014). Sources of bias may include sonication or other shearing methods, PCR amplification – especially when the starting DNA amount is low – and coverage bias, which mainly arises from genome sequence duplications, repetitive elements, and paralogues (Wheeler et al., 2012).

Normalization can be particularly critical when comparing samples with different magnitudes of the specific target. A clear example is cells treated with HDACi, which produce a much higher signal compared to controls if a ChIP-seq experiment is performed to evaluate H3K27ac or other histone acetylations.

Several ChIP-seq normalization methods have been developed in recent years. The main methods, upon which further implementations could be divided into two main categories, intra-sample normalization and inter-sample normalization and are:

- Intra-sample normalization:
  - Normalization by read-depth normalization (CPM, RPGC, RPKM)
  - Normalization by background
- Inter-sample normalization:
  - Normalization by distribution

- Spike-in normalization
- TMM (edgeR)
- RLE (DESeq2)

Read-depth normalization, or library size normalization, is a commonly used method to compare samples by scaling their reads to the total number of sequenced reads, but it requires only small differences between the mapped reads of different samples. There are different algorithms that involve the read-depth to scale the samples and the most popular like CPM, RPGC and RPKM are implemented in widely used tools for ChIP-seq analysis as Deeptools (Ramírez et al., 2016). Normalization by background (i.e. the input sample) assumes that background noise, represented by the total counts in the non-peak regions, is similar between samples (Colando et al., 2025; Steinhauser et al., 2016). These methods are good for exploratory analysis or for data visualization, but they cannot be used for differential testing since they do not take into account, for instance, antibody efficiency differences among samples or library composition.

In contrast, normalization through distribution, like quantile normalization, offers a more robust comparison between samples because it considers the differences in the distribution of reads among samples (Colando et al., 2025; Nair et al., 2012). Spike-in normalization has emerged to address the limitations of other normalization methods. Spike-in control involves the addition of an equal amount of external DNA (e.g. from Drosophila or yeast) to the samples, allowing all experimental steps to be monitored (Colando et al., 2025; Patel et al., 2024). Researchers have developed new normalization methods to overcome some limitations and improve analysis (Patel et al., 2024). An advancement of the spike-in method is the dual-spike-in normalization approach for ChIP-seq (ChIP wrangler), which uses spike-in chromatin from two different species. This increases sample quality and minimises possible errors, as the two species have different features (genome size, GC content, etc.) (Patel et al., 2025). Methods for normalization to reveal global changes in histone modifications without spike-in have been developed (Jin et al., 2020). An interesting normalization tool is based on housekeeping genes or genes with unchanged expression between samples. These normalizations require RNA-seq data and assume that, on average, no differences in ChIP-seq signals should be observed in the regulatory regions of genes whose expression levels are constant across samples (Allhoff et al., 2016; Polit et al., 2021). For differential testing, the most widely used normalisation methods are borrowed from RNA-seq analysis, such as DESeq2 and edgeR (Love et al., 2014; Robinson et al., 2010). DESeq2 uses a method called Relative Log Expression (RLE), while edgeR uses the Trimmed Mean of M-values (TMM). After peak calling, reads can be counted within the peaks. The assumption, especially for sharp histone marks or transcription factors, is that the number of reads counted within peaks follows a negative binomial distribution, where most peaks do not change between conditions, similar to RNA-seq gene-level counts. Thus, they can be normalised in the same way. Specific tools developed for differential analysis of ChIP-seq data, such as DiffBind (Stark & Brown, 2011), efficiently implement these normalisation strategies. According to a recent study, DiffBind performs optimally when peaks are sharp, but its accuracy declines with broad peaks. In contrast, a third tool, bdgdiff (MACS2), handles broad and heterogeneous peak architectures more effectively (Eder & Grebien, 2022). For these reasons, in our laboratory, DiffBind is preferentially used for analysing sharp acetylation peaks, while bdgdiff is applied to more heterogeneous regions.

5. Concluding remarks

ChIP-seq offers several advantages. Firstly, the data are clean, with high resolution and sensitivity, and low background noise, providing precise information on protein-DNA interaction. In addition, ChIP-seq enables coverage of the entire genome with a relatively small amount of material (Massie & Mills, 2012; Park, 2009).

The main disadvantages of ChIP-seq are its cost and time requirements, although both have decreased significantly with the widespread adoption of the technique. Furthermore, the success of ChIP-seq is closely linked to the quality of antibodies and sonication, which are parameters that must be carefully evaluated. Data analysis for ChIP-seq also requires specific expertise in bioinformatics and statistics (Park, 2009).

Acknowledgments

This work was supported by research grants from AIRC under IG 2021—ID. 26200 project; from Interreg Italia-Osterreich ITAT11-018 SENECA; from CIB "Approcci biotecnologici di frontiera per la salute, per l'ambiente e l'industria"; from MUR-PRIN Bando 2022 Prot. 20228A7JM7 "Targeting class IIa HDACs to reset super-enhancers activity in cancer cells: towards selective epigenetic therapies in the 3D reality".

References

Allhoff, M., Seré, K., F. Pires, J., Zenke, M., & G. Costa, I. (2016). Differential peak calling of ChIP-seq signals with replicates with THOR. *Nucleic Acids Research*, gkw680. https://doi.org/10.1093/nar/gkw680.

Ausio, J. (2006). Histone variants–the structure behind the function. *Briefings in Functional Genomics and Proteomics, 5*(3), 228–243. https://doi.org/10.1093/bfgp/ell020.

Barski, A., Cuddapah, S., Cui, K., Roh, T.-Y., Schones, D. E., Wang, Z., ... Zhao, K. (2007). High-resolution profiling of histone methylations in the human genome. *Cell, 129*(4), 823–837. https://doi.org/10.1016/j.cell.2007.05.009.

Brancolini, C. (2025). Class IIa HDACs are important signal transducers with unclear enzymatic activities. *Biomolecules, 15*(8), 1061. https://doi.org/10.3390/biom15081061.

Brancolini, C., Gagliano, T., & Minisini, M. (2022). HDACs and the epigenetic plasticity of cancer cells: Target the complexity. *Pharmacology & Therapeutics, 238*, 108190. https://doi.org/10.1016/j.pharmthera.2022.108190.

Clocchiatti, A., Florean, C., & Brancolini, C. (2011). Class IIa HDACs: From important roles in differentiation to possible implications in tumourigenesis. *Journal of Cellular and Molecular Medicine, 15*(9), 1833–1846. https://doi.org/10.1111/j.1582-4934.2011.01321.x.

Colando, S., Schulz, D., & Hardin, J. (2025). Selecting ChIP-seq normalization methods from the perspective of their technical conditions. *Briefings in Bioinformatics, 26*(4), https://doi.org/10.1093/bib/bbaf431.

Creyghton, M. P., Cheng, A. W., Welstead, G. G., Kooistra, T., Carey, B. W., Steine, E. J., ... Jaenisch, R. (2010). Histone H3K27ac separates active from poised enhancers and predicts developmental state. *Proceedings of the National Academy of Sciences, 107*(50), 21931–21936. https://doi.org/10.1073/pnas.1016071107.

Cricchi, E., Bertozzo, A., Minisini, M., & Brancolini, C. (2025). Genomic and epigenomic plasticity in the hypoxic environment. *DNA (Mary Ann Liebert, Inc.), 5*(2), 22. https://doi.org/10.3390/dna5020022.

Davey, C. A., Sargent, D. F., Luger, K., Maeder, A. W., & Richmond, T. J. (2002). Solvent mediated interactions in the structure of the nucleosome core particle at 1.9Å resolution. *Journal of Molecular Biology, 319*(5), 1097–1113. https://doi.org/10.1016/S0022-2836(02)00386-8.

Delcuve, G. P., Khan, D. H., & Davie, J. R. (2012). Roles of histone deacetylases in epigenetic regulation: Emerging paradigms from studies with inhibitors. *Clinical Epigenetics, 4*(1), 5. https://doi.org/10.1186/1868-7083-4-5.

Di Giorgio, E., Gagliostro, E., & Brancolini, C. (2015). Selective class IIa HDAC inhibitors: Myth or reality. *Cellular and Molecular Life Sciences, 72*(1), 73–86. https://doi.org/10.1007/s00018-014-1727-8.

Eckschlager, T., Plch, J., Stiborova, M., & Hrabeta, J. (2017). Histone deacetylase inhibitors as anticancer drugs. *International Journal of Molecular Sciences, 18*(7), 1414. https://doi.org/10.3390/ijms18071414.

Eder, T., & Grebien, F. (2022). Comprehensive assessment of differential ChIP-seq tools guides optimal algorithm selection. *Genome Biology, 23*(1), 119. https://doi.org/10.1186/s13059-022-02686-y.

Egan, B., Yuan, C.-C., Craske, M. L., Labhart, P., Guler, G. D., Arnott, D., ... Trojer, P. (2016). An alternative approach to ChIP-Seq normalization enables detection of genome-wide changes in histone H3 lysine 27 trimethylation upon EZH2 inhibition. *PLoS One, 11*(11), e0166438. https://doi.org/10.1371/journal.pone.0166438.

Haberland, M., Montgomery, R. L., & Olson, E. N. (2009). The many roles of histone deacetylases in development and physiology: Implications for disease and therapy. *Nature Reviews. Genetics, 10*(1), 32–42. https://doi.org/10.1038/nrg2485.

Helin, K., & Dhanak, D. (2013). Chromatin proteins and modifications as drug targets. *Nature, 502*(7472), 480–488. https://doi.org/10.1038/nature12751.

Iyer, V. R., Horak, C. E., Scafe, C. S., Botstein, D., Snyder, M., & Brown, P. O. (2001). Genomic binding sites of the yeast cell-cycle transcription factors SBF and MBF. *Nature, 409*(6819), 533–538. https://doi.org/10.1038/35054095.

Jayathilaka, N., Han, A., Gaffney, K. J., Dey, R., Jarusiewicz, J. A., Noridomi, K., ... Chen, L. (2012). Inhibition of the function of class IIa HDACs by blocking their interaction with MEF2. *Nucleic Acids Research, 40*(12), 5378–5388. https://doi.org/10.1093/nar/gks189.

Jin, H., Kasper, L. H., Larson, J. D., Wu, G., Baker, S. J., Zhang, J., & Fan, Y. (2020). ChIPseqSpikeInFree: A ChIP-seq normalization approach to reveal global changes in histone modifications without spike-in. *Bioinformatics (Oxford, England), 36*(4), 1270–1272. https://doi.org/10.1093/bioinformatics/btz720.

Joseph, F. M., & Young, N. L. (2023). Histone variant-specific post-translational modifications. *Seminars in Cell & Developmental Biology, 135*, 73–84. https://doi.org/10.1016/j.semcdb.2022.02.012.

Kaya-Okur, H. S., Wu, S. J., Codomo, C. A., Pledger, E. S., Bryson, T. D., Henikoff, J. G., ... Henikoff, S. (2019). CUT&Tag for efficient epigenomic profiling of small samples and single cells. *Nature Communications, 10*(1), 1930. https://doi.org/10.1038/s41467-019-09982-5.

Kidder, B. L., Hu, G., & Zhao, K. (2011). ChIP-Seq: Technical considerations for obtaining high-quality data. *Nature Immunology, 12*(10), 918–922. https://doi.org/10.1038/ni.2117.

Klemm, S. L., Shipony, Z., & Greenleaf, W. J. (2019). Chromatin accessibility and the regulatory epigenome. *Nature Reviews. Genetics, 20*(4), 207–220. https://doi.org/10.1038/s41576-018-0089-8.

Lee, T. I., Johnstone, S. E., & Young, R. A. (2006). Chromatin immunoprecipitation and microarray-based analysis of protein location. *Nature Protocols, 1*(2), 729–748. https://doi.org/10.1038/nprot.2006.98.

Love, M. I., Huber, W., & Anders, S. (2014). Moderated estimation of fold change and dispersion for RNA-seq data with DESeq2. *Genome Biology, 15*(12), 550. https://doi.org/10.1186/s13059-014-0550-8.

Luger, K., Mäder, A. W., Richmond, R. K., Sargent, D. F., & Richmond, T. J. (1997). Crystal structure of the nucleosome core particle at 2.8 Å resolution. *Nature, 389*(6648), 251–260. https://doi.org/10.1038/38444.

Luo, Y. N., Liang, Y., Wu, W., & Wang, L. (2026). Evolutionarily conserved regulation and functions of H1 linker histones in development. *Trends in Genetics*. https://doi.org/10.1016/j.tig.2025.12.003.

Massie, C.E., Mills, I.G. (2012). Mapping protein–DNA interactions using ChIP-sequencing. 157–173. https://doi.org/10.1007/978-1-61779-376-9_11.

Meyer, C. A., & Liu, X. S. (2014). Identifying and mitigating bias in next-generation sequencing methods for chromatin biology. *Nature Reviews. Genetics, 15*(11), 709–721. https://doi.org/10.1038/nrg3788.

Minisini, M., Di Giorgio, E., Kerschbamer, E., Dalla, E., Faggiani, M., Franforte, E., ... Brancolini, C. (2022). Transcriptomic and genomic studies classify NKL54 as a histone deacetylase inhibitor with indirect influence on MEF2-dependent transcription. *Nucleic Acids Research, 50*(5), 2566–2586. https://doi.org/10.1093/nar/gkac081.

Nair, N. U., Sahu, A. Das, Bucher, P., & Moret, B. M. E. (2012). ChIPnorm: A statistical method for normalizing and identifying differential regions in histone modification ChIP-seq libraries. *PLoS One, 7*(8), e39573. https://doi.org/10.1371/journal.pone.0039573.

O'Neill, L. (2003). Immunoprecipitation of native chromatin: NChIP. *Methods (San Diego, Calif.), 31*(1), 76–82. https://doi.org/10.1016/S1046-2023(03)00090-2.

Orlando, D. A., Chen, M. W., Brown, V. E., Solanki, S., Choi, Y. J., Olson, E. R., ... Guenther, M. G. (2014). Quantitative ChIP-Seq normalization reveals global modulation of the epigenome. *Cell Reports, 9*(3), 1163–1170. https://doi.org/10.1016/j.celrep.2014.10.018.

Park, P. J. (2009). ChIP–seq: Advantages and challenges of a maturing technology. *Nature Reviews. Genetics, 10*(10), 669–680. https://doi.org/10.1038/nrg2641.

Patel, L. A., Cao, Y., Mendenhall, E. M., Benner, C., & Goren, A. (2024). The wild west of spike-in normalization. *Nature Biotechnology, 42*(9), 1343–1349. https://doi.org/10.1038/s41587-024-02377-y.

Patel, L., Cao, Y., Xu, T., Modolo, E., Dishon, T., Zhang, L., ... Goren, A. (2025). Improved spike-in normalization clarifies the relationship between active histone modifications and transcription. *bioRxiv*. https://doi.org/10.1101/2025.11.25.690627.

Polit, L., Kerdivel, G., Gregoricchio, S., Esposito, M., Guillouf, C., & Boeva, V. (2021). CHIPIN: ChIP-seq inter-sample normalization based on signal invariance across transcriptionally constant genes. *BMC Bioinformatics, 22*(1), 407. https://doi.org/10.1186/s12859-021-04320-3.

Ramírez, F., Ryan, D. P., Grüning, B., Bhardwaj, V., Kilpert, F., Richter, A. S., ... Manke, T. (2016). DeepTools2: A next generation web server for deep-sequencing data analysis. *Nucleic Acids Research, 44*(W1), W160–W165. https://doi.org/10.1093/nar/gkw257.

Robinson, M. D., McCarthy, D. J., & Smyth, G. K. (2010). EdgeR: A bioconductor package for differential expression analysis of digital gene expression data. *Bioinformatics (Oxford, England), 26*(1), 139–140. https://doi.org/10.1093/bioinformatics/btp616.

Skene, P. J., & Henikoff, S. (2017). An efficient targeted nuclease strategy for high-resolution mapping of DNA binding sites. *ELife, 6*. https://doi.org/10.7554/eLife.21856.

R. Stark G. Brown DiffBind differential binding analysis of ChIP-Seq peak data. R Package Version 2011. https://bioconductor.org/packages/release/bioc/vignettes/DiffBind/inst/doc/DiffBind.pdf.

Steinhauser, S., Kurzawa, N., Eils, R., & Herrmann, C. (2016). A comprehensive comparison of tools for differential ChIP-seq analysis. *Briefings in Bioinformatics*, bbv110. https://doi.org/10.1093/bib/bbv110.

Villar-Garea, A., & Imhof, A. (2006). The analysis of histone modifications. *Biochimica et Biophysica Acta (BBA) - Proteins and Proteomics, 1764*(12), 1932–1939. https://doi.org/10.1016/j.bbapap.2006.08.009.

Wheeler, T. J., Clements, J., Eddy, S. R., Hubley, R., Jones, T. A., Jurka, J., ... Finn, R. D. (2012). Dfam: A database of repetitive DNA based on profile hidden Markov models. *Nucleic Acids Research, 41*(D1), D70–D82. https://doi.org/10.1093/nar/gks1265.

Wu, D., Wang, L., & Huang, H. (2021). Protocol to apply spike-in ChIP-seq to capture massive histone acetylation in human cells. *STAR Protocols, 2*(3), 100681. https://doi.org/10.1016/j.xpro.2021.100681.

Zhang, T., Zhang, Z., Dong, Q., Xiong, J., & Zhu, B. (2020). Histone H3K27 acetylation is dispensable for enhancer activity in mouse embryonic stem cells. *Genome Biology, 21*(1), 45. https://doi.org/10.1186/s13059-020-01957-w.

CHAPTER TWO

Methods for the kinetic analysis of time-dependent HDAC6 inhibitors

Edoardo Cellupica*, **Gianluca Fossati, Christian Steinkühler, and Gianluca Caprini**
Italfarmaco Group, New Drug Incubator Department, Cinisello Balsamo, Italy
*Corresponding author. e-mail address: e.cellupica@italfarmacogroup.com

Contents

1. Introduction 18
2. HDAC6 Activity assay 19
   2.1 Reagents 20
   2.2 Buffers and solutions 21
   2.3 Equipment 21
3. Assay setup 22
   3.1 Procedure for progress-curve analysis 22
   3.2 Procedure for jump dilution assay 23
4. Data analysis 24
   4.1 Progress-curve analysis 24
   4.2 IC_{50} determination 25
   4.3 Mechanism interpretation based on k_{obs} 25
   4.4 Jump dilution assay analysis 25
5. Representative results 26
6. Notes 28
7. Summary 29
Appendix A. Control for compound-dependent quenching/enhancement (deacetylated product standard ± compound) 29
References 30

Abstract

The class IIb histone deacetylase HDAC6 is responsible for the deacetylation of multiple cytoplasmic substrates, including α-tubulin and Miro-1, and has emerged as an important therapeutic target in neuropathies. In this context, several HDAC6 inhibitors effective in neuropathic models, exhibit slow-binding and slow-release behavior, thereby requiring kinetic approaches that extend beyond commonly used endpoint assays, which can lead to misinterpretation of inhibitor potency and mechanism of action. Notably, compounds with slow dissociation kinetics may offer improved therapeutic properties in neuropathic conditions, as sustained target

Methods in Enzymology, Volume 733
ISSN 0076-6879, https://doi.org/10.1016/bs.mie.2026.05.038

engagement can translate into prolonged improvement of axonal transport and cytoskeletal dynamics. In this chapter, we outline a comprehensive methodology for the study of HDAC6 inhibition and the identification of slow-binding and slow-dissociating inhibitors.

1. Introduction

Within the Zn^{2+}-dependent histone deacetylase family, HDAC6 is a structurally and mechanistically distinct enzyme acting mainly in the cytoplasm (Dompierre et al., 2007; Haggarty et al., 2003). It contains two catalytic domains (CD1 and CD2) and, unlike class I HDACs, primarily deacetylates non-histone substrates such as α-tubulin and Miro-1 (Dompierre et al., 2007; Haggarty et al., 2003; Kalinski et al., 2019). Through the activity on these two substrates, HDAC6 directly links deacetylation to microtubule-dependent processes, including axonal transport and mitochondrial trafficking (Dompierre et al., 2007; Kalinski et al., 2019).

From a mechanistic enzymology perspective, HDAC6 inhibitors often display slow-binding behavior (Cellupica et al., 2022, 2024; König et al., 2023; Motlová et al., 2023). Time-dependent inhibition requires explicit kinetic treatment to correctly define inhibitor potency and mechanism of action (Cellupica et al., 2023; Ehmann et al., 2012; Knockenhauer & Copeland, 2023; Madsen & Olsen, 2016). Accordingly, rigorous characterization of HDAC6 inhibitors requires a mechanistic enzymology framework rather than reliance on endpoint assays alone, since the latter can lead to substantially misleading structure–activity relationships (SAR) and may result in the inadvertent exclusion of promising inhibitor candidates (Cellupica et al., 2023). Time-course experiments are essential to identify slow-binding behavior and to extract kinetically meaningful parameters, including residence time and apparent kinetic rate constants (Ehmann et al., 2012; Knockenhauer & Copeland, 2023; Madsen & Olsen, 2016).

Finally, from a translational perspective, HDAC6 inhibition has been extensively investigated in models of chemotherapy-induced peripheral neuropathy (CIPN), where inhibition of HDAC6 enzymatic activity has been associated with preservation of axonal integrity and neuronal function (Krukowski et al., 2017). In paclitaxel-induced peripheral neuropathy, the DFMO-based selective HDAC6 inhibitor ITF6475 showed neuroprotective effects, preserving peripheral nerve function and structure in addition to reducing mechanical allodynia (Cavaletti et al., 2025). Protective effects were also observed in a cisplatin-induced peripheral neurotoxicity model (Cavaletti et al., 2026). These findings support the continued exploration of HDAC6 as a therapeutic target in disease contexts where sustained regulation of cytoskeletal dynamics and intracellular transport is beneficial by using long-acting inhibitors with prolonged residence time, including difluoromethyl-1,3,4-oxadiazoles (DFMOs) (Cellupica et al., 2022, 2024; König et al., 2023; Motlová et al., 2023).

In this work, ITF6475 is used as a representative HDAC6 inhibitor to illustrate kinetic analysis workflows under conditions that capture time-dependent inhibition.

In summary, HDAC6 and DFMOs constitute a representative enzymatic system for studying time-dependent inhibition and residence time effects (Cellupica et al., 2022, 2024; Ehmann et al., 2012; Knockenhauer & Copeland, 2023; König et al., 2023; Madsen & Olsen, 2016; Motlová et al., 2023). The methodologies described in this chapter provide a framework for rigorous kinetic characterization of HDAC6 inhibitors, enabling mechanistic discrimination of binding modes and supporting the development of compounds with well-defined kinetic profiles (Ehmann et al., 2012; Knockenhauer & Copeland, 2023; Madsen & Olsen, 2016).

2. HDAC6 Activity assay

Fluor-de-Lys Green was chosen as the commercial synthetic fluorogenic substrate to monitor HDAC activity. The two-step deacetylase activity assay is based on HDAC-mediated deacetylation of lysine residues, which renders the amide bond susceptible to trypsin cleavage, resulting in the release of a fluorophore (Fig. 1). Therefore, the deacetylase activity can be readily detected as an increase of intensity of the 535 nm light emitted upon excitation with 485 nm light.

Other coumarin-based substrates are commercially available and have been widely used in the past, but Fluor de Lys Green is preferred because it emits in the green range, reducing interference from compound autofluorescence and improving the signal-to-noise ratio compared with traditional blue-emitting coumarin-based substrates.

The assay is discontinuous and can be performed either by measuring product formation at a fixed time point or by monitoring its time course. The

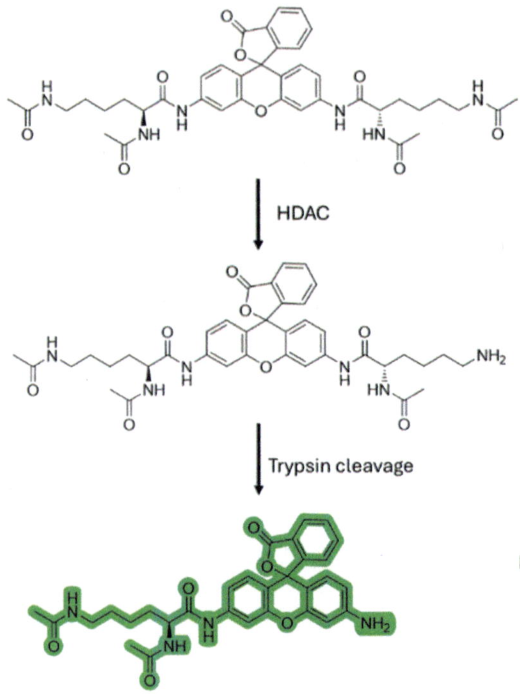

Fig. 1 Reaction scheme of the Fluor de Lys Green-based HDAC activity assay and formation of the rhodamine 110 dye indicator to quantify deacetylase activity.

latter approach, which is the one described in this work, can reveal slow-binding and/or slow-dissociating inhibitors that might otherwise be missed in fixed-time formats, thus preventing the generation of misleading data about inhibitor potency, stoichiometry and mechanism of action. Moreover, the time-course format allows measurements of the kinetic rate constants.

The time-course assay described in this study can be readily applied to other HDAC isozymes with minor modifications to account for differences in substrate concentration, additives, and buffer composition.

2.1 Reagents

- Tris base (Sigma-Aldrich, T4661)
- HCl (Supelco, 1.09057)
- NaCl (Sigma-Aldrich, 71376)
- KCl (Sigma-Aldrich, P9333)
- $MgCl_2$ (Sigma-Aldrich, M4880)
- Tween−20 (Sigma-Aldrich, P9416)
- Glycerol (EMD Millipore, 356352)

- Bovine serum albumin (BSA; Sigma-Aldrich, A7030)
- Tris(2-carboxyethyl)phosphine (TCEP; Sigma-Aldrich, C4706)
- DMSO (Sigma-Aldrich, D2650)
- Fluor de Lys Green substrate (Enzo Life Sciences, BML-KI572)
  - used from a concentrated stock in DMSO (see "Buffers and solutions")
- Fluor de Lys Developer I (Enzo Life Sciences, BML-KI105)
- Trichostatin A (TSA; Sigma-Aldrich, T8552), 2 mM in DMSO
- Test inhibitor: ITF6475 (20 mM stock solution in DMSO) (Cavaletti et al., 2025)

2.2 Buffers and solutions

All solutions are prepared using ultrapure water and analytical grade reagents.

- **Assay Buffer 1 (AB1).** 50 mM Tris, 130 mM NaCl, 0.05% Tween−20, and 10 % glycerol. The pH was adjusted to 8.0 with 1 M HCl, and the buffer was filtered and stored in a Pyrex bottle at 4 °C. Immediately before use, supplement with 0.5 mM TCEP from stock solutions prepared in water (500 mM) and 1 mg/mL BSA in AB1 (20 mg/mL).
- **Assay Buffer 2 (AB2).** 50 mM Tris/HCl, pH 8.0, 137 mM NaCl, 2.7 mM KCl, and 1 mM MgCl2.
- **Stop/developing solution (S/D).** Prepared by diluting Fluor de Lys Green Developer I 200-fold in AB2 supplemented with 2 μM TSA (freshly added from a 2 mM stock solution in DMSO).
- **Substrate.** Fluor de Lys Green substrate: concentrated stock in DMSO (examples used: 200x stock for time-course inhibition assays; 50 mM stock for jump dilution)
- **Enzyme.** Human GST-HDAC6 (full-length, N-terminal GST) (BPS Bioscience, Cat. 50006). Note: The enzyme should be diluted to prepare an enzyme working solution in AB1 at a concentration of 10−50 ng/μL, allowing convenient aliquot volumes to be added to the activity assays. This working solution is stable at −80 °C and may be stored for future experiments after being aliquoted into small volumes to allow only 2–3 freeze-thaw cycles.

2.3 Equipment

- 96-well flat-bottom, white half-area plates (PerkinElmer, 6005569)
- Microcentrifuge tubes (1.5–2.0 mL)
- Pipettors and electronic repeater (e.g., Eppendorf Repeater E3 multipette)
- Temperature-controlled incubator

- Plate reader suitable for excitation at 485 nm and emission fluorescence measurement at535 nm (e.g., Wallac 1420 Victor2/VICTOR Multilabel Plate Reader)
- Software for curve fitting and IC_{50} analysis (e.g. GraphPad Prism v10.2.2)

3. Assay setup

ITF6475 was used as a representative compound for the analyses described below.

3.1 Procedure for progress-curve analysis

The appropriate amounts of enzyme and substrate should be determined in a series of preliminary assays (see Notes 1 and 2):

- HDAC6 final concentration: 190 pM
- Fluor de Lys Green substrate final concentration: 3 μM (close to K_m: 3.4 μM under these experimental conditions)

Procedure:

1. Prepare complete AB1 sufficiently in advance to equilibrate to 25 °C.
2. For each condition, aliquot substrate from 200x stock in DMSO (600 μM) into 1.5 mL test tube, using 2.5 μL for a final reaction volume of 500 μL or 5 μL for a final reaction volume of 1000 μL.
3. Add inhibitor from 200x stocks in DMSO to each vial (add DMSO only for no-inhibitor controls). 2.5 μL for a final reaction volume of 500 μL or 5 μL for a final reaction volume of 1000 μL. Mix (vortex briefly) and spin down. Incubate at 25 °C.
4. In parallel, dilute enzyme working solution into AB1 to reach the desired final catalytic concentration upon mixing. Equilibrate enzyme solution 10–15 min at 25 °C.
5. Prepare a blank sample containing all components of the reaction mixture except the enzyme and the inhibitor. The enzyme does not contribute to the background signal, whereas the inhibitor (particularly at high concentrations) may, in some cases, interfere with the fluorescent signal (see Appendix A).

   Reaction initiation and sampling:

1. At time zero, start the reaction by adding the enzyme solution (or AB1 for blank sample) to each test tube containing substrate plus inhibitor or DMSO. 495 μL for a final reaction volume of 500 μL or 990 μL for a

final reaction volume of 1000 µL. Final DMSO concentration is equal to 1%. Using electronic pipettes facilitates enzyme addition. Mixing is performed by gently inverting the tube up and down.

2. At various reaction times, withdraw 110 µL or 60 µL from each reaction mix.
3. Quench/develop by mixing each aliquot with an equal volume of Stop/Developer solution.
4. Incubate quenched samples at room temperature for 30 min in the dark.
5. Transfer 100 µL of each developed sample in duplicate or singlicate to the plate.
6. Read fluorescence emission at 535 nm upon excitation at 485 nm.

3.2 Procedure for jump dilution assay

The jump dilution assay is a rapid, large-fold dilution of a pre-equilibrated enzyme–inhibitor complex formed under saturating inhibitor conditions. The dilution sharply reduces the inhibitor concentration to a level at which full recovery of enzyme activity is expected if the inhibition is reversible. Monitoring activity recovery over time provides insight into inhibitor reversibility, dissociation kinetics (k_{off}) and residence time.

Preparation of working solutions:

- Pre-incubation mixes were prepared to a final DMSO concentration of 1 % by mixing enzyme in AB1 with an equal volume of AB1 containing 2% DMSO (±inhibitor).
- Substrate solution for jump dilution step: 50 µM Fluor de Lys Green in AB1, 1 % DMSO, equilibrated at 25 °C.
- HDAC6 working solution:15 ng/µL GST-HDAC6 in AB1. After 500x jump dilution, final enzyme concentration is 15 pg/µL (93 pM).
- Inhibitor working solutions: ITF6475 25 µM in DMSO; then diluted to 500 nM in AB1 (final DMSO 2 %) to generate 250 nM inhibitor in the 6 µL pre-incubation mix.

Enzyme–inhibitor pre-incubation:

1. Prepare pre-incubation mixes (total 6 µL each) in AB1 supplemented with final 1 % DMSO:
   - Control (no inhibitor): 3 µL HDAC6 (15 ng/µL) + 3 µL AB1 with 2 % DMSO
   - Inhibitor (ITF6475): 3 µL HDAC6 (15 ng/µL) + 3 µL inhibitor (in the case of ITF6475 we use 500 nM → 250 nM in pre-incubation)
   - Blank (no enzyme): 3 µL AB1 + 3 µL AB1 with 2% DMSO

2. Incubate pre-incubation mixtures for 60 min at 25 °C.
3. Transfer 3 μL of each pre-incubation mixture into a fresh 2 mL tube (to be used for the jump dilution step).

   Jump dilution and initiation of post-jump reaction:

1. Initiate the reaction by adding 1500 μL of substrate solution (50 μM Fluor de Lys Green) pre-equilibrated at 25 °C to 3 μL of each pre-incubation mixture (500 × dilution).
2. The post-jump reaction proceeds under substrate-saturating conditions ($S \gg K_m$) and at inhibitor concentrations expected to allow near-complete recovery of activity at steady state. Final conditions after jump dilution (as reported):
   - 1 % DMSO in AB1.
   - 50 μM Fluor de Lys Green substrate.
   - 15 pg/μL GST-HDAC6 (93 pM).
   - 0 or 0.5 nM ITF6475.

   Sampling, quenching/development, and readout:

1. Stop reactions at 10, 20, 30, 60, 90, 120, 150, 210, 240, 270, 300, 330, and 360 min by withdrawing 110 μL and mixing 1:1 with S/D solution.
2. Transfer 100 μL to a 96-well plate in duplicate.
3. Read fluorescence after 30 min development (Ex 485 nm/Em 535 nm).

4. Data analysis

4.1 Progress-curve analysis

The background-corrected fluorescence values (proportional to product formed) are plotted versus reaction time (progress curves).

- Fast-binding inhibitors: product formation is linear; determine the initial velocity (v_i) as ΔF/min by interpolating the experimental values in the linear part.
- Slow-binding inhibitors: the reaction transitions from an initial velocity (v_i) to a steady-state velocity (v_s). Fit the full progress curve to Eq. 1 to obtain v_i, v_s, and k_{obs}:

$$P = v_s \cdot t + \frac{(v_i - v_s)\left[1 - e^{-k_{obs} \cdot t}\right]}{k_{obs}} \tag{1}$$

where P is the product concentration (here expressed as fluorescence); v_i is the initial reaction velocity (as ΔF/min); v_s is the steady-state velocity (ΔF/min); t is the reaction time (min); k_{obs} (min^{-1}) is the observed rate constant for conversion from the initial velocity (of the free enzyme) to the steady-state velocity when equilibrium is reached. When the initial velocity values (v_i) appeared to be similar to that of the enzyme in the absence of inhibitor, the initial velocity (v_i) was often set to the value obtained in control reactions in the absence of inhibitor to improve the fitting quality.

4.2 IC_{50} determination

The initial velocity (v_i) for fast-on inhibitors or the steady-state velocity (v_s) for slow-on inhibitors was used to calculate the residual activity as a percentage relative to the value obtained in the absence of inhibitor. Such percental residual activity (y) as a function of inhibitor concentration (x) was fitted with Eq. 2 to obtain the IC_{50} value and the Hill coefficient (h):

$$y = \frac{100}{1 + \left(\frac{x}{IC_{50}}\right)^h} \tag{2}$$

4.3 Mechanism interpretation based on k_{obs}

Plot the calculated k_{obs} versus inhibitor concentration and fit to:

- One-step mechanism (apparent linear dependence):

$$k_{obs} = k_{on} \cdot [I] + k_{off} \tag{3}$$

The linear fit of the k_{obs} plot provides estimates of the kinetic rate constants: the dissociation rate constant (k_{off}), obtained from the y-intercept, and the apparent association rate constant (k_{on}), derived from the slope.

- Two-step mechanism (hyperbolic dependence):

$$k_{obs} = k_{off} + \frac{k_{on} \cdot [I]}{K_i + [I]} \tag{4}$$

Should the k_{off} value be too small, as observed for very slowly dissociating inhibitors (e.g., DFMO compounds), it cannot be accurately determined from the y-intercept of these plots because it is indistinguishable from zero. In such cases, the jump dilution assay is the most appropriate method for determining the k_{off} value.

4.4 Jump dilution assay analysis

The background-corrected fluorescence versus time data obtained after jump dilution are used to quantify recovery of HDAC6 activity upon rapid dilution of inhibitor concentration. If the inhibitor dissociates rapidly, the progress curve should be linear. If the inhibitor dissociates slowly, the recovery of activity will be curvilinear and data are fit using Eq. 1. For jump dilutions, the k_{obs} value describes the transition from the initial inhibited reaction to the steady-state and, under conditions of minimal inhibitor rebinding, this rate constant reflects the k_{off} value. For very slow-dissociating compounds, it is often not feasible to measure full recovery of activity within a reaction time frame that preserves enzyme stability. In such cases, the steady-state rate (v_s) can be fixed to the value obtained from control reactions in the absence of inhibitor, thereby improving the quality of the fit and enabling a more accurate estimation of the dissociation rate constant (k_{off}). The residence time (RT) is defined as the reciprocal of k_{off}.

5. Representative results

Using the approach and methodology described above, the results obtained with the DFMO-based inhibitor ITF6475 are detailed below.

Progress-curve analysis revealed a time-dependent inhibition (Fig. 2, left), consistent with a slow onset of inhibition. A replot of the calculated k_{obs} as a function of [ITF6475] yielded a straight line with slope corresponding to a k_{on} value equal to $4.17 \pm 0.02 \times 10^5$ M^{-1} min^{-1} and y-intercept corresponding to k_{off}, which was not distinguishable from zero (Fig. 2, right). The linear plot supports a simple one-step slow-binding mechanism, although it does not rule out the possibility of a kinetically unobservable induced-fit process.

As anticipated from k_{off} values approaching zero in the kinetic analysis, ITF6475 exhibited very slow recovery of activity following jump dilution (Fig. 3), consistent with extremely slow inhibitor dissociation. Kinetic analysis yielded a k_{off} of $0.79 \pm 0.06 \times 10^{-3}$ min^{-1}, corresponding to a residence time of 1264 min (~21.1 h).

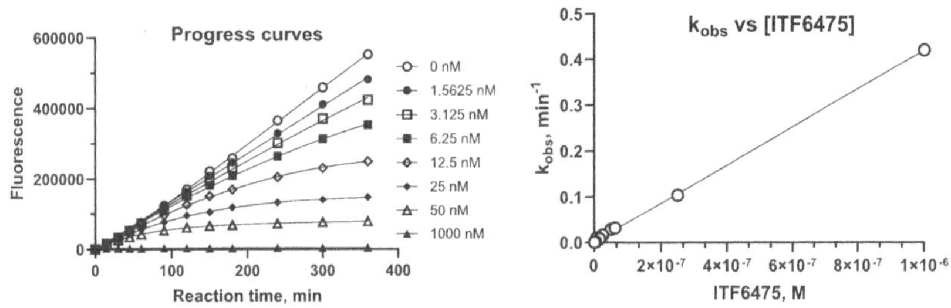

Fig. 2 The time-dependent inhibition of HDAC6 by ITF6475. Left: Representative reaction progress curves for HDAC6 in the presence of increasing concentrations of ITF6475, fitted using Eq. 1. **Right:** Replot of the corresponding k_{obs} values for HDAC6 inhibition. The experimental dataset includes the values derived from the left panel together with additional measurements not shown in the figure.

Finally, because inhibition is time-dependent, potency was quantified using steady-state velocities (v_s) derived from progress-curve fitting, yielding an IC_{50} of 4.3 nM for ITF6475 (Fig. 4). Assuming a competitive inhibition model, this corresponds to a K_i of 2.3 nM, which is in excellent agreement with the K_i calculated from kinetic parameters as the k_{off}/k_{on} ratio (1.9 nM).

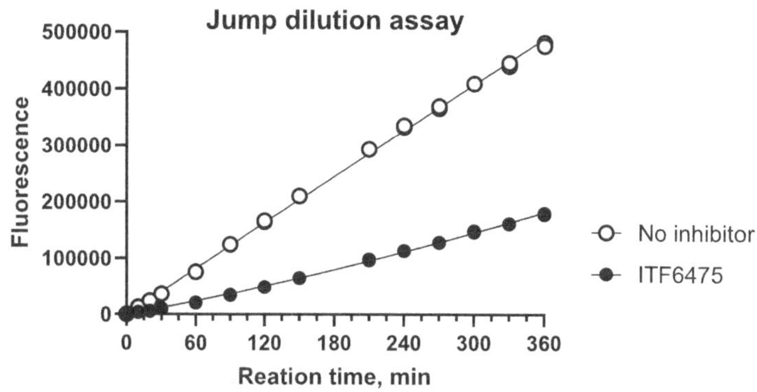

Fig. 3 Jump-dilution assay supports long residence time of ITF6475 on HDAC6. HDAC6 was incubated in the absence (○) or in the presence of 250 nM ITF6475 (●) for 60 min. The solutions were then rapidly diluted 500-fold by addition of 50 μM Fluor-de-Lys Green substrate in assay buffer.

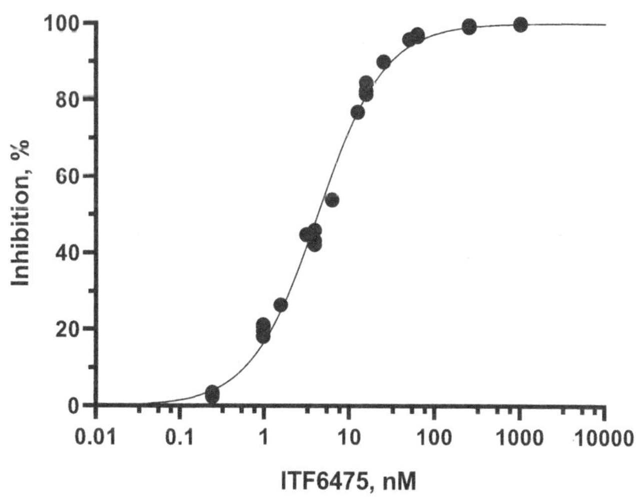

Fig. 4 ITF6475 is a potent HDAC6 inhibitor. Percent of inhibition was calculated from steady-state velocities (v_s) extracted by nonlinear regression of ITF6475 progress curves (Fig. 2), normalized to the no-inhibitor control and fitted to Eq. 2. The experimental dataset includes the values derived from experiment in Fig. 2 together with additional measurements not shown.

6. Notes

1. The amount of enzyme used in the assay at a given substrate concentration depends on the specific activity of the enzyme batch. The effect of enzyme titration on the observed reaction velocity should therefore be determined experimentally for a specific batch, and the enzyme concentration used for inhibition studies should be selected within the linear portion of the curve, while still providing a robust fluorescence signal.
2. For inhibition studies, the Fluor de Lys Green substrate was used at a concentration close to K_m whenever feasible. In order to derive dissociation constants from IC_{50} values, the K_m value must be known. Where the inhibition mechanism is consistent with a competitive model, cross-isozyme comparisons may be facilitated by converting IC_{50} values to K_i using the Cheng–Prusoff relationship: $K_i = \frac{IC_{50}}{1 + [S] / K_m}$.
3. The presence of both TCEP and BSA in the assay mixture is essential to maintain HDAC6 stability for up to 6−7 h during the reaction, which is the typical reaction time required to detect very slow-binding inhibitors such as DFMO compounds.
4. Strict temperature control at 25 °C must be maintained throughout pre-incubation, reaction, and sampling.
5. Classify inhibitors as fast- vs slow-binding from progress-curve shape before choosing v_i or v_s for downstream IC_{50} analysis.
6. Sampling time-point spacing should be in sufficient number to resolve curvature for slow-binding behavior and to define the true linear window (where applicable).
7. For jump dilution assays, ensure the dilution factor (500 × used here) reduces inhibitors to a concentration expected to yield near-complete recovery at steady state.
8. Developer/stop solution. Prepare the developer dilution fresh and protect developed samples from light during incubation. The optimal developer concentration may need to be adjusted depending on the fluorogenic substrate and signal range (e.g., higher developer concentration may be required for some substrates).
9. Optical interference (compound autofluorescence, quenching, inner-filter effects) is common in screening or early SAR, where compounds are often tested at higher concentrations, and can bias apparent inhibition in fluorogenic HDAC assays; see Appendix A for a detailed control protocol using a deacetylated-product standard under assay-matched readout conditions.

7. Summary

HDAC6 enzymatic inhibition is well known to exert neuroprotective effects, largely through the preservation of axonal cytoskeletal integrity and related transport processes. Furthermore, HDAC6 KO mice are more refractory to CIPN (Ma et al., 2019). In this context, an extended target residence time may underline durable neuroprotective activity in models of chemotherapy-induced peripheral neuropathy, including paclitaxel-evoked painful neuropathy, in which ITF6475 — the compound used in this work — has demonstrated in vivo efficacy (Cavaletti et al., 2025, 2026). Overall, the methodologies described herein provide a robust framework for the kinetic characterization of HDAC6 inhibitors and are expected to be broadly applicable, supporting the identification and optimization of compounds with favorable kinetic profiles and enhanced therapeutic potential in neuropathic conditions.

Appendix A. Control for compound-dependent quenching/enhancement (deacetylated product standard ± compound)

Purpose.

To quantify compound-dependent optical interference in the Fluor de Lys Green readout (fluorescence quenching or enhancement) by measuring the impact of test compounds on a deacetylated-product standard curve under assay-matched readout conditions. This control is recommended for early screening/SAR, where compounds are often tested at higher concentrations (up to 1 mM); it was not required for the representative ITF6475 datasets shown in this chapter.

Reagents and solutions.

- Deacetylated product standard: FLUOR DE LYS®-Green Standard (Enzo Life Science BML-KI605; 1 mM in DMSO).
- Standard working stocks (100× in DMSO): prepare 2.5, 5, 10 μM (100×). Final concentrations in the 100 μL mix: 25, 50, 100 nM, respectively.
- Test compounds: prepare 100× stocks in DMSO.
- AB1: as defined in the main Methods (final DMSO in the assay-matched simulation: 2%).
- Stop/Developer solution (S/D): as defined in the main Methods.

Procedure.

1. For each compound (or compound concentration), label three microcentrifuge tubes (one per standard level: 2.5, 5, 10 μM 100×). Prepare matched DMSO-only control tubes.
2. Add 1 μL of 100 × compound stock to each tube (or 1 μL DMSO for controls).
3. Add 1 μL of the corresponding 100 × deacetylated-product standard (2.5, 5, or 10 μM).
4. Add 98 μL of AB1 pre-equilibrated at 25 °C (final volume 100 μL; final DMSO 2%). Mix gently.
5. Incubate 45 min at 25 °C.

6. Add an equal volume of Stop/Developer solution (1:1, v/v). Mix.
7. Incubate 30 min at room temperature in the dark.
8. Transfer 95 μL of each sample in duplicate to a 96-well plate and read fluorescence (Ex 485 nm/Em 535 nm).

Data analysis

1. Plot fluorescence versus final deacetylated-product concentration (25, 50, 100 nM).
2. Fit each condition by linear regression to obtain slopes:
   - $Slope_{DMSO}$ (vehicle control)
   - $Slope_{I}$ (compound condition)
3. Calculate the slope ratio: $R = \frac{Slope_I}{Slope_{DMSO}}$
4. Report quenching or enhancement:
   - If $R < 1$: % quenching = $100 \times (1 - R)$
   - If $R > 1$: % enhancement = $100 \times (R-1)$

Interpretation/acceptance criteria.

- A significant change in standard-curve slope indicates compound-dependent optical interference affecting the fluorescence readout. We suggest using the slope ratio R as a calibration factor according to the following quenching: ($R < 1$); enhancement: ($R > 1$).
- Equivalent "percent" adjustment. If inhibition is calculated directly as a percentage from fluorescence and a single quenching/enhancement value is used, the measured percent inhibition can be adjusted by subtracting the quenching contribution (or subtracting a negative value in case of enhancement), consistent with the sign convention above.
- Acceptance criteria. Corrections should only be applied when (i) the standard curve remains linear in the relevant range, and (ii) the interference metric is consistent across the assay window; otherwise, compounds should be flagged for follow-up using an orthogonal readout.

References

Cavaletti, G., Canta, A., Chiorazzi, A., Ballarini, E., Rodriguez, V., Scuteri, A., ... Licandro, S. A. (2026). Protective effect of new histone deacetylase 6 inhibitors in a cisplatin-induced peripheral neurotoxicity murine model. *PAIN Reports, 11*(2), e1395. https://doi.org/10.1097/PR9.0000000000001395.

Cavaletti, G., Canta, A., Chiorazzi, A., Pozzi, E., Carozzi, V., Meregalli, C., ... Licandro, S. A. (2025). ITF6475, a new histone deacetylase 6 inhibitor, prevents painful neuropathy induced by paclitaxel. *Toxics, 13*(9), 767. https://doi.org/10.3390/toxics13090767.

Cellupica, E., Caprini, G., Cordella, P., Cukier, C., Fossati, G., Marchini, M., ... Steinkühler, C. (2022). Difluoromethyl-1,3,4-oxadiazoles are slow-binding substrate analog inhibitors of histone deacetylase 6 with unprecedented isotype selectivity. *The Journal of Biological Chemistry, 299*(1), 102800. https://doi.org/10.1016/j.jbc.2022.102800.

Cellupica, E., Caprini, G., Fossati, G., Mirdita, D., Cordella, P., Marchini, M., ... Vanoni, M. A. (2023). The importance of the "time factor" for the evaluation of inhibition mechanisms: The case of selected HDAC6 inhibitors. *Biology (Basel), 12*(8), 1049. https://doi.org/10.3390/biology12081049.

Cellupica, E., Gaiassi, A., Rocchio, I., Rovelli, G., Pomarico, R., Sandrone, G., ... Vergani, B. (2024). Mechanistic and structural insights on difluoromethyl-1,3,4-oxadiazole inhibitors of HDAC6. *International Journal of Molecular Sciences, 25*(11), 5885. https://doi.org/10.3390/ijms25115885.

Dompierre, J. P., Godin, J. D., Charrin, B. C., Cordelières, F. P., King, S. J., Humbert, S., & Saudou, F. (2007). Histone deacetylase 6 inhibition compensates for the transport deficit in Huntington's disease by increasing tubulin acetylation. *The Journal of Neuroscience, 27*(13), 3571–3583. https://doi.org/10.1523/JNEUROSCI.0037-07.2007.

Ehmann, D. E., Jahić, H., Ross, P. L., Gu, R. F., Hu, J., Kern, G., ... Fisher, S. L. (2012). Avibactam is a covalent, reversible, non-β-lactam β-lactamase inhibitor. *Proceedings of the National Academy of Sciences of the United States of America, 109*(29), 11663–11668. https://doi.org/10.1073/pnas.1205073109.

Haggarty, S. J., Koeller, K. M., Wong, J. C., Grozinger, C. M., & Schreiber, S. L. (2003). Domain-selective small-molecule inhibitor of histone deacetylase 6 (HDAC6)-mediated tubulin deacetylation. *Proceedings of the National Academy of Sciences of the United States of America, 100*(8), 4389–4394. https://doi.org/10.1073/pnas.0430973100.

Kalinski, A. L., Kar, A. N., Craver, J., Tosolini, A. P., Sleigh, J. N., Lee, S. J., ... Twiss, J. L. (2019). Deacetylation of Miro1 by HDAC6 blocks mitochondrial transport and mediates axon growth inhibition. *The Journal of Cell Biology, 218*(6), 1871–1890. https://doi.org/10.1083/jcb.201702187.

Knockenhauer, K. E., & Copeland, R. A. (2023). The importance of binding kinetics and drug-target residence time in pharmacology. *British Journal of Pharmacology, 181*(21), 4103–4116. https://doi.org/10.1111/bph.16104.

König, B., Watson, P. R., Reßing, N., Cragin, A. D., Schäker-Hübner, L., Christianson, D. W., & Hansen, F. K. (2023). Difluoromethyl-1,3,4-oxadiazoles are selective, mechanism-based, and essentially irreversible inhibitors of histone deacetylase 6. *Journal of Medicinal Chemistry, 66*(19), 13821–13837. https://doi.org/10.1021/acs.jmedchem.3c01345.

Krukowski, K., Ma, J., Golonzhka, O., Laumet, G. O., Gutti, T., van Duzer, J. H., ... Kavelaars, A. (2017). HDAC6 inhibition effectively reverses chemotherapy-induced peripheral neuropathy. *Pain, 158*(6), 1126–1137. https://doi.org/10.1097/j.pain.0000000000000893.

Ma, J., Trinh, R. T., Mahant, I. D., Peng, B., Matthias, P., Heijnen, C. J., & Kavelaars, A. (2019). Cell-specific role of histone deacetylase 6 in chemotherapy-induced mechanical allodynia and loss of intraepidermal nerve fibers. *Pain, 160*(12), 2877–2890. https://doi.org/10.1097/j.pain.0000000000001667.

Madsen, A. S., & Olsen, C. A. (2016). A potent trifluoromethyl ketone histone deacetylase inhibitor exhibits class-dependent mechanism of action. *MedChemComm, 7*(5), 823–831. https://doi.org/10.1039/c5md00451a.

Motlová, L., Šnajdr, I., Kutil, Z., Andris, E., Ptáček, J., Novotná, A., ... Bařinka, C. (2023). Comprehensive mechanistic view of the hydrolysis of oxadiazole-based inhibitors by histone deacetylase 6 (HDAC6). *ACS Chemical Biology, 18*(7), 1594–1610. https://doi.org/10.1021/acschembio.3c00212.

CHAPTER THREE

Synthesis and functional characterisation of small molecule activator of p300/CBP lysine acetyltransferase activity: Implications in spinal injury repair

Amrish Rai[a,b], Prajjval Mishra[c], Akash Kumar Singh[c,d], Priyanka Pandey[c], and Tapas K. Kundu[c,d,*]

[a]Division of Neuroscience and Ageing Biology, CSIR-Central Drug Research Institute, Lucknow, India
[b]Academy of Scientific and Innovative Research (AcSIR), Ghaziabad, India
[c]Molecular Biology and Genetics Unit, Transcription and Disease Laboratory, Jawaharlal Nehru Centre for Advanced Scientific Research, Jakkur, Bengaluru, India
[d]Neuroscience Unit, Jawaharlal Nehru Centre for Advanced Scientific Research, Bangalore, India
*Corresponding author. e-mail address: tapas@jncasr.ac.in

Contents

1. Introduction 34
2. Synthesis and characterization of the KAT activator for sufficient HAT activation 37
   2.1 Synthesis of TTK21 37
   2.2 Synthesis of glucose-derived carbon nanosphere (CSP) 39
   2.3 Conjugation of CSP to TTK21 41
   2.4 Biological characterization of CSP-TTK21 43
3. KAT activator molecule: Implication in spinal injury repair 46
   3.1 Creation of mid-thoracic dorsal hemisection model of spinal cord injury 46
   3.2 Oral administration of KAT activator CSP-TTK21 after spinal cord injury 48
   3.3 Assessment of functional recovery after spinal cord injury (locomotor test) 49
   3.4 Studying the induction of acetylation after spinal cord injury 51
Acknowledgments 52
References 53

Abstract

p300/CBP is a lysine acetyltransferase enzyme that modifies the lysine residue of histones by adding acetyl groups from acetyl-CoA. Dysregulation of p300/CBP KAT activity has been implicated in several neurological disorders. Pharmacological activation of p300/CBP via small-molecule provides a promising strategy to restore acetylation balance and promote neuronal repair. Our laboratory has reported that the p300/CBP-specific small molecule KAT

Methods in Enzymology, Volume 733
ISSN 0076-6879, https://doi.org/10.1016/bs.mie.2026.05.027

activator TTK21 (N-[4-chloro-3-(trifluoromethyl)phenyl]-2-propoxy-benzamide), when conjugated to a glucose-derived carbon nanosphere (CSP), efficiently crossed the plasma membrane and induced acetylation. Surprisingly, its systemic administration in mice resulted in increased acetylation levels in the frontal cortex and hippocampus. Furthermore, it induced axon regeneration and improved functional recovery after spinal cord injury. Interestingly, oral administration of CSP-TTK21 also induced acetylation and promoted motor recovery in a rat model of spinal cord injury. In this chapter, we illustrate a method for the chemical synthesis of TTK21, glucose-derived carbon nanosphere (CSP), and their conjugate (CSP-TTK21). In addition, we systematically describe the methodologies for the chemical and biological characterization of the KAT activator, including the evaluation of histone acetylation in vitro and in vivo. Lastly, we present detailed experimental procedures to assess the therapeutic efficacy of CSP-TTK21 in models of spinal cord injury. Collectively, this chapter provides comprehensive protocols and practical guidance for the characterization and biological evaluation of p300/CBP activators for spinal cord injury repair.

1. Introduction

Acetylation is one of the most prevalent posttranslational epigenetic modifications that occurs on histones and non-histone proteins. Histone acetylation regulates a variety of physiological processes and has a profound impact on diseases, such as neurodegeneration and cancer (Kimura et al., 2005). It is mainly mediated by lysine acetyltransferases (KATs), which work by catalyzing the transfer of acetyl group from acetyl-CoA to the lysine residues on the N-terminal of histones. Within the KAT superfamily, p300 and CBP (CREB-binding protein) are highly homologous, multidomain proteins in terms of both structure and function as transcriptional co-activators and acetyltransferases (Valor et al., 2013). These enzymes play crucial roles in regulating gene expression, cell proliferation, and normal embryonic development. During the early developmental stage, complete knockout of either p300 or CBP in mice is embryonically lethal (Yao et al., 1998). p300/CBP primarily acetylates histones H3 (K9, K14, K27) and H4 (K5, K8). Dysregulation of their KAT activity has been linked to several neurological disorders, such as Rubinstein-Taybi Syndrome (RTS), Huntington's disease (HD), Alzheimer's disease (AD), and Parkinson's disease (PD) (Dekker & Haisma, 2009; Singh & Kundu, 2026; Singh et al., 2018). Recently, CBP and its KAT activity have also been implicated in spinal cord injury (SCI) and autism (Singh et al., 2024, 2025).

Given the central role of p300/CBP in epigenetic regulation, several restorative strategies have been proposed to reestablish the dysregulated lysine acetylation and p300/CBP KAT activity. Although many effective histone deacetylase inhibitors (HDACi) are known, their lack of specificity and side effects limit their long-term therapeutic application. Moreover, one study showed that at least one functional copy of the CBP gene is essential for HDAC inhibitors to perform and HDAC inhibition was ineffective under a complete CBP knockout condition (Abel & Zukin, 2008; Vecsey et al., 2007). Thus, using an HDAC inhibitor to induce acetylation cannot simply replace HAT activation, displaying the role of lysine acetyltransferase enzymes. Thus, the alternative approach could be direct KAT activation, which serves a similar aim and may offer a more specific therapeutic approach. The discovery of small-molecule activators of lysine acetyltransferases (KATs) represented a transformative advance in epigenetic drug discovery (Simon et al., 2016). Notably, directly targeting KAT enzymes not only activates the targeted enzymatic function but also other related functions that are not generally rescued by HDAC inhibition. Hence, KAT activation is a more efficient strategy than inhibiting HDAC enzymes in this pathological context (Fig. 1).

Mature neurons in the mammalian central nervous system regenerate minimally after spinal cord injury (SCI) (Sofroniew, 2018). The connections between the spinal cord and the brain are disrupted, resulting in the loss of sensory motor functions (He & Jin, 2016). This failure of axonal regeneration after SCI is attributed to the presence of growth-inhibitory molecules and the absence of growth-promoting molecules at the injury site (Schwab & Bartholdi, 1996). Notably, in a maturing neuron, transcriptional mechanisms

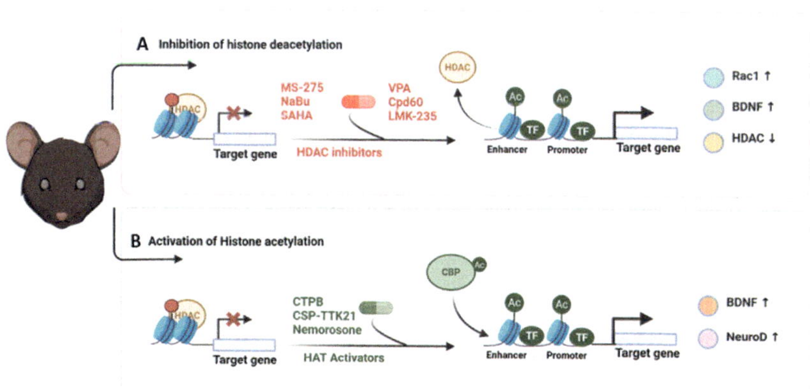

Fig. 1 Small molecule modulators of acetylation. (A) Use of HDAC inhibitors to inhibit the histone deacetylation, thereby inducing acetylation on chromatin. (B) Use of HAT activators to directly activate the histone acetyltransferase activity, thereby increasing histone acetylation.

play an essential role in inducing the expression of regeneration-associated genes, but these mechanisms fail after injury. This failure of transcriptional changes creates a roadblock to axon regeneration. Although the mechanisms underlying this transcriptional regulation are not well understood, epigenetic mechanisms are known to contribute to these effects (Bird, 2007). One of the best-known epigenetic mechanisms is the reversible acetylation (Kouzarides, 2007). Therefore, targeting epigenetic regulators such as p300/CBP to reactivate pro-regenerative transcriptional programs has emerged as a promising therapeutic strategy.

Our laboratory discovered the first small-molecule modulators of the p300 acetyltransferase by chemically modifying the natural HAT inhibitor anacardic acid, leading to the discovery of CTPB (N-(4-chloro-3-trifluoromethyl-phenyl)-2-ethoxy-6-pentadecyl-benzamide; Fig. 2). CTPB is a potent and selective activator of p300 KAT activity; however, it is cell-impermeable. When conjugated to a glucose-derived carbon nanosphere (CSP), it can enter cells and induce H3 acetylation (Balasubramanyam et al., 2003; Selvi et al., 2008). Another study showed that CTPB behaves as an allosteric activator and its binding induces conformational changes in p300 (Mantelingu et al., 2007). This work provided one of the earliest examples of allosteric regulation of a chromatin-modifying enzyme by a small molecule. It established a new method that has now been widely exploited in epigenetic drug discovery and design. Further, in need of a more stable and potent KAT activator, TTK21 (N-[4-chloro-3-(trifluoromethyl) phenyl]-2-propoxy-benzamide) was synthesized from salicylic acid. Similar to the parent compound CTPB, TTK21 also displayed poor cell permeability, but upon conjugation to CSP, it efficiently crossed the plasma membrane and induced histone acetylation.

Furthermore, when CSP-TTK21 was systemically injected into adult C57BL/6J mice, increased acetylation was observed on histones (H2B and H3) in the frontal cortex and hippocampus, indicating that CSP-TTK21 crosses the

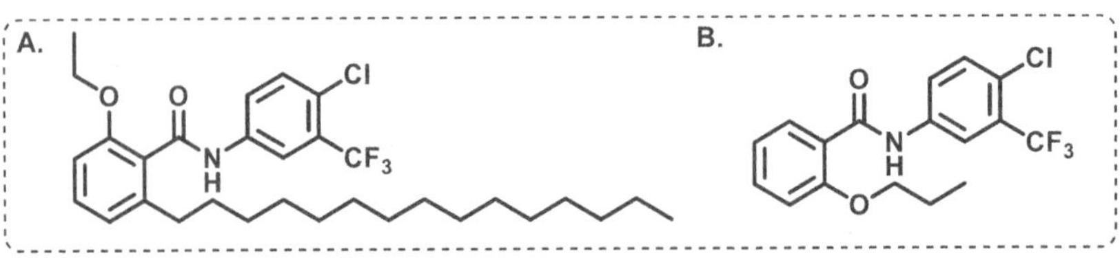

Fig. 2 Chemical structures: (A) CTPB (N-(4-chloro-3-trifluoromethyl-phenyl)-2-ethoxy-6-pentadecyl-benzamide). (B) TTK21 (N-[4-chloro-3-(trifluoromethyl) phenyl]-2-propoxy-benzamide).

blood-brain barrier and reaches various parts of the brain. CSP-TTK21 was also tested for two key hippocampal functions: hippocampal neurogenesis and spatial memory. Notably, CSP-TTK21 was found to promote the formation of DCX-positive neurons in the subgranular zone of DG. Molecular studies have demonstrated that CSP-TTK21 increases mRNA levels of NeuroD1 and BDNF, which are essential for neuronal differentiation (Chatterjee et al., 2013; Smitha et al., 2024). In another study, selective KAT activation of p300/CBP by CSP-TTK21 restored spatial memory and plasticity deficits in a tauopathy mouse model of Alzheimer's disease (Chatterjee et al., 2018). p300/CBP are the master regulators of gene expression and are known to induce regenerative growth of DRG neurons (Mahar & Cavalli, 2018). It was found that the pharmacological activation of CBP via intraperitoneal (IP) administration of CSP-TTK21 induced axon regeneration and sprouting and facilitated functional recovery in both acute and chronic rodent models of spinal cord injury (SCI) (Hutson et al., 2019; Muller et al., 2022). CSP-TTK21 was shown to enhance neuronal activity via epigenetic modifications, particularly by inducing CBP-mediated histone acetylation, thereby increasing expression of regeneration-associated genes, enhancing regenerative potential, and promoting functional recovery after SCI. Our study, which involved oral delivery of CSP-TTK21, also demonstrated comparable effects to IP administration in increasing the expression of regeneration-associated genes and effectively contributed to the repair of spinal cord injuries (Singh et al., 2024).

Here, we provide detailed laboratory methods for the design and synthesis of the p300/CBP KAT activator TTK21 and its glucose-derived carbon nanosphere conjugate. We described methods for chemical and biological characterization. Furthermore, we have given the methodologies for developing a preclinical model of spinal injury, which is very useful for testing molecules of this class to develop better therapeutics for efficient spinal injury repair.

2. Synthesis and characterization of the KAT activator for sufficient HAT activation

2.1 Synthesis of TTK21

2.1.1 Equipment

Glassware (beakers, flasks, measuring cylinder and dropper), magnetic stirrer with hot plate, reflux condenser, rotary evaporator, separating funnel, litmus, and filter paper.

2.1.2 Reagents

Salicylic acid, iodopropane, potassium carbonate, sodium hydroxide, thionyl chloride, 4-chloro-3-(trifluoromethyl)aniline, triethylamine, and sodium sulphate. **Solvents-** Ethyl acetate*, tetrahydrofuran* (THF), dichloromethane* (DCM), methanol*, dimethylformamide* (DMF), hexane and chloroform (*anhydrous).

2.1.3 Procedure

1. In a round-bottom flask, add salicylic acid (7.24 mmol, 1 equiv.), K_2CO_3 (21.7 mmol, 3 equiv.), and 1-iodopropane (21.7 mmol, 3 equiv.) in dimethylformamide (5 ml).
2. Heat the reaction mixture at 80 °C for 24 h and monitor the reaction using thin-layer chromatography (TLC).
3. Upon completion, pour the reaction mixture into ice-cold water and extract it with ethyl acetate (X3).
4. Combine all the organic layers, dry over sodium sulphate, and evaporate under vacuum to get the product in 84% yield.
5. The product can be used in the next step without further purification (Fig. 3, Step 1).
6. In a round-bottom flask, add aqueous sodium hydroxide (1 equiv.) to the solution of crude product of step 1 (2-propoxysalicylic acid, 1 equiv.) in methanol at 0 °C (keep 3:1, Methanol: Water).
7. Let the reaction mixture stir at 0 °C for 30 min and then at room temperature for 4–5 h, and monitor the reaction using TLC.

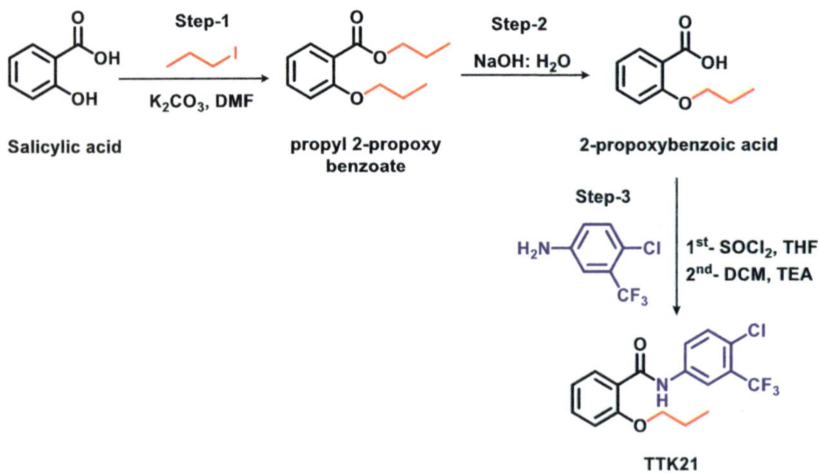

Fig. 3 3-Step synthetic procedure of TTK21 from salicylic acid.

8. Upon completion, pour the reaction mixture into cold water and neutralize using 3 M HCl, finally extract with dichloromethane (X3).
9. Combine all the organic layers, dry over sodium sulphate and concentrate under vacuum to get the product in 89 % yield.
10. The product can be used in the next step without further purification (Fig. 3, Step 2).
11. In a round-bottom flask, dissolve 2-propoxysalicylic acid (2.77 mmol, 1 equiv.) in 5 ml of THF at 0 °C, and add thionyl chloride (6.94 mmol, 2.5 equiv.) dropwise, followed by a few drops of dimethylformamide as a catalyst to the solution.
12. Transfer and heat the reaction mixture at 45 °C for 6–8 h. Upon completion, evaporate the solvent under vacuum to obtain 2-propoxysalicyl chloride.
13. In another round-bottom flask, take 4-chloro-3-(trifluoromethyl) aniline (3.33 mmol, 1.2 equiv.), dissolve in dichloromethane, and add triethylamine (3.33 mmol, 1 equiv.) dropwise.
14. Stir the reaction mixture at room temperature for 30 min. Further add this resulting solution dropwise to the 2-propoxysalicyl chloride dissolved in dichloromethane and stir the reaction mixture in a sealed system for 3–4 h.
15. Upon completion, pour the reaction mixture into cold water and then extract with dichloromethane (X3).
16. Combine all the organic layers, dry over sodium sulphate, and concentrate under vacuum. Recrystallise from the dichloromethane-hexane mixture (1:10) to yield the pure product as a white solid in 92 % yield (Fig. 3, Step 3).

All the synthesized compounds were identified and characterized by ^{1}H NMR, ^{13}C NMR, and mass (HRMS) spectroscopy (Fig. 4).

2.1.4 Notes

Use thionyl chloride carefully as it is highly toxic and corrosive.

2.2 Synthesis of glucose-derived carbon nanosphere (CSP)

2.2.1 Equipment

Hydrothermal autoclave, hot air oven, glass petri dish, crucibles, funnel, dishes, and filtration flask.

2.2.2 Reagents

D-(+) glucose. **Solvents-** triply distilled (TD) water, ethanol, and acetone.

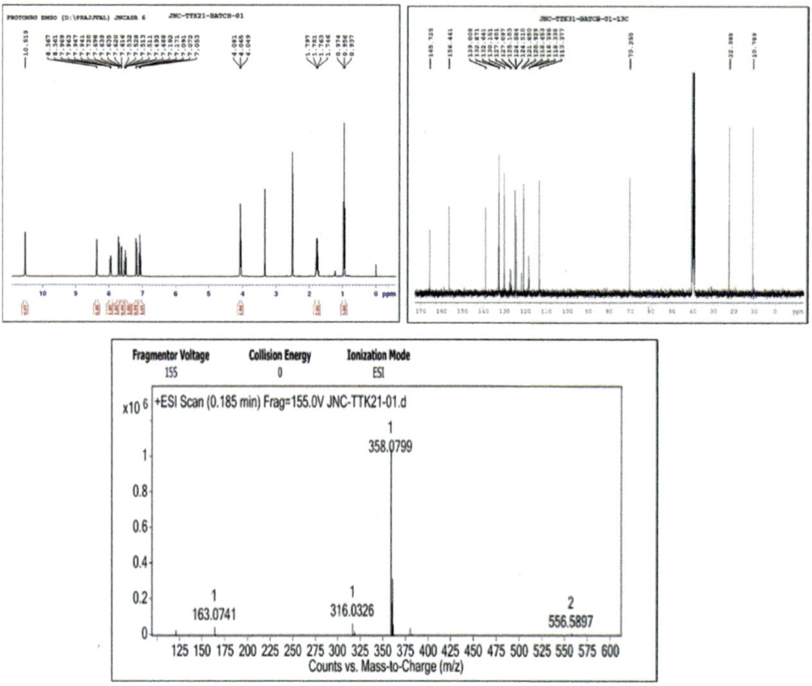

Fig. 4 ^{1}H and ^{13}C NMR of TTK21 in DMSO-d_6 and HRMS data.

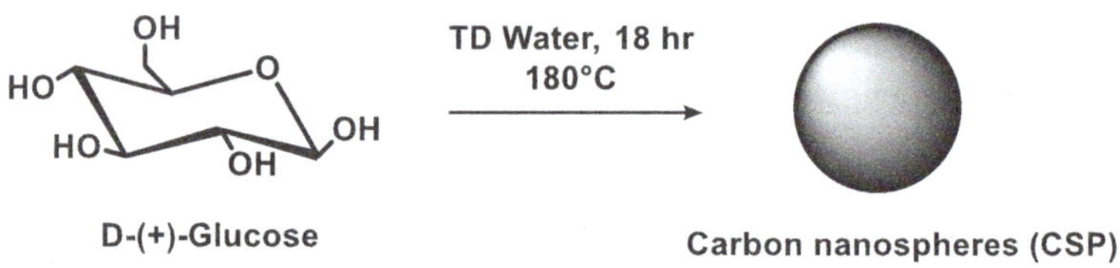

Fig. 5 Synthetic procedure of CSP from D-(+) glucose.

2.2.3 Procedure

1. Dissolve 5 g of D-(+) glucose in 50 ml of TD water to form a clear solution (Fig. 5).
2. Place it in a 60 ml Teflon-lined, sealed stainless-steel autoclave and maintain the reaction mixture at 180 °C for 18 h in a hot air oven.
3. Filter the reaction mixture using a sintered crucible or a Buchner funnel.
4. Wash the obtained crude product with water (X3) and ethanol (X3), and finally wash with acetone to remove maximum water content.
5. Dry the solid in the oven at 50 °C for 4 h to get the product (CSP) as brownish-black powder Fig. 5.

To determine the surface morphology and nanosphere diameter, a scanning electron microscope (SEM) was used (Fig. 6).

2.2.4 Notes

Wash and keep CSP carefully as it is highly light-sensitive. Storage- 4 °C.

2.3 Conjugation of CSP to TTK21

2.3.1 Equipment

magnetic stirrer, glass dropper, filtration flask.

2.3.2 Reagents

CSP, TTK-21, thionyl chloride ($SOCl_2$), 4-(Dimethylamino)pyridine (DMAP). **Solvent-** Dichloromethane, dimethylformamide, and water.

2.3.3 Procedure

1. Add dropwise $SOCl_2$ (1 equiv.) to a suspension of 100 mg of CSP in DCM, followed by a few drops of DMF.
2. Let the reaction mixture stir at room temperature for 8–9 h.
3. Add a solution of 100 mg TTK21 dissolved in DCM and DMAP (0.1 equiv.) dropwise to the CSP suspension and stir for 8−9 h at room temperature.
4. Filter the reaction mixture using a Buchner funnel and wash with cold water and DCM.
5. Finally, test the filtrate for the absence of TTK21 using TLC.
6. Dry the CSP-TTK21 conjugate at 50 °C until a constant weight is achieved Fig. 7.

To determine the surface morphology, nanosphere diameter, and conjugation, a scanning electron microscope (SEM) and EDAX were used. Figs. 8 and 9.

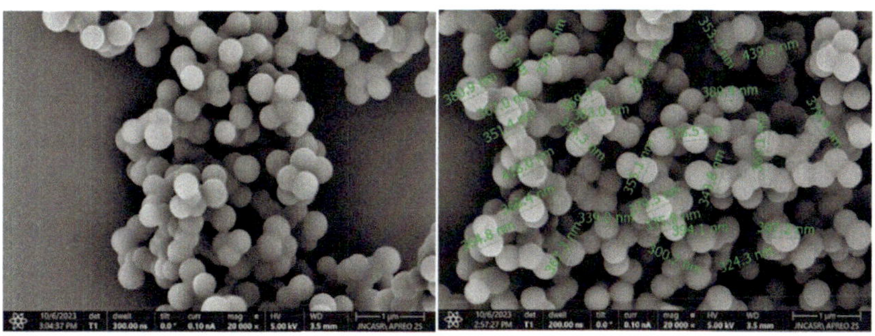

Fig. 6 SEM images of CSP.

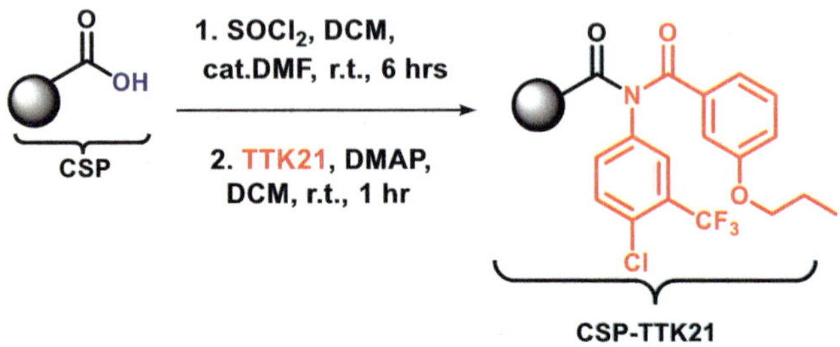

Fig. 7 Synthetic procedure for the conjugation of CSP and TTK21.

Fig. 8 (A). SEM image of CSP-01 after conjugation. (B). Mapping region of CSP-TTK21. (C). EDAX of CSP-TTK21 (Presence of F in EDAX analysis confirms conjugation of TTK21 with CSP). **Note-** Wash and keep CSP-TTK21 carefully as it is highly light-sensitive. Storage- 4 °C.

Fig. 9 (A). TTK21 (B). CSP (C). CSP-TTK21 (D). Ultrasonic Bath/Sonicator (used for washing CSP) (E). hot air oven (for synthesising CSP) (F). filtration setup (G). Teflon-lined hydrothermal autoclave (H). magnetic stirrer with a hot plate.

2.4 Biological characterization of CSP-TTK21

2.4.1 Equipment

Intraperitoneal injection supplies (insulin syringes, sterile needles), CO_2 chamber or approved euthanasia setup, Cryostat, Confocal laser-scanning microscope, Cell culture incubator (37 °C, 5 % CO_2), Gel electrophoresis and blotting apparatus, Sonicator, Eppendorf tubes, and Centrifuge.

2.4.2 Reagents

SH-SY5Y human neuroblastoma cells, Dulbecco's Modified Eagle Medium (DMEM) or DMEM/F-12, Fetal bovine serum (FBS), Penicillin–streptomycin, CSP and CSP-TTK21 compounds, Phosphate-buffered saline (PBS), Laemmli sample buffer (2× or 4×), Protease inhibitor cocktail, SDS-PAGE reagents, PVDF or nitrocellulose membrane, Phosphate-buffered saline with Tween-20 (PBST), Blocking reagent (5 % non-fat dry milk or BSA), Enhanced chemiluminescence (ECL) detection reagents, Paraformaldehyde (PFA), 4 % in PBS, Sucrose solutions (10 %, 20 %, 30 %), Cryoprotectant or OCT embedding medium, Normal serum for blocking (species-matched to secondary antibody), Triton X-100, Fluorescent secondary antibodies (species-specific), DAPI or equivalent nuclear counterstain, Mounting medium (anti-fade), Acetylated-H3K14ac antibody (1:2000; in-house raised) – immunoblotting, Total H3 histone antibody (1:10,000; in-house raised) – immunoblotting loading control, Acetylated-H3K14ac antibody (1:250; in-house raised) – immunohistochemistry.

2.4.3 Procedure

2.4.3.1 *In-vitro* characterization in SH-SY5Y cells

1. Maintain SH-SY5Y cells in complete growth medium (DMEM or DMEM/F-12 supplemented with 10 % FBS and 1 % penicillin–streptomycin) at 37 °C in 5 % CO_2.
2. Plate cells at appropriate density (e.g., 60–70 % confluence) one day prior to treatment.
3. Treat cells with CSP or CSP-TTK21 at the desired working concentration for 24 h. Include vehicle-treated cells as controls.
4. At the end of treatment, wash cells once with ice-cold PBS.
5. Lyse cells directly in culture dishes using Laemmli sample buffer supplemented with protease inhibitors.
6. Scrape cells thoroughly and transfer lysates to microcentrifuge tubes.
7. Boil samples at 95 °C for 5 min to denature proteins.
8. Briefly centrifuge to collect condensate and store at −20 °C or proceed immediately to SDS-PAGE.
9. Separate equal volumes of lysates by SDS-PAGE.
10. Transfer proteins onto PVDF or nitrocellulose membranes using standard wet or semi-dry transfer methods.
11. Block membranes in 5 % non-fat milk or BSA in PBST for 1 h at room temperature.
12. Incubate membranes overnight at 4 °C with acetylated-H3K14ac antibody (1:2000) diluted in blocking buffer.
13. Wash membranes 3 × 5 min with PBST.
14. Incubate with appropriate HRP-conjugated secondary antibody for 1 h at room temperature.
15. Wash membranes and detect signal using ECL.
16. Strip membranes if required and re-probe with total H3 antibody (1:10,000) as a loading control (Fig. 10A).

2.4.3.2 *In vivo* CSP/CSP-TTK21 administration and characterization

1. Prepare CSP or CSP-TTK21 freshly in sterile PBS suitable for IP injection.
2. Inject mice intraperitoneally with a single dose of 20 mg/kg body weight.
3. Monitor animals daily for general health and behaviour.
4. Three days post-injection, euthanize animals according to institutional ethical guidelines.

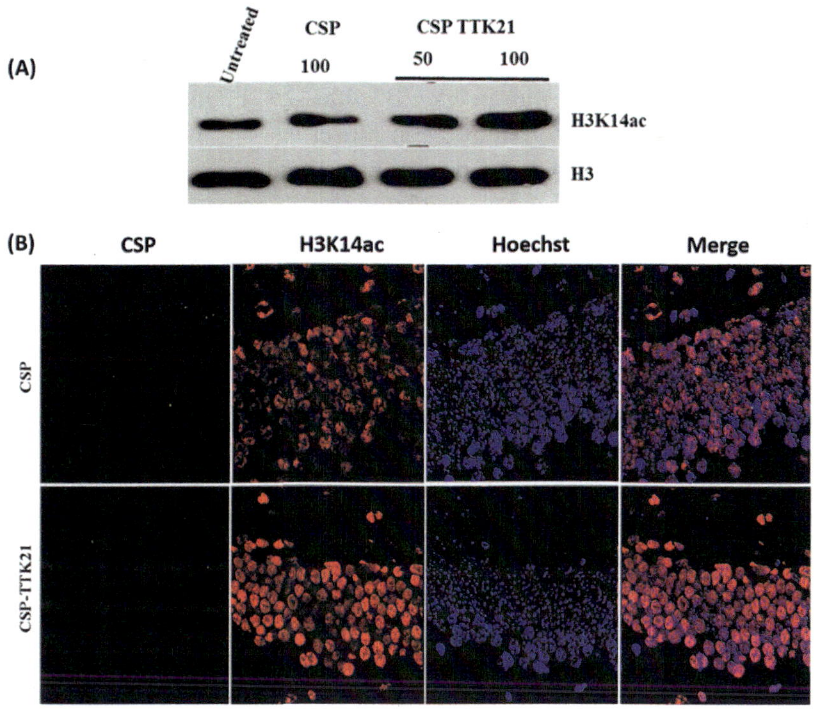

Fig. 10 (A) in vitro and (B) in vivo characterisation of CSP-TTK21 activity. *Adapted from Singh et al., Aging Cell, 2025*[8].

Brain Tissue Collection and Processing

1. Perfuse mice transcardially with 50 ml PBS followed by 30 ml of 4 % PFA.
2. Dissect brains and post-fix in 4 % PFA for 12–16 h at 4 °C.
3. Cryoprotect brains sequentially in 10 % and 20 % sucrose solutions for 24 h each, followed by 30 % sucrose solution until the brain sank to the bottom of the Falcon tube (approx. 6−7 h).
4. Embed tissue in OCT solution and keep at −20 °C for 15 min for proper embedding. Section using a cryostat (20–40 μm thickness).
5. Wash free-floating or mounted sections in 1X PBS.
6. Permeabilize and block sections in 1X PBS containing 0.3 % Triton X-100 and 5 % BSA for 1 h at room temperature.
7. Incubate sections with acetylated-H4K12 antibody (1:250) overnight at 4 °C.
8. Wash Section 3 × 10 min in 1X PBS.

9. Incubate with fluorescent secondary antibody (1:250) for 1–2 h at room temperature in the dark.
10. Wash Section 3 × 10 min in 1X PBS, add DAPI 2 min before the end of the second wash to counterstain nuclei.
11. Mount sections using anti-fade mounting media for long-term storage. 70% glycerol can also be used, but the fluorescence intensity will decrease over time.
12. Acquire images using a confocal microscope with identical settings across experimental groups (Fig. 10B).
13. Collect z-stacks where appropriate to capture nuclear acetylation signals.
14. Quantify fluorescence intensity using image analysis software, normalizing to background and/or cell number.

2.4.4 Notes

1. The immunoblotting process needs equal protein levels through total H3 measurement, which serves as the loading control.
2. The antibody dilutions that are provided as optimal for in-house raised antibodies need modification when testing different antibody batches.
3. During imaging, use the same imaging parameters across samples for a specific antibody; this will allow for direct quantitative comparisons between their results. Samples processed with different antibodies might need further modifications in the imaging parameters; that's totally acceptable as long as the parameters are the same across all the samples.
4. The most effective time point for detecting persistent changes in brain tissue histone acetylation occurs three days after the injection.

3. KAT activator molecule: Implication in spinal injury repair

3.1 Creation of mid-thoracic dorsal hemisection model of spinal cord injury

3.1.1 Animals

1. Animal Species – Sprague Dawley (SD) male rats of 100–120 g of weight.
2. Animal Housing – Pathogen-free environment, temperature 22 ± 2 °C, 12/12 h light/dark cycle with free access to food and water.

3.1.2 Equipment

Surgicals and injections – Syringes (25-gauge 1-inch needles), scissors, pointed forceps, scalpel, microscissor, tissue forceps, bone nipper, forceps, needle holder, Vicryl, and silk sutures.

3.1.3 Reagents

Antibiotics (Enrofloxacin), Analgesics (Meloxicam) and Saline (0.9 %).

3.1.4 Procedure

A mid-thoracic dorsal hemisection surgical model of spinal cord injury is useful, handy model system that we established in a rodent species (SD rats). The schematic shows the step-by-step procedures – Fig. 11.

1. Place the animals in the surgery area 1 h prior to the procedure starting. Clean the surgical area with alcohol, then place the sterile surgical equipment in it.
2. Anesthetize the animals with a mixture solution of 35 mg/kg body weight of ketamine and 8 mg/kg body weight of xylazine. Check for the anaesthetic depth by manually pinching the hindlimbs; a well anaesthetised animal shows a breathing response but no limb movement.
3. Shave the skin over the thoracic area using a hair trimmer and clean the skin surface with alcohol wipes.
4. Perform the dorsal laminectomy (i.e., partial vertebral laminectomy) to expose the laminae T8–T10. Briefly, incise the skin, shift apart the superficial fat, and dissect the muscle tissue to expose laminae T8–T10. Perform the T9 laminectomy and remove the meningeal layer at the midline longitudinally.

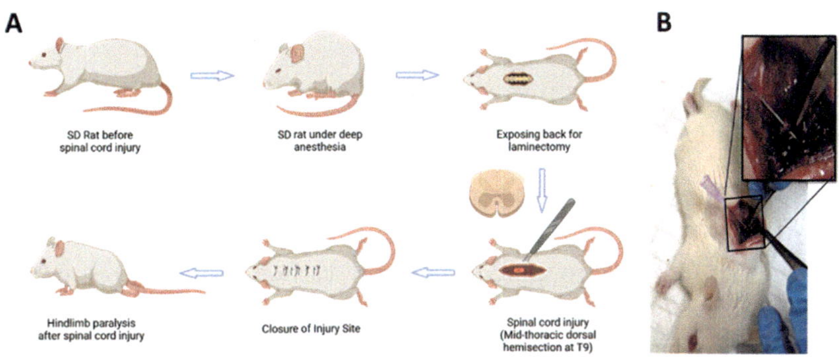

Fig. 11 Mid-thoracic dorsal hemisection spinal cord injury model generation in SD rats. (A) Schematic showing the spinal cord injury model generation in SD rats. (B) Mid-thoracic dorsal hemisection at T9 levels.

5. Perform the mid-thoracic dorsal hemisection spinal injury at the T9 level with the help of an injection needle (24G) and a microscissor by cutting the dorsal half area of the spinal cord.
6. Confirm the status of injury by observing the animal. A well-injured animal shows a small jump, hindlimbs go to extension, and the tail whips back and forth multiple times.
7. Close the detached muscles and skin using Vicryl and silk suture, and place the animals back in their home cage.
8. Postoperatively, treat animals with an antibiotic, i.e., enrofloxacin (BD 10–20 mg/kg B. Wt. for 7 days) and an analgesic (BD 2 mg/kg B. Wt. meloxicam + 60 mg/kg B. Wt. paracetamol for 7 days).

3.1.5 Notes

1. It is essential to check the animal daily for general health, mobility within the cage, wounds, swellings, infections, and autophagia of toes. Animals should be given 0.9% saline (10 ml/kg body weight) orally twice daily for 3 days and once daily from day 4 to 7 after surgery. This will protect the animal from dehydration.
2. It is essential that animals receive twice-daily bladder expression for the first week after surgery and once daily until needed. This will prevent animals from developing urinary infections.

3.2 Oral administration of KAT activator CSP-TTK21 after spinal cord injury

3.2.1 Equipment

Stainless steel curved animal feeding cannula (20G, 50 mm), Syringe (1 ml), and Ultrasonic bath.

3.2.2 Reagents

A fresh new batch of CSP and CSP-TTK21, with prior testing of both chemical and biological properties, and Autoclaved MQ.

3.2.3 Procedure

The API (TTK21) is formulated as a nanosphere-based preparation by covalent conjugation to glucose-derived carbon nanospheres (CSP). This formulation enables the drug to cross the blood-brain barrier and exert its activity in a controlled-release manner. For rats, the efficacious dose of CSP-TTK21 via the oral route is 10 mg/Kg body weight; CSP can be used as a vehicle control. A suspension of both CSP and CSP-TTK21 (2 mg/ml) can be prepared by dispersing the compound in

water (MQ) using an ultrasonic bath for 5 min. Further, the doses of CSP and CSP-TTK21 can be administered to the animals based on their weight, as per the study plan.
The oral delivery of CSP and CSP-TTK21 procedure includes –.

1. Assign the animals randomly into two groups – for treatment with KAT activator (CSP-TTK21) or vehicle control (carbon nanosphere, CSP) (Rats – 10 mg/kg body weight, orally once per week).
2. During the oral gavage, scruff the animal in a way that the forelimbs extend laterally and the head is gently extended backward, aligning with the animal's body.
3. Introduce the feeding cannula into the mouth of the animal near the esophagus, without any forceful action.
4. Deliver the CSP and CSP-TTK21 solution slowly. Take precautions to prevent reflux or aspiration.
5. Observe the animal for 5–10 min post gavage for any sign of distress or procedure-related complications.

3.2.4 *Notes*

1. In our past studies, animals received the first oral dose of CSP-TTK21 4 h after spinal cord injury to achieve a clinically feasible delay in treatment. Then they received it once a week thereafter for 5 weeks.
2. In our acute toxicity studies, no adverse effects with the oral administration of CSP-TTK21 were observed.

3.3 Assessment of functional recovery after spinal cord injury (locomotor test)

The locomotor test was performed to assess the animals' movement. This test uses an open-field arena with infrared beams to measure the sensory-motor function of the animal after spinal cord injury and treatment with the KAT activator CSP-TTK21 (Fig. 12).

3.3.1 *Equipment*

Open field square arena (45 (W) x 45 (D) x 40 (H) cm), Photoelectric actimeter (Acti-Track System, Panlab), and Tissue paper.

3.3.2 *Reagents*

10 % Alcohol.

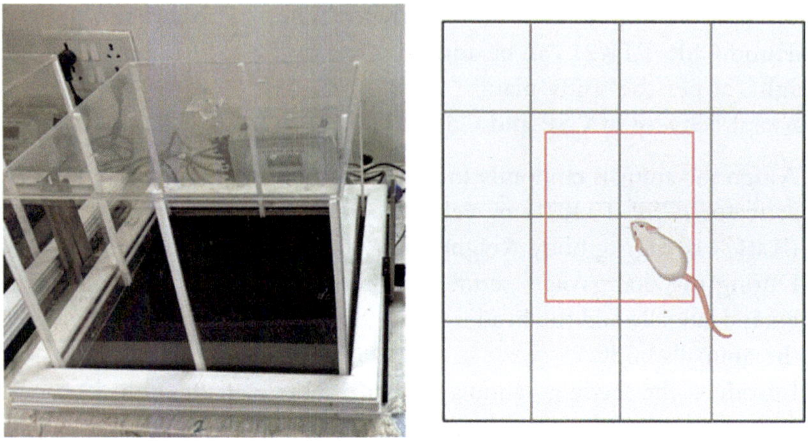

Fig. 12 Open field test. The instrument used to test locomotor activity in animals.

3.3.3 Procedure

1. Firstly, prepare the testing room and testing apparatus. Use at least four activity chambers of measurement (45 (W) x 45 (D) x 40 (H) cm).
2. Clean the open field arena with a 10% ethanol prior to use and in between the subsequent tests. This will remove the odor cues from the previous test animal.
3. Use a tracking system and software (Panlab) for tracking and analysis to evaluate animals' movement.
4. Place the animals in the testing room to acclimate for a minimum of 1 h before starting the experiment.
5. Remove the animal from the home cage by gently grasping it and placing it in the middle of the activity chamber while immediately starting the software to begin tracking the animal's activity.
6. Allow uninterrupted exploration of the test animal in the activity chamber for 30 30-minute period, during which the ActiTrack software will record both horizontal and vertical activity of the animal.
7. On the completion of the testing period, pick up the animal gently and return it to its homecage.
8. Before introducing the next animal, clean the floor and walls of the activity chamber with 10% ethanol and wipe down with a clean paper towel to avoid odor interference from the previous animal.

3.3.4 Notes

While looking at the functional recovery and locomotor improvements, the two aspects of open field activity can be recorded: (1) Horizontal activity, which includes the total distance traveled by the animal during the testing period, and (2) Vertical activity, which includes rearing behavior by counting the number of events animal stand using both hind limbs in a vertical upright position.

3.4 Studying the induction of acetylation after spinal cord injury

3.4.1 Equipment

Cryostat or freezing microtome, Microscope slides, Confocal microscope, rocking platform/shaker, light-protected containers, Slide storage boxes wrapped in foil (dark), and Zeiss LSM 510 META and a Zeiss LSM 810 system.

3.4.2 Reagents

Polyfreeze OCT media, nuclear stain (DAPI), 1X PBS (pH 7.4), Permeabilisation solution: 0.3 % Triton X-100 in PBS, Blocking/antibody diluent: 5 % BSA in 0.3 % Triton X-100 in 1X PBS, Primary antibody (Anti-H4K12ac, Anti-H3K27ac). Example dilutions: 1:500–1:1000, Fluorescent secondary antibodies: species-specific, highly cross-absorbed, Alexa fluors, Dilution: 1:500–1:1000.

3.4.3 Procedure

1. Perform the cryosectioning of brain and spinal cord tissue using a cryotome (Leica system). Collect the sections in an appropriate buffer (PBS 1X).
2. Permeabilize the sections by incubating them in 0.3 % Triton X-100 in PBS for 30 min at room temperature.
3. Perform the blocking of tissue sections by incubating it in blocking buffer (5 % BSA + 0.3 % Triton X-100 in PBS) for 1 h at room temperature.
4. Prepare primary antibody in antibody diluent (blocking buffer or PBS + 1–2 % BSA + 0.1 % Triton X-100). Typical dilutions tested: 1:500 for Anti-H4K12ac and Anti-H3K27ac antibodies.
5. Incubate tissue sections with primary antibody overnight at 4 °C in a light-protected chamber with gentle rocking.
6. After primary antibody incubation, wash the tissue Section 3 times, 10 min each, in PBS (or PBS + 0.1 % Triton X-100) at RT with gentle agitation.

7. Incubate the tissue sections with Alexa fluor-tagged secondary antibody diluted in antibody diluent (Recommended dilution 1:500) for 1–2 h at RT.
8. For the nuclear staining, use DAPI during the washing steps for 10 min, gently on a rocking shaker.
9. Perform the confocal imaging using a confocal scanning laser microscopy Zeiss (LSM 510 META and a Zeiss LSM 810 system) and Zen imaging software.

3.4.4 Notes

1. It is essential to take an appropriate thickness of tissue sections (recommended thickness 20−40 uM μm) for the immunostaining.
2. Select appropriate laser lines, detector settings, and objective lenses for each fluorophore. Imaging parameters should be kept constant across experimental groups to enable a reliable comparison of fluorescence intensity and signal distribution. All images should be captured under identical acquisition settings to minimize variability.

3.4.5 Summary

The chapter provides a comprehensive methodological framework for the development and evaluation of p300/CBP lysine acetyltransferase (KAT) activators for candidates who desire to explore epigenetic mechanisms involved in spinal cord injury and neuroregeneration. A detailed procedure is provided for the chemical synthesis and structural characterization of TTK21, glucose-derived carbon nanospheres (CSP), and their conjugation protocol. In addition, standardized methods for the quantitative evaluation of in vitro and in vivo histone acetylation are described. The chapter further outlines reproducible methods for establishing a dorsal hemisection spinal cord injury model, administering CSP-TTK21, and assessing functional recovery through behavioural and molecular analyses. Collectively, this chapter provides methodological insight into the development and evaluation of small-molecule p300/CBP KAT activators in spinal cord injury models.

Acknowledgments

This work was supported by the Jawaharlal Nehru Center for Advanced Scientific Research (JNCASR), CSIR-Central Drug Research Institute, Lucknow, J C Bose Fellowship, Department of Science and Technology (DST), India (grant no. JBR/2020/000029), and core grant of SERB (grant no. CRG/2021/000356) to TKK.

References

Abel, T., & Zukin, R. S. (2008). Epigenetic targets of HDAC inhibition in neurodegenerative and psychiatric disorders. *Current Opinion in Pharmacology, 8*, 57–64.

Balasubramanyam, K., Swaminathan, V., Ranganathan, A., & Kundu, T. K. (2003). Small molecule modulators of histone acetyltransferase p300. *The Journal of Biological Chemistry, 278*, 19134–19140.

Bird, A. (2007). Perceptions of epigenetics. *Nature, 447*, 396–398.

Chatterjee, S., Cassel, R., Schneider-Anthony, A., Merienne, K., Cosquer, B., Tzeplaeff, L., ... Eswaramoorthy, M. (2018). Reinstating plasticity and memory in a tauopathy mouse model with an acetyltransferase activator. *EMBO Molecular Medicine, 10*, e8587.

Chatterjee, S., Mizar, P., Cassel, R., Neidl, R., Selvi, B. R., Mohankrishna, D. V., ... Boutillier, A. L. (2013). A novel activator of CBP/p300 acetyl-transferases promotes neurogenesis and extends memory duration in adult mice. *The Journal of Neuroscience, 33*, 10698–10712.

Dekker, F. J., & Haisma, H. J. (2009). Histone acetyl transferases as emerging drug targets. *Drug Discovery Today, 14*, 942–948.

He, Z., & Jin, Y. (2016). Intrinsic control of axon regeneration. *Neuron, 90*, 437–451.

Hutson, T. H., Kathe, C., Palmisano, I., Bartholdi, K., Hervera, A., De Virgiliis, F., ... Barraud, Q. (2019). Cbp-dependent histone acetylation mediates axon regeneration induced by environmental enrichment in rodent spinal cord injury models. *Science Translational Medicine, 11*, eaaw2064.

Kimura, A., Matsubara, K., & Horikoshi, M. (2005). A decade of histone acetylation: Marking eukaryotic chromosomes with specific codes. *Journal of Biochemistry, 138*, 647–662.

Kouzarides, T. (2007). Chromatin modifications and their function. *Cell, 128*, 693–705.

Mahar, M., & Cavalli, V. (2018). Intrinsic mechanisms of neuronal axon regeneration. *Nature Reviews. Neuroscience, 19*, 323–337.

Mantelingu, K., Kishore, A. H., Balasubramanyam, K., Kumar, G. V., Altaf, M., Swamy, S. N., ... Kundu, T. K. (2007). Activation of p300 histone acetyltransferase by small molecules altering enzyme structure: probed by surface-enhanced Raman spectroscopy. *The Journal of Physical Chemistry. B, 111*, 4527–4534.

Muller, F., De Virgiliis, F., Kong, G., Zhou, L., Serger, E., Chadwick, J., ... Kundu, T. K. (2022). CBP/p300 activation promotes axon growth, sprouting, and synaptic plasticity in chronic experimental spinal cord injury with severe disability. *PLoS Biology, 20*, e3001310.

Schwab, M. E., & Bartholdi, D. (1996). Degeneration and regeneration of axons in the lesioned spinal cord. *Physiological Reviews, 76*, 319–370.

Selvi, B. R., Jagadeesan, D., Suma, B. S., Nagashankar, G., Arif, M., Balasubramanyam, K., ... Kundu, T. K. (2008). Intrinsically fluorescent carbon nanospheres as a nuclear targeting vector: Delivery of membrane-impermeable molecule to modulate gene expression in vivo. *Nano Letters, 8*, 3182–3188.

Simon, R. P., Robaa, D., Alhalabi, Z., Sippl, W., & Jung, M. (2016). KATching-up on small molecule modulators of lysine acetyltransferases. *Journal of Medicinal Chemistry, 59*, 1249–1270.

Singh, A. K., & Kundu, T. K. (2026). Landscape of protein post-translational modifications in age-related neurodegenerative disorders: Implications in protein aggregation, altered cellular processes, and therapeutics. *ACS Omega.*

Singh, A. K., Halder-Sinha, S., Clement, J. P., & Kundu, T. K. (2018). Epigenetic modulation by small molecule compounds for neurodegenerative disorders. *Pharmacological Research, 132*, 135–148.

Singh, A. K., Joshi, I., Reddy, N. M. N., Purushotham, S. S., Eswaramoorthy, M., Vasudevan, M., ... Kundu, T. K. (2025). Epigenetic modulation rescues neurodevelopmental deficits in Syngap1(+/-) mice. *Aging Cell, 24*, e14408.

Singh, A. K., Rai, A., Joshi, I., Reddy, D. N., Guha, R., Alka, K., ... Clement, J. P. (2024). Oral administration of a specific p300/CBP lysine acetyltransferase activator induces synaptic plasticity and repairs spinal cord injury. *ACS Chemical Neuroscience, 15*, 2741–2755.

Smitha, A. S., Singh, A. K., PR, J. L., Bhatt, R., Mishra, P., Eswaramoorthy, M., ... Kundu, T. K. (2024). p300/CBP KATs are critical for maturation and differentiation of adult neural progenitors. *ACS Chemical Biology, 11*, 2345–2358.

Sofroniew, M. V. (2018). Dissecting spinal cord regeneration. *Nature, 557*, 343–350.

Valor, L. M., Viosca, J., Lopez-Atalaya, J. P., & Barco, A. (2013). Lysine acetyltransferases CBP and p300 as therapeutic targets in cognitive and neurodegenerative disorders. *Current Pharmaceutical Design, 19*, 5051–5064.

Vecsey, C. G., Hawk, J. D., Lattal, K. M., Stein, J. M., Fabian, S. A., Attner, M. A., ... Abel, T. (2007). Histone deacetylase inhibitors enhance memory and synaptic plasticity via CREB: CBP-dependent transcriptional activation. *The Journal of Neuroscience, 27*, 6128–6140.

Yao, T. P., Oh, S. P., Fuchs, M., Zhou, N. D., Ch'ng, L. E., Newsome, D., ... Eckner, R. (1998). Gene dosage-dependent embryonic development and proliferation defects in mice lacking the transcriptional integrator p300. *Cell, 93*, 361–372.

CHAPTER FOUR

Chromatin immunoprecipitation assay for analyzing histone H3-mediated regulation of copper homeostasis

Sakshi Singh[a,1], Rakesh Kumar Sahu[b,1], and Raghuvir Singh Tomar[c,*]
[a]Division of Molecular and Cellular Biology, Eunice Kennedy Shriver National Institute of Child Health and Human Development, National Institutes of Health, Bethesda, MD, United States
[b]Laboratory of Biochemistry and Molecular Biology, National Cancer Institute, National Institutes of Health, Bethesda, MD, United States
[c]Department of Biological Sciences, Indian Institute of Science Education and Research Bhopal, Madhya Pradesh, India
*Corresponding author. e-mail address: rst@iiserb.ac.in

Contents

1. Introduction 56
2. Before you begin 58
   2.1 Yeast strains and growth conditions 58
   2.2 Buffer recipes 58
3. Key resources table 59
4. Materials and equipment 61
5. Step-by-step method details 62
   5.1 Assays of copper tolerance in yeast strains 62
   5.2 Analysis of metallothionine gene induction 63
   5.3 Chromatin immunoprecipitation assay to investigate the mechanisms of metallothionein gene induction 64
6. Quantification and statistical analysis 71
7. Expected outcomes 72
8. Advantages and limitations 73
9. Optimization and troubleshooting 73
10. Concluding remarks 74
Acknowledgements 75
Author contributions 75
References 75

Abstract

Histone tails undergo various post-translational modifications that aid in recruiting effector proteins involved in diverse transcriptional responses. Our previous work

[1] These authors contributed equally

Methods in Enzymology, Volume 733
ISSN 0076-6879, https://doi.org/10.1016/bs.mie.2026.05.030

explored the role of histone H3 N-terminal residues and their modifications on the transcriptional regulation of the *CUP1* metallothionein gene during cellular copper stress. To pinpoint critical histone residues, we used complementary methods, including growth assays and gene expression kinetics. We screened a library of synthetic histone H3 and H4 mutants for copper response defects by assessing their sensitivity to copper stress and quantifying *CUP1* transcript levels via real-time PCR. We further investigated the mechanisms of impaired *CUP1* transcription by assessing the recruitment of Ace1, a copper-sensing transcription factor that activates *CUP1* in response to high intracellular copper, and of TATA-binding protein (TBP), using chromatin immunoprecipitation (ChIP) assays. Our results showed that mutations in the H3 N-terminal tail impair the recruitment of Ace1 and TBP to the *CUP1* promoter, causing defects in its induction. Collectively, these findings highlight the significance of histone residues and their modifications in regulating metal homeostasis. This chapter outlines the methods used to study histone-mediated regulation of copper homeostasis, with a focus on the chromatin immunoprecipitation assay.

1. Introduction

Copper is an essential micronutrient that serves as a cofactor for essential metabolic enzymes involved in various biological pathways, including mitochondrial respiration, free radical detoxification, protein trafficking, and cellular signaling (Askwith et al., 1994; Baker et al., 2017; Culotta et al., 1995; Horn & Barrientos, 2008; Kim et al., 2008; Krishnamoorthy et al., 2016). Intracellular copper levels must be tightly regulated to prevent deficiency or toxic accumulation, which is controlled by copper homeostatic mechanisms. This regulation involves a coordinated system of copper uptake, intracellular chaperoning, targeted delivery, utilization, and detoxification (Chen et al., 2022; Kim et al., 2008; Zhu et al., 2024). Disruption of these homeostatic processes impairs cellular and organismal physiology, leading to developmental disorders, metabolic diseases, and copper-related pathologies, including Menkes disease, Wilson's disease, and neurodegenerative disorders (Chen et al., 2022; Kim et al., 2008; Zhu et al., 2024).

Copper uptake requires copper in its reduced form (Cu^{+1}), catalyzed by reductases (Fre1/Fre2) in budding yeast *Saccharomyces cerevisiae* (Georgatsou et al., 1997; Hassett & Kosman, 1995). This reduced copper is then transported into the cell by high-affinity transporters, specifically Ctr1 and Ctr3 (Dancis et al., 1994; Pena et al., 2000). Once internalized, copper binds to chaperone proteins such as Ccs1, Atx1, and Cox17, which deliver it to its cellular destinations (Culotta et al., 1997; Glerum et al., 1996; Kim et al., 2008; Pufahl et al., 1997). Cysteine-rich metallothionein proteins (MTs) buffer intracellular copper by sequestering excess ions (Ecker et al., 1986; Fogel & Welch, 1982; Winge et al., 1985), preventing toxic effects such as protein misfolding and oxidative stress (Singh, Sahu, & Tomar, 2021; Singh, Sahu, Sugathan, et al., 2021). In budding yeast, Cup1 and Crs5 are the primary MTs responsible for copper detoxification (Culotta et al., 1994; Ecker et al., 1986; Fogel & Welch, 1982; Jensen et al., 1996; Winge et al., 1985), and their expression is regulated by the transcription factor Ace1 (Thiele, 1988; Welch et al., 1989). Ace1 senses intracellular copper levels by binding Cu^{+1} ions via its cysteine-rich domain, triggering a conformational change that activates transcription of MTs and other copper stress-responsive genes (Buchman et al., 1989; Fürst et al., 1988; Gross et al., 2000; Szczypka & Thiele, 1989). During copper starvation, the transcription factor Mac1 detects low copper levels and stimulates copper uptake by increasing the expression of copper importers, reductases, and chaperones (Graden & Winge, 1997; Gross et al., 2000; Jungmann et al., 1993; Martins et al., 1998; Yamaguchi-Iwai et al., 1997).

S. cerevisiae serves as an excellent model for studying transcriptional regulatory mechanisms, due to its genetic tractability and conservation with higher eukaryotes. The *CUP1* gene is widely used to investigate the dynamic regulation of chromatin organization during gene expression (Peng et al., 2015; Shen et al., 2001; Wimalarathna et al., 2012). Because nucleosomes act as barriers to transcription, chromatin remodeling is necessary during the activation of several stress-responsive genes (Erkina et al., 2010; Li & Reese, 2001; Qiu et al., 2016; Rawal et al., 2018, 2022; Sahu et al., 2021; Tomar et al., 2009)—this process displaces, evicts, or mobilizes nucleosomes, creating nucleosome-depleted regions and allowing transcriptional machinery to access DNA (Clapier et al., 2017; Sahu et al., 2020). Histone modifications and transcription factors influence the recruitment of the chromatin remodeling enzymes to target gene promoters, thereby facilitating gene activation or repression (Ford et al., 2008; Li et al., 2007; Peterson & Logie, 2000; Sahu et al., 2021, 2024; Sharma et al., 2003; Yudkovsky et al., 1999). Histone residues, therefore, play a critical role in various stress-responsive transcriptional programs (Kouzarides, 2007; Singh, Sahu, & Tomar, 2021; Thakre et al., 2019, 2020, 2021; Weiner et al., 2012). When intracellular copper levels are high, the *CUP1* gene exhibits transcriptional bursting, which depends on the short, repetitive binding of Ace1 and the rapid, repetitive chromatin remodeling mediated by the RSC (Remodeling the Structure of Chromatin) complex (Karpova et al., 2008; Mehta et al., 2018; Poorey et al., 2013; Shen et al., 2001; Wimalarathna et al., 2012). Our previous study demonstrates the role of histone H2A, H3, and H4 tail residues, as well as their post-translational modifications, in the copper sensitivity and transcriptional induction of the *CUP1* gene under elevated intracellular copper levels (Singh, Sahu, & Tomar, 2021; Singh, Sahu, Sugathan, et al., 2021). We found that modifiable lysine residues of histone H3, such as K23, K27, and K36, along with other residues in its N-terminal tail, are essential for *CUP1* induction (Singh, Sahu, & Tomar, 2021). Mutations of these residues or truncations of the H3 tail severely impair *CUP1* induction by affecting the recruitment of Ace1 and TBP to the regulatory sites of the *CUP1* promoter. Consequently, these mutants exhibit severe copper-induced proteotoxicity and fail to survive under sustained copper stress (Acharjee et al., 2025; Singh, Sahu, & Tomar, 2021). This chapter details approaches for identifying copper-sensitive histone mutants through growth assays, followed by quantifying *CUP1* induction kinetics using real-time quantitative PCR (RT-qPCR). We then describe the procedure for conducting a chromatin immunoprecipitation (ChIP) assay to examine Ace1 or TBP binding at the *CUP1* promoter during induction, and include key precautionary and optimization notes to enhance experimental throughput.

2. Before you begin

2.1 Yeast strains and growth conditions

Saccharomyces cerevisiae strains containing either wild-type or mutant H3/H4 are from the Dharmacon library (YSC5106), which are isogenic to *MATa his3Δ200 leu2Δ0 lys2Δ0 trp1Δ63 ura3Δ0 met15Δ0 can1::MFA1pr-HIS3 hht1-hhf1::NatMX4 hht2-hhf2::[HHTS-HHFS]*-URA3* (Dai et al., 2008). Ace1-myc–tagged yeast strains for the ChIP assay were created by integrating a 13-myc tag at the 3' end of *ACE1*, using the pFA-6a-13myc plasmid (Longtine et al., 1998).

Cells should be cultured at 30 °C with shaking at 200 rpm in synthetic complete (SC) medium composed of 0.18 % of amino acids, adenine and uracil mix, 0.17 % yeast nitrogen base, 0.5 % ammonium sulfate, and 2 % glucose. For the drop test, use SC-Agar plates (2 %) and incubate the cells at 30 °C using a platform incubator.

2.2 Buffer recipes

(i) ChIP-FA Lysis buffer: 50 mM HEPES, pH 8, 150 mM NaCl, 2 mM EDTA, 1 % Triton X-100, 0.1 % sodium deoxycholate, and 0.1 % sodium dodecyl sulfate (SDS)

(ii) ChIP-Low salt buffer: 20 mM HEPES, pH 8, 0.1 % Triton X-100, 2 mM EDTA, 0.1 % SDS, and 150 mM NaCl
(iii) ChIP-High salt buffer: 20 mM HEPES, pH 8, 0.1 % Triton X-100, 2 mM EDTA, 0.1 % SDS, and 500 mM NaCl
(iv) ChIP-LiCl buffer: 100 mM Tris-Cl, pH 7.5, 0.5 M LiCl, 1 % NP−40, and 1 % sodium deoxycholate
(v) 1X TE: 100 mM Tris-Cl, pH 8, and 100 mM EDTA
(vi) ChIP-Elution Buffer: 10 mM Tris-Cl pH 8, 1 mM EDTA, and 1 % SDS
(vii) 0.5 % bovine serum albumin (BSA): Dissolve BSA in 1X phosphate-buffered saline (PBS), filter-sterilize with a 0.22 μm syringe filter, and store at 4 °C.
(viii) TAE Buffer, pH 8: 40 mM Tris, 20 mM acetic acid, and 1 mM EDTA

3. Key resources table

Reagent or Resource	Source	Identifier
Antibodies		
Anti-myc antibody (9E10)	Thermo Fisher Scientific	MA1-980
Anti-TBP	Joseph C. Reese Lab	N/A
Chemicals, Peptides, and Recombinant Proteins		
Copper (II) chloride dihydrate ($CuCl_2.2H_2O$)	Millipore Sigma	307483
Sodium dodecyl sulfate (SDS) solution (10 %)	Millipore Sigma	71736
Phenol solution (pH 4.3)	Millipore Sigma	P4682
Phenol solution (pH 8.0)	Millipore Sigma	P4557
Chloroform ($CHCl_3$)	Millipore Sigma	319988
Isoamyl alcohol	Millipore Sigma	8.22255
Ethyl alcohol (≥99.45 %, for molecular biology)	Millipore Sigma	E7023
KAPA SYBR FAST qPCR Master Mix (2X) Universal	Millipore Sigma	KK4618

Bovine Serum Albumin (BSA)	Millipore Sigma	A8022
Formaldehyde solution 37 % (stabilized with about 10 % methanol)	Millipore Sigma	8.18708
Glycine	Millipore Sigma	G7126
cOmplete, Mini, EDTA-free Protease Inhibitor Cocktail (PIC)	Millipore Sigma	11836170001
Phenylmethanesulfonyl fluoride (PMSF)	Millipore Sigma	P7626
HEPES sodium salt	Millipore Sigma	H7006
Sodium chloride (NaCl)	Millipore Sigma	S9888
Lithium chloride (LiCl)	Millipore Sigma	310468
Sodium deoxycholate	Millipore Sigma	D6750
DNase/RNase-Free Deionized Water	Thermo Fisher Scientific	A57775
Proteinase-K Solution, ChIP grade (stock 20 mg/mL)	Thermo Fisher Scientific	26160
RNase-A (stock 20 mg/mL)	Thermo Fisher Scientific	12091021
Glycogen (20 mg/mL)	Thermo Fisher Scientific	R0561
GlycoBlue Coprecipitant (15 mg/mL)	Thermo Fisher Scientific	AM9515
Dynabead Protein G for Immunoprecipitation	Thermo Fisher Scientific	10004D
EDTA (0.5 M), pH 8.0	Thermo Fisher Scientific	AM9260G
NP−40	Thermo Fisher Scientific	85124
Triton X−100	Thermo Fisher Scientific	28313
1 M Tris-HCI Buffer, pH 7.5	Thermo Fisher Scientific	15567027
1 M Tris-HCI Buffer, pH 8	Thermo Fisher Scientific	15568025
GeneRuler DNA Ladders (100 bp plus)	Thermo Fisher Scientific	SM0323
iScript cDNA synthesis kit	BioRad	1708891
Experimental Models: Organisms/Strains		
Saccharomyces cerevisiae; isogenic H3/H4 mutant library strains, Genotype: *MATa his3Δ200 leu2Δ0 lys2Δ0 trp1Δ63 ura3Δ0 met15Δ0 can1::MFA1pr-HIS3 hht1-hhf1::NatMX4 hht2-hhf2::[HHTS-HHFS]*-URA3*	Dharmacon	YSC5106

Oligonucleotides		
CUP1 Fr-CCAATGCCAATGTGGT AGCTG	Millipore Sigma	N/A
CUP1 Rv- CATTTCCCAGAGCAG CATGAC	Millipore Sigma	N/A
ACT1 Fr- TCGTCGGTAGACCAAG ACAC	Millipore Sigma	N/A
ACT1 Rv- TTCTTCTGGGGCAAC TCTCA	Millipore Sigma	N/A
CUP1-P2 Fr- TTCACCACCCTTTA TTTCAGGCT	Millipore Sigma	N/A
CUP1-P2 Rv- TGGAACGGTTCAG CGGAAAA	Millipore Sigma	N/A
CUP1-P3 Fr- TTTTCCGCTGAACC GTTCCA	Millipore Sigma	N/A
CUP1-P3 Rv- TTTATGTGATGATT GATTGA	Millipore Sigma	N/A
ACE1-MYC tag Fr-GCTATACAAAC AAGTATCTTCGCATAACTCAC ATTCACAACGGATCCCCGGGT TAATTAA	Millipore Sigma	N/A
ACE1-MYC tag Rv- AATATATGTA TATGTATATTTATATAGATGT CCAGATGCAGAATTCGAGCTC GTTTAAAC	Millipore Sigma	N/A
Software and Algorithms		
Prism 10.6.1 software	Graph Pad	N/A

4. Materials and equipment

1. Petri dishes
2. Glass tubes and flasks (500 mL)
3. 96-well flat-bottom plates (Nunc)
4. Centrifuge tubes (15 and 50 mL)
5. 15 mL Bioruptor TPX tubes (Diagenode)
6. Microcentrifuge tubes (MCTs, Axygen- RNase and DNase free)

7. Screw cap MCTs
8. QIAquick PCR Purification Kit
9. Glass beads 0.5 mm (acid-washed)
10. Syringe filter 0.22 μm
11. Syringe 20 mL
12. Platform incubator
13. Shaker incubator
14. Vortex mixer
15. Multivortexor
16. Rotator
17. Magnetic rack
18. Refrigerated Centrifuge
19. Cold cabinet (4 °C)
20. 4 °C Refrigerator
21. −20 °C freezer
22. −80 °C freezer
23. Thermal mixer
24. Water Bath
25. Scanner
26. Plate reader (Biotek Instruments Inc, USA)
27. Nano-drop Spectrophotometer
28. Real-Time PCR machine
29. Diagenode Bioruptor sonicator
30. Agarose Gel Electrophoresis System
31. Gel Documentation System (LAS-mini-400)

5. Step-by-step method details

5.1 Assays of copper tolerance in yeast strains

Copper is an essential micronutrient, required in trace amounts for optimal cellular functions; however, excessive accumulation is cytotoxic (Chen et al., 2022). Here, we describe growth assays to test the sensitivity of any strain towards copper stress by the drop test, colony-forming units, and liquid kinetics assay.

5.1.1 Drop test assay and colony-forming assay to determine the copper dosage

1. Grow the yeast strains overnight in SC media at 30 °C on shaking conditions at 200 rpm.

2. Conduct a drop test analysis by preparing 10-fold serial dilutions of each strain. Spot 3 μL of both wild-type (WT) and mutant strains onto freshly prepared SC agar plates, both untreated and treated with varying concentrations of $CuCl_2 \cdot 2H_2O$ (0.5, 1, and 1.5 mM).
3. Incubate the plates at 30 °C and monitor the growth by capturing images at 48 and 72 h.
4. Following the selection of copper-sensitive strains, carry out a Colony-Forming Unit (CFU) assay to evaluate their viability across different concentrations of $CuCl_2 \cdot 2H_2O$. To determine percent viability, use the formula: (number of viable cells / total number of cells) × 100.
5. The optimum dose for subsequent experiments was determined by assessment of the viability of the mutants. For our assays, the optimal dose was 0.5 mM, at which more than 70 % of the cells remained viable after 30 min of copper treatment.

Note: Cell viability should not be severely affected by the treatment. After exposure to any concentration of copper, at least 70 % of cells should remain viable.

5.1.2 Growth kinetics analysis

Validate the growth of WT and mutant strains to assess mutant sensitivity under different conditions (untreated or $CuCl_2$-treated) by performing growth kinetics assays in liquid SC media. Dilute the overnight primary cultures to an OD_{600} of 0.2 in fresh media and treat with $CuCl_2$ as required. After treatment, dispense 200 μL of each culture into 96-well flat-bottom plates. Monitor growth using a plate reader at 30 °C with continuous orbital shaking and recording OD_{600} every 30 min for up to 48 h. Subtract the media blank OD_{600} from the sample readings and plot the corrected OD_{600} values against time.

5.2 Analysis of metallothionine gene induction

To analyze the *CUP1* induction, allow the cells to grow until mid-log phase (OD_{600} range between 0.8 and 1) in SC media at 30 °C before adding 0.5 mM $CuCl_2 \cdot 2H_2O$ to the culture. To monitor the *CUP1* induction kinetics (Peña et al., 1998), harvest the cells at different time points (0, 10, 20, 30, and 40 min) after induction. Perform total RNA extraction using the hot-phenol method as described earlier (Schmitt et al., 1990; Singh & Hinnebusch, 2025), and use 1 μg of RNA for cDNA preparation. Use standard protocols for real-time PCR. Normalize *CUP1* expression to that of the internal control gene *ACT1*, and calculate the relative fold change in gene expression using the $2^{-\Delta\Delta CT}$ method (Livak & Schmittgen, 2001). The results can then be visualized using Prism software. We noticed that *CUP1* induction peaks at about 25–30 fold after 20 min in WT cells treated with 0.5 mM $CuCl_2 \cdot 2H_2O$, while several histone mutants, such as H3 (K23Q, K27R, K36Q, Δ5–16, Δ13–16, Δ13–28, Δ25–28, Δ28–31, Δ29–32), fail to induce *CUP1* up to WT levels (Singh, Sahu, & Tomar, 2021). This suggests that the N-terminal tail residues of histone H3 are critical for efficient *CUP1* MT induction in response to increased intracellular copper accumulation.

Note: Set up the cell culture in a 250 mL or 500 mL flask. When the cells reach mid-log phase, remove 10 mL of culture as the uninduced sample. Then, treat the remaining cells (about 50–60 mL) with 0.5 mM $CuCl_2 \cdot 2H_2O$ and collect around 10 mL at various time points. This approach reduces sample variability and improves reproducibility. Immediately freeze the harvested cell pellets at −80 °C to prevent any changes in *CUP1* levels. Avoid any heat shock during culturing or harvesting.

5.3 Chromatin immunoprecipitation assay to investigate the mechanisms of metallothionein gene induction

To investigate the role of histone residues in regulating *CUP1* transcription, we performed Ace1 and TBP ChIP-qPCR analyses on copper-sensitive histone mutants that display reduced *CUP1* expression compared to wild-type cells (Singh, Sahu, & Tomar, 2021). A ChIP assay consists of several key steps (Fig. 1): proteins are crosslinked to DNA, chromatin is fragmented, and immunoprecipitation is performed using a specific antibody. The crosslinks are then reversed to purify the DNA for analysis. DNA regions bound by the protein can be identified using techniques such as RT-qPCR or next-generation sequencing (ChIP-seq).

Day 1: Culturing, crosslinking, and harvesting cells

1. Inoculate cells in 10 mL of SC-selection media before beginning the experiment. Incubate the preculture overnight at 30 °C, shaking at 200 rpm.
2. Regrow the cells from the preculture (with a starting OD_{600} of 0.1 or 0.2) in two flasks containing 100 mL of SC selection media until they reach mid-log phase (OD_{600} = 0.8). Induce cells in one flask by adding 0.5 mM $CuCl_2 \cdot 2H_2O$ (treated), then incubate for 10 min. Leave the other flask untreated.

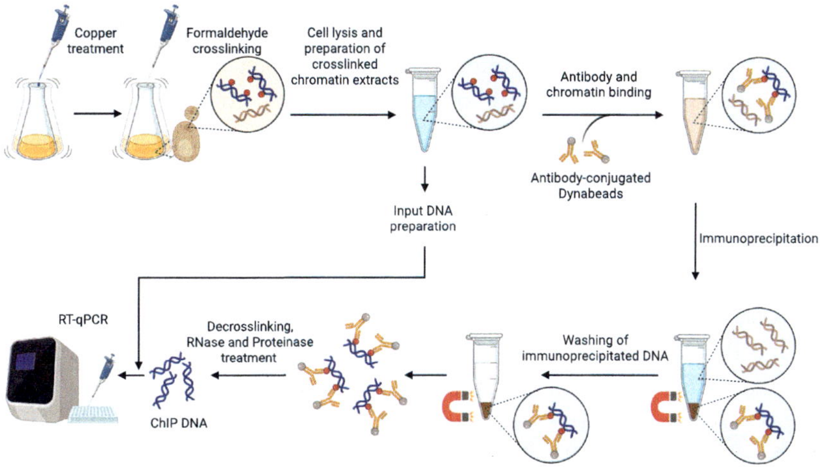

Fig. 1 Schematic illustrating the workflow of a chromatin immunoprecipitation assay. *Created in BioRender. Singh, S. (2025) https://BioRender.com/8hd1689.*

Note: Based on the kinetics of *CUP1* expression observed over a 40-minute time course, the 10-minute time point was identified as optimal for induction and was therefore maintained for ChIP experiments.

3. Crosslink both copper-treated and untreated cells by adding 2.7 mL of 37 % formaldehyde stock directly to the culture medium to achieve a final concentration of 1 %. Incubate the cells at 25 °C with constant shaking for 15 min.
4. Add 6 mL of freshly prepared 2.5 M glycine to quench excess cross-linking agents. Incubate at 25 °C with constant shaking for 5 min.
5. Transfer the cells to two 50 mL centrifuge tubes and spin the cells at 3000 rpm for 5 min at 4 °C.
6. Resuspend the cells in pre-chilled 1X PBS and transfer them to two 1.5 mL microcentrifuge tubes. Wash the cells twice by adding PBS and centrifuging at 3000 rpm for 5 min at 4°C. Carefully remove the PBS and store the resulting cell pellets at −80 °C until lysis and sonication.

Note: In this protocol, 100 mL of cells were pelleted into two 1.5 mL microcentrifuge tubes, and the lysis and sonication steps were optimized accordingly. You may pellet 100 mL of cells in one MCT, adjusting lysis and sonication steps as needed.

Day 2: Cell lysis and preparation of crosslinked chromatin extracts

7. Prepare 50 mL of lysis buffer by adding 1 tablet of protease inhibitor cocktail (PIC) and 500 μL of 100 mM PMSF to the ChIP FA lysis buffer.

   Note: PMSF has a very short half-life and therefore must be added just before starting the lysis step.
8. Take the pellets out of the −80 °C freezer and thaw them on ice for 5 min.
9. Resuspend each cell pellet in 200 μL of FA lysis buffer containing PIC and PMSF. Add an equal volume of 0.5 mm glass beads to each pellet. Ensure that no glass beads remain on the tube rim, as they may cause leakage during vortexing.
10. Perform mechanical lysis by briefly vortexing the sample for 30 s, then placing it on ice for 30 s. Repeat this cycle up to 5 times.
11. Add 200 μL FA lysis buffer containing PIC and PMSF to each sample and repeat step 10.
12. Again, add 200 μL FA lysis buffer containing PIC and PMSF to each sample and place the tubes on a multi-vortexer and shake them continuously at maximum speed at 4 °C for 30 min.
13. After shaking, transfer the tubes to ice. Puncture tube bottoms using a red-hot needle, place them on new MCTs, and centrifuge at 2000 rpm for 30 s at 4 °C to collect cell extracts.
14. Collect the respective cell extracts from two MCTs in a 15 mL centrifuge tube and add 0.6 mL FA Lysis buffer containing PIC and PMSF, bringing the final volume to 1.8 mL.

    Note: For efficient sonication, 15 mL Bioruptor TPX tubes by Diagenode can be used.
15. Fill pre-chilled distilled water up to the marked line in the Diagenode Bioruptor. Arrange the centrifuge tubes on the tube holder. Set the sonicator to deliver 5 pulses (30 s ON, 30 s OFF) at maximum amplitude. Sonicate the samples for 3 rounds (a total of 15 cycles).

    Note: (i) Sonication cycles should be optimized according to the number of cells. A greater number of cells requires more sonication cycles.

    (ii) To prevent the denaturation of chromatin samples, keep the sonicator's water bath at approximately 4 °C throughout the procedure. After every five cycles, remove the water and refill the bath with chilled distilled water to maintain the temperature. In newer sonicators, the water cooler units are attached to circulate cold water through the sonication bath, so changing the water after every five cycles is not necessary in these models.

16. After sonication, split the cell extracts into two MCTs and centrifuge at 13,000 rpm for 30 min at 4 °C to clarify the lysate, and then transfer the supernatant to fresh tubes.
17. Remove 200 μL of cell extract for input DNA preparation and store the remaining lysate at −80 °C. For input DNA preparation, perform the decrosslinking and DNA extraction steps as described below.
18. Add 200 μL of elution buffer and 10 μL of Proteinase-K (from a 20 mg/mL stock) to 200 μL of cell extract. Mix well and incubate at 65 °C overnight in a water bath for decrosslinking.

Day 3: Input DNA preparation

19. On the following day, remove the tubes and allow the samples to cool to room temperature. Centrifuge the tubes to collect the condensates, then add 400 μL of phenol-chloroform-isoamyl alcohol (PCIAA, 25:24:1).
20. Vortex the tubes and centrifuge at 13,000 rpm for 15 min at room temperature.
21. Transfer the supernatant to new MCTs, then add 200 μL of chloro form-isoamyl alcohol (CIAA, 24:1).
22. Vortex and centrifuge at 13,000 rpm for 15 min at room temperature.
23. Transfer the supernatant into fresh MCTs. Add 1 mL of 100 % ethanol to each tube, then keep the samples at −80 °C for 30–60 min.
24. Centrifuge the tubes at 13,000 rpm for 5 min at 4 °C and carefully remove the ethanol.
25. Wash the pellet with 70 % ethanol by spinning at 13,000 rpm for 5 min at 4 °C, and air-dry the pellets.
26. Dissolve the pellet in 100 μL of 1X TE buffer. Add 1 μL of RNase-A (from a 20 mg/mL stock), and incubate at 37 °C for 1 h. This DNA will serve as the input DNA. Measure the DNA concentration of each sample using a Nanodrop instrument to determine the chromatin volume required for immunoprecipitation. Alternatively, the DNA can be purified using a QIAquick PCR Purification Kit, eluted in 100 μL of elution buffer, and quantified.
27. Check the size of sheared chromatin on 2 % agarose gel by loading 20 μL of Input DNA along with the 100 bp DNA ladder on the gel. Ideally, the resulting DNA fragments should range between 100 and 600 base pairs (bp) (Fig. 2).

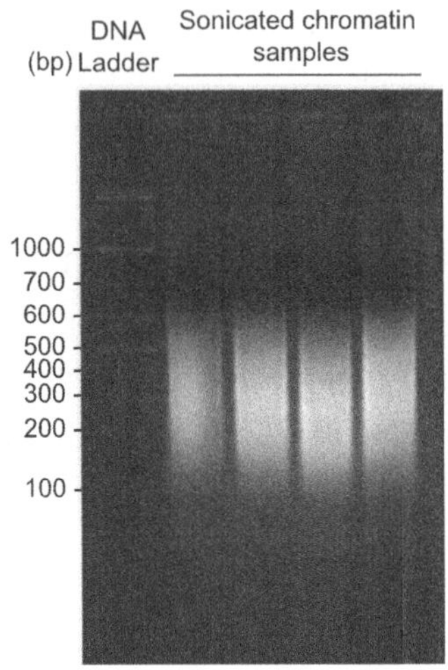

Fig. 2 Agarose gel electrophoresis image of chromatin samples sonicated with the Diagenode Bioruptor.

Day 4: Immunoprecipitation (IP)

28. Thoroughly resuspend the Dynabeads on a rotator before use to ensure that the bead slurry is evenly mixed. Make a small cut in the pipette tip, and transfer the total volume of beads needed—30 μL of bead slurry per sample—into an MCT.

29. Carefully remove the storage solution by placing the beads on a magnetic stand for 1–2 min.

30. Remove the tube from the magnet. Add 1 mL of 0.5 % BSA, then wash the beads on a rotator at 4 °C for 1 min. Repeat the wash two more times.

Note: BSA should be freshly prepared in 1X PBS and filter sterilized.

31. Resuspend the beads in 0.5 % BSA, adjusting the total volume based on the number of IP samples (use 50 μL per IP sample).

32. Add 2−4 μg of the antibody of interest per IP (e.g., for Myc and TBP) to the beads.

33. Incubate the mixture on a rotator at 4 °C for 2 h. The incubation time can be adjusted depending on the antibody's affinity for specific beads (protein A or protein G Dynabeads).

Note: Here, the beads required for all IPs were washed together in a single MCT and incubated with the specific antibody in one tube. Mixing and incubating in a single tube minimizes sample variation.

34. Remove the supernatant by placing the beads on a magnetic stand for 1–2 min.
35. Add 1 mL of 0.5 % BSA to the antibody-conjugated beads and wash them on a rotator at 4 °C for 1 min. Repeat this wash one additional time.
36. Resuspend the beads in 0.5 % BSA, based on the number of IP samples. Next, transfer equal aliquots of beads (50 μL) into new MCTs for each sample.
37. Add 2.5–5 μg of sonicated chromatin to the antibody-conjugated beads, based on the DNA quantification from step 26. If necessary, adjust the volume to 250 μL with FA lysis buffer (containing PIC and PMSF). Incubate the mixture overnight at 4°C in a cold room or cold cabinet, using a rotator set to low speed.

Day 5: Washing of Immunoprecipitated chromatin

38. Place the tubes on the magnetic rack for 1–2 min to allow the beads to separate. Carefully remove the unbound lysate using a pipette. Add 1 mL of low salt buffer (containing PIC and PMSF) to the chromatin-bound beads, then wash by rotating the tubes at 4 °C for 10 min at low rpm.
39. Place the tubes on magnetic stand for 1-2 min and carefully remove the buffer. Repeat the low salt buffer washing step.
40. Wash the beads once with 1 mL of high-salt buffer (containing PIC and PMSF) for 10 min at 4 °C.
41. Wash the beads once using 1 mL of LiCl buffer for 10 min on rotation at 4 °C.
42. Wash the beads once using 1 mL TE pH 8 for 10 min on rotation at 4 °C.
43. Remove the TE by placing the tubes on the magnetic rack. Gently centrifuge the beads at 1000 rpm for 15 s, then place them back on the magnetic rack and remove any remaining traces of TE.
44. Add 150 μL of elution buffer to the washed chromatin-bound beads. Place the tubes on a thermal mixer set to 65 °C and shake at 500–1000 rpm for 30 min.

    Note: To prevent sample spillage and evaporation during elution at 65 °C on the thermal mixer, wrap the MCT caps with parafilm.
45. Place the tubes on the magnetic rack and transfer the supernatant to a new set of screw cap MCTs.

46. Repeat the elution by adding 100 μL of elution buffer to the beads and incubating at 65 °C on a thermal mixer for 30 min with shaking.
47. Afterwards, place the tubes on the magnetic rack and carefully collect the supernatant into the corresponding screw cap MCTs.
48. To decrosslink, place the tubes in a water bath at 65 °C and incubate overnight.

Day 6: RNase-A and Proteinase-K treatment and ChIP DNA isolation

49. The next day, remove the tubes from water bath, gently spin the tubes to collect the condensate and allow the samples to cool to room temperature. Add 250 μL of 1X TE and 5 μL of RNase-A (from a 20 mg/mL stock) to each sample. Briefly vortex and spin the mixtures, then incubate at 37 °C in a water bath for 2 h.
50. Add 10 μL of Proteinase K (from a 20 mg/mL stock solution) to each sample, briefly vortex, spin down, and incubate at 55 °C for 2 h.
51. After treating the samples with RNase-A and Proteinase-K, proceed with DNA isolation. Add 500 μL of PCIAA (25:24:1) to each tube, gently vortex, and centrifuge at 13,000 rpm for 15 min at room temperature.
52. Transfer the supernatant to new MCTs. Add glycogen to a final concentration of 2 mg/mL, followed by 1 mL of 100 % ethanol. Place the tubes at −80 °C for 30 min.

    Note: Instead of glycogen, you may add 1.2 μL of GlycoBlue co-precipitant (15 mg/mL stock) during ethanol precipitation. GlycoBlue colors the DNA pellet blue, making it easier to see and reducing the risk of pellet loss during washing.
53. Centrifuge the tubes at 13,000 rpm for 10 min at 4 °C.
54. Wash the pellet with 70 % ethanol by centrifuging at 13,000 rpm for 10 min at 4 °C. Carefully handle and air-dry the pellet after centrifugation.
55. Dissolve the DNA pellet in 50 μL DNase-free ultrapure water and store the Chip DNA at −20 °C.

Precautions:

1. Prepare all buffers fresh and store them at 4 °C during the experiment.
2. To prevent chromatin degradation, keep samples at 4 °C or on ice during all immunoprecipitation and washing steps.
3. Limit sonication to 4–5 cycles at a time, pausing for 10−15 min before repeating. This prevents machine overheating and sample denaturation. During any pause in sonication, be sure to keep chromatin samples on ice.

4. Always store Dynabeads at 4 °C to prevent denaturation. Avoid freezing the Dynabeads, as it is not recommended.
5. Be careful when dispensing Dynabeads into MCTs. Add an equal amount of beads to each MCT to avoid variations in pulldown.
6. While washing the beads with buffer, take care not to lose any beads. Allow the beads to completely settle on the magnetic rack before removing the supernatant.

6. Quantification and statistical analysis

1. Quantification of ChIP enrichment by RT-qPCR:
   i. To normalize and accurately assess enrichment of target DNA in ChIP samples compared to the starting material (input), dilute each input DNA to create a consistent baseline. Make serial dilutions in the linear range—1/200, 1/400, 1/600, 1/800, 1/1000, and 1/1200—using 0.1X TE buffer.
   ii. Perform a semiquantitative PCR with 1 μL from each dilution, using control primers (*ACT1*) in a 10 μL PCR reaction (28–30 cycles), to identify the optimal dilution for ChIP-qPCR.
   iii. By analyzing the gel electrophoresis, determine an input dilution for each sample that is not saturated and approximately equivalent to the rest of the samples.
   iv. Perform RT qPCR on ChIP-DNA and the corresponding diluted Input-DNA using specific target oligonucleotides.
   v. ChIP enrichment will be normalized to input DNA and represented as Enrichment/Input (%). To calculate this, use the formula:

   $$\Delta Ct = Ct_{\text{ChIP}} - [Ct_{\text{Input}} - \log_{E} (\text{Input dilution factor})],$$

   where Ct is the cycle threshold and E is the specific primer efficiency (2 for 100 % efficiency). Enrichment/Input (%) is then given as $E^{-\Delta Ct}$ (or $2^{-\Delta Ct}$ for oligos with 100 % efficiency). Values can be visualized using GraphPad Prism software.
2. To ensure reproducibility, include at least three biological replicates for each ChIP experiment. Assess statistical significance using the unpaired Mann-Whitney *t*-test for two-sample comparisons. For multiple pairwise comparisons, use one-way or two-way ANOVA tests. Statistical analyses and data visualization can be performed using GraphPad Prism.

7. Expected outcomes

Our prior study identified certain histone residues and modifications essential for inducing *CUP1* MT, which helps cells tolerate elevated intracellular copper (Singh, Sahu, & Tomar, 2021). Mutations in the N-terminal tails of histone H3 (K23Q, K27R, K36Q, Δ5–16, Δ13–16, Δ13–28, Δ25–28, Δ28–31, and Δ29–32) resulted in decreased *CUP1* expression and rendered cells unable to withstand copper accumulation or copper-induced protein misfolding. We hypothesized that such mutations affect *CUP1* transcription by altering the binding of Ace1 and TBP at the *CUP1* promoter. To test this, we performed Ace1 ChIP at the *CUP1* promoter in copper-sensitive histone mutants. Ace1 binding sites were determined from the literature, indicating that Ace1 binds to the upstream activation sequence (UAS) region of the *CUP1* promoter (Buchman et al., 1990; Huibregtse et al., 1989; Karin et al., 1984; Shen et al., 2002). To assess Ace1 occupancy at the *CUP1* locus, we designed primers targeting two regions relative to the transcription start site (TSS): distal (−182 to −332, *CUP1-P2*) and proximal (−1 to −202, *CUP1-P3*) (key resources table). Fig. 3 presents a schematic of the *CUP1* locus, including the UAS, Ace1-binding sites, and TATA boxes.

To measure pre-initiation complex assembly, we examined TBP binding at the TATA box sequence in the *CUP1* promoter. The *CUP1* promoter region P3 (−1 to −202) contains two TATA boxes (Karin et al., 1984), so higher TBP binding is expected at this locus compared to P2, which lacks TBP binding sites. Based on our previous findings (Singh, Sahu, & Tomar, 2021), Ace1 binding at the *CUP1* UAS regions and TBP occupancy at the *CUP1* TATA boxes are expected to decrease in the H3K23Q, K36Q, Δ5–16, Δ13–28, Δ25–28, and Δ28–31 mutants compared to H3WT cells, correlating with reduced *CUP1* transcription.

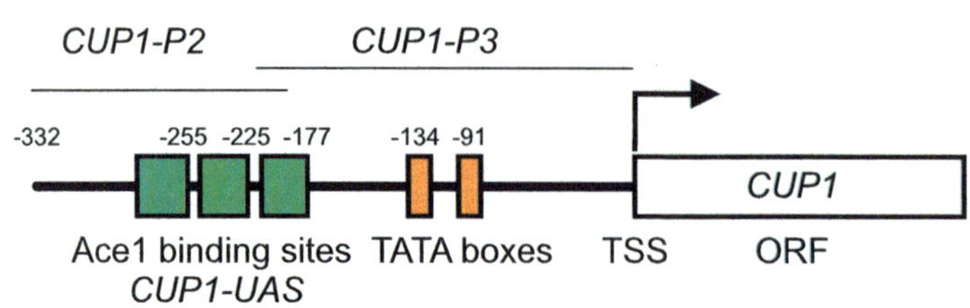

Fig. 3 Schematic of the *CUP1* locus showing the transcription regulatory sites.

8. Advantages and limitations

ChIP is an efficient and powerful assay for measuring specific DNA–protein interactions *in vivo*, using selective antibodies against the protein of interest. It is a rapid and quantitative method with several advantages. For example, ChIP can be used to compare the presence of specific proteins at various loci, map histone modifications at genomic regions of interest, or quantify transcription factor binding at inducible gene promoters after specific stimulation. However, because ChIP relies on antibodies to isolate specific proteins and involves mechanical lysis of chromatin samples, it can sometimes introduce misleading artifacts. Therefore, results should be interpreted with caution, and various internal controls should be employed when analyzing ChIP data.

A positive control, such as a well-characterized genomic region previously defined in the literature, should be included to confirm that the ChIP protocol is working. Including Negative controls, such as primers targeting intergenic regions that should not show enrichment, helps estimate background signal and noise. Additionally, each immunoprecipitation should be normalized to the input DNA amount for each strain and condition. To minimize experimental limitations, the ChIP assay protocol should integrate careful antibody selection, appropriate controls, use of biological replicates, optimization of shearing protocols, and rigorous downstream analysis.

9. Optimization and troubleshooting

1. Transcriptional measurements of stress-inducible genes are highly sensitive; any changes in growth conditions or exposure to external stress can significantly affect transcript levels. Yeast cells exhibit an integrated stress response (ISR), and minor alterations in nutritional conditions or temperature may trigger the ISR. Therefore, it is crucial to exercise caution during cell growth and induction to ensure consistent experimental outcomes.
2. For fast-cycling transcription factors, selecting the optimal induction time point is crucial. Run a time-course experiment to determine when the maximum binding events are captured.
3. Successful cell lysis can be visualized under a microscope. Take a 10 μL sample before and after lysis to observe the nucleus versus whole cells.

4. Before proceeding with the ChIP assay, it is recommended to check the bulk levels of the protein of interest using a Western blot assay. A reduction in protein levels can lead to decreased genomic binding and ChIP bias.
5. Crosslinking time is critical, as over-crosslinking can introduce artifacts into the ChIP reaction. Optimize this step carefully to avoid both under- and over-crosslinking. Also, monitor the quenching reaction duration.
6. Optimize chromatin sonication based on your starting material. Initially, use WT or control strains to test different sonication cycles.
7. Specificity of the antibody is critical in ChIP assays. Whenever possible, use ChIP-grade or previously validated antibodies. If these are not available, tag the target protein with common epitopes (like Myc, GFP, FLAG, His, or TAP) so that you can use commercially available ChIP-grade antibodies against these tags for immunoprecipitation.
8. Various beads are available for immunoprecipitation. Choose Dynabead Protein A or G based on your immunoglobulin species and subclass, or use antibody-conjugated Dynabeads directly.
9. Washing steps are critical to reduce background signals. Increase wash stringency or frequency if background noise persists.
10. Normalize each IP sample to its respective input DNA to correct for fragmentation or pulldown bias.

10. Concluding remarks

Histone post-translational modifications are key regulators of transcription, orchestrating gene expression through the recruitment of chromatin remodelers and transcriptional coactivators to specific promoters. Chromatin-remodeling enzymes equipped with histone reader domains recognize these histone marks, enabling nucleosome repositioning and histone displacement that facilitate transcriptional activation. Perturbations in these modifications—whether through loss or excessive accumulation—can disrupt pre-initiation complex assembly and impair proper transcriptional control. The experimental workflow and chromatin immunoprecipitation assay detailed in this chapter provide a framework for investigating the mechanistic roles of histone modifications in transcriptional regulation, particularly under stress-induced conditions.

Acknowledgements

We thank Joseph C. Reese for providing us with the anti-TBP antibody. This work was supported by funds from the Council of Scientific and Industrial Research, Government of India [CSIR grant no. 38(1468)/18/EMR-II] to R.S.T. and intramural funds from IISER Bhopal.

Author contributions

S.S. and R.K.S. optimized the protocol and conducted the experiments. S.S., R.K.S., and R.S.T. wrote the manuscript.

References

Acharjee, S., Pal, R., Anand, S., Thakur, P., Anjana, V., Singh, R., ... Tomar, R. S. (2025). Mutations in histones dysregulate copper homeostasis leading to defect in Sec61-dependent protein translocation mechanism in *Saccharomyces cerevisiae*. *The Journal of Biological Chemistry, 301*(2), 108163. https://doi.org/10.1016/j.jbc.2025.108163.

Askwith, C., Eide, D., Van Ho, A., Bernard, P. S., Li, L., Davis-Kaplan, S., ... Kaplan, J. (1994). The *FET3* gene of *S. cerevisiae* encodes a multicopper oxidase required for ferrous iron uptake. *Cell, 76*(2), 403–410. https://doi.org/10.1016/0092-8674(94)90346-8.

Baker, Z. N., Cobine, P. A., & Leary, S. C. (2017). The mitochondrion: A central architect of copper homeostasis. *Metallomics: Integrated Biometal Science, 9*(11), 1501–1512. https://doi.org/10.1039/c7mt00221a.

Buchman, C., Skroch, P., Dixon, W., Tullius, T. D., & Karin, M. (1990). A single amino acid change in CUP2 alters its mode of DNA binding. *Molecular and Cellular Biology, 10*(9), 4778–4787. https://doi.org/10.1128/mcb.10.9.4778-4787.1990.

Buchman, C., Skroch, P., Welch, J., Fogel, S., & Karin, M. (1989). The *CUP2* gene product, regulator of yeast metallothionein expression, is a copper-activated DNA-binding protein. *Molecular and Cellular Biology, 9*(9), 4091–4095. https://doi.org/10.1128/mcb.9.9.4091-4095.1989.

Chen, L., Min, J., & Wang, F. (2022). Copper homeostasis and cuproptosis in health and disease. *Signal Transduction and Targeted Therapy,* 7(1), 378. https://doi.org/10.1038/s41392-022-01229-y.

Clapier, C. R., Iwasa, J., Cairns, B. R., & Peterson, C. L. (2017). Mechanisms of action and regulation of ATP-dependent chromatin-remodelling complexes. *Nature Reviews. Molecular Cell Biology, 18*(7), 407–422. https://doi.org/10.1038/nrm.2017.26.

Culotta, V. C., Howard, W. R., & Liu, X. F. (1994). *CRS5* encodes a metallothionein-like protein in *Saccharomyces cerevisiae*. *The Journal of Biological Chemistry, 269*(41), 25295–25302.

Culotta, V. C., Joh, H. D., Lin, S. J., Slekar, K. H., & Strain, J. (1995). A physiological role for *Saccharomyces cerevisiae* copper/zinc superoxide dismutase in copper buffering. *The Journal of Biological Chemistry, 270*(50), 29991–29997. https://doi.org/10.1074/jbc.270.50.29991.

Culotta, V. C., Klomp, L. W., Strain, J., Casareno, R. L., Krems, B., & Gitlin, J. D. (1997). The copper chaperone for superoxide dismutase. *The Journal of Biological Chemistry, 272*(38), 23469–23472. https://doi.org/10.1074/jbc.272.38.23469.

Dai, J., Hyland, E. M., Yuan, D. S., Huang, H., Bader, J. S., & Boeke, J. D. (2008). Probing nucleosome function: A highly versatile library of synthetic histone H3 and H4 mutants. *Cell, 134*(6), 1066–1078. https://doi.org/10.1016/j.cell.2008.07.019.

Dancis, A., Yuan, D. S., Haile, D., Askwith, C., Eide, D., Moehle, C., ... Klausner, R. D. (1994). Molecular characterization of a copper transport protein in *S. cerevisiae*: An unexpected role for copper in iron transport. *Cell, 76*(2), 393–402. https://doi.org/10.1016/0092-8674(94)90345-x.

Ecker, D. J., Butt, T. R., Sternberg, E. J., Neeper, M. P., Debouck, C., Gorman, J. A., & Crooke, S. T. (1986). Yeast metallothionein function in metal ion detoxification. *The Journal of Biological Chemistry, 261*(36), 16895–16900.

Erkina, T. Y., Zou, Y., Freeling, S., Vorobyev, V. I., & Erkine, A. M. (2010). Functional interplay between chromatin remodeling complexes RSC, SWI/SNF and ISWI in regulation of yeast heat shock genes. *Nucleic Acids Research, 38*(5), 1441–1449. https://doi.org/10.1093/nar/gkp1130.

Fogel, S., & Welch, J. W. (1982). Tandem gene amplification mediates copper resistance in yeast. *Proceedings of the National Academy of Sciences of the United States of America, 79*(17), 5342–5346. https://doi.org/10.1073/pnas.79.17.5342.

Ford, J., Odeyale, O., & Shen, C.-H. (2008). Activator-dependent recruitment of SWI/SNF and INO80 during INO1 activation. *Biochemical and Biophysical Research Communications, 373*(4), 602–606. https://doi.org/10.1016/j.bbrc.2008.06.079.

Fürst, P., Hu, S., Hackett, R., & Hamer, D. (1988). Copper activates metallothionein gene transcription by altering the conformation of a specific DNA binding protein. *Cell, 55*(4), 705–717. https://doi.org/10.1016/0092-8674(88)90229-2.

Georgatsou, E., Mavrogiannis, L. A., Fragiadakis, G. S., & Alexandraki, D. (1997). The yeast Fre1p/Fre2p cupric reductases facilitate copper uptake and are regulated by the copper-modulated Mac1p activator. *The Journal of Biological Chemistry, 272*(21), 13786–13792. https://doi.org/10.1074/jbc.272.21.13786.

Glerum, D. M., Shtanko, A., & Tzagoloff, A. (1996). Characterization of COX17, a yeast gene involved in copper metabolism and assembly of cytochrome oxidase. *The Journal of Biological Chemistry, 271*(24), 14504–14509. https://doi.org/10.1074/jbc.271.24.14504.

Graden, J. A., & Winge, D. R. (1997). Copper-mediated repression of the activation domain in the yeast Mac1p transcription factor. *Proceedings of the National Academy of Sciences of the United States of America, 94*(11), 5550–5555. https://doi.org/10.1073/pnas.94.11.5550.

Gross, C., Kelleher, M., Iyer, V. R., Brown, P. O., & Winge, D. R. (2000). Identification of the copper regulon in *Saccharomyces cerevisiae* by DNA microarrays. *The Journal of Biological Chemistry, 275*(41), 32310–32316. https://doi.org/10.1074/jbc.M005946200.

Hassett, R., & Kosman, D. J. (1995). Evidence for Cu(II) reduction as a component of copper uptake by *Saccharomyces cerevisiae. The Journal of Biological Chemistry, 270*(1), 128–134. https://doi.org/10.1074/jbc.270.1.128.

Horn, D., & Barrientos, A. (2008). Mitochondrial copper metabolism and delivery to cytochrome c oxidase. *IUBMB Life, 60*(7), 421–429. https://doi.org/10.1002/iub.50.

Huibregtse, J. M., Engelke, D. R., & Thiele, D. J. (1989). Copper-induced binding of cellular factors to yeast metallothionein upstream activation sequences. *Proceedings of the National Academy of Sciences of the United States of America, 86*(1), 65–69. https://doi.org/10.1073/pnas.86.1.65.

Jensen, L. T., Howard, W. R., Strain, J. J., Winge, D. R., & Culotta, V. C. (1996). Enhanced effectiveness of copper ion buffering by CUP1 metallothionein compared with CRS5 metallothionein in *Saccharomyces cerevisiae. The Journal of Biological Chemistry, 271*(31), 18514–18519. https://doi.org/10.1074/jbc.271.31.18514.

Jungmann, J., Reins, H. A., Lee, J., Romeo, A., Hassett, R., Kosman, D., & Jentsch, S. (1993). MAC1, a nuclear regulatory protein related to Cu-dependent transcription factors is involved in Cu/Fe utilization and stress resistance in yeast. *The EMBO Journal, 12*(13), 5051–5056. https://doi.org/10.1002/j.1460-2075.1993.tb06198.x.

Karin, M., Najarian, R., Haslinger, A., Valenzuela, P., Welch, J., & Fogel, S. (1984). Primary structure and transcription of an amplified genetic locus: The *CUP1* locus of yeast. *Proceedings of the National Academy of Sciences of the United States of America, 81*(2), 337–341. https://doi.org/10.1073/pnas.81.2.337.

Karpova, T. S., Kim, M. J., Spriet, C., Nalley, K., Stasevich, T. J., Kherrouche, Z., ... McNally, J. G. (2008). Concurrent fast and slow cycling of a transcriptional activator at an endogenous promoter. *Science (New York, N. Y.), 319*(5862), 466–469. https://doi.org/10.1126/science.1150559.

Kim, B.-E., Nevitt, T., & Thiele, D. J. (2008). Mechanisms for copper acquisition, distribution and regulation. *Nature Chemical Biology, 4*(3), 176–185. https://doi.org/10.1038/nchembio.72.

Kouzarides, T. (2007). Chromatin modifications and their function. *Cell, 128*(4), 693–705. https://doi.org/10.1016/j.cell.2007.02.005.

Krishnamoorthy, L., Cotruvo, J. A., Chan, J., Kaluarachchi, H., Muchenditsi, A., Pendyala, V. S., ... Chang, C. J. (2016). Copper regulates cyclic-AMP-dependent lipolysis. *Nature Chemical Biology, 12*(8), 586–592. https://doi.org/10.1038/nchembio.2098.

Li, B., Carey, M., & Workman, J. L. (2007). The role of chromatin during transcription. *Cell, 128*(4), 707–719. https://doi.org/10.1016/j.cell.2007.01.015.

Li, B., & Reese, J. C. (2001). Ssn6-Tup1 regulates *RNR3* by positioning nucleosomes and affecting the chromatin structure at the upstream repression sequence. *The Journal of Biological Chemistry, 276*(36), 33788–33797. https://doi.org/10.1074/jbc.M104220200.

Livak, K. J., & Schmittgen, T. D. (2001). Analysis of relative gene expression data using real-time quantitative PCR and the 2(-Delta Delta C(T)) Method. *Methods (San Diego, Calif.), 25*(4), 402–408. https://doi.org/10.1006/meth.2001.1262.

Longtine, M. S., McKenzie, A., Demarini, D. J., Shah, N. G., Wach, A., Brachat, A., ... Pringle, J. R. (1998). Additional modules for versatile and economical PCR-based gene deletion and modification in *Saccharomyces cerevisiae. Yeast (Chichester, England), 14*(10), 953–961. https://doi.org/10.1002/(SICI)1097-0061(199807)14:103.0.CO;2-U.

Martins, L. J., Jensen, L. T., Simon, J. R., Keller, G. L., & Winge, D. R. (1998). Metalloregulation of *FRE1* and *FRE2* homologs in *Saccharomyces cerevisiae. The Journal of Biological Chemistry, 273*(37), 23716–23721. https://doi.org/10.1074/jbc.273.37.23716.

Mehta, G. D., Ball, D. A., Eriksson, P. R., Chereji, R. V., Clark, D. J., McNally, J. G., & Karpova, T. S. (2018). Single-molecule analysis reveals linked cycles of RSC chromatin remodeling and Ace1p transcription factor binding in yeast. *Molecular Cell, 72*(5), 875–887.e9. https://doi.org/10.1016/j.molcel.2018.09.009.

Peña, M. M., Koch, K. A., & Thiele, D. J. (1998). Dynamic regulation of copper uptake and detoxification genes in *Saccharomyces cerevisiae. Molecular and Cellular Biology, 18*(5), 2514–2523. https://doi.org/10.1128/MCB.18.5.2514.

Pena, M. M., Puig, S., & Thiele, D. J. (2000). Characterization of the *Saccharomyces cerevisiae* high affinity copper transporter Ctr3. *The Journal of Biological Chemistry, 275*(43), 33244–33251. https://doi.org/10.1074/jbc.M005392200.

Peng, B., Williams, T. C., Henry, M., Nielsen, L. K., & Vickers, C. E. (2015). Controlling heterologous gene expression in yeast cell factories on different carbon substrates and across the diauxic shift: A comparison of yeast promoter activities. *Microbial Cell Factories, 14*, 91. https://doi.org/10.1186/s12934-015-0278-5.

Peterson, C. L., & Logie, C. (2000). Recruitment of chromatin remodeling machines. *Journal of Cellular Biochemistry, 78*(2), 179–185. https://doi.org/10.1002/(sici)1097-4644(20000801)78:23.0.co;2-e.

Poorey, K., Viswanathan, R., Carver, M. N., Karpova, T. S., Cirimotich, S. M., McNally, J. G., ... Auble, D. T. (2013). Measuring chromatin interaction dynamics on the second time scale at single-copy genes. *Science (New York, N. Y.), 342*(6156), 369–372. https://doi.org/10.1126/science.1242369.

Pufahl, R. A., Singer, C. P., Peariso, K. L., Lin, S. J., Schmidt, P. J., Fahrni, C. J., ... O'Halloran, T. V. (1997). Metal ion chaperone function of the soluble Cu(I) receptor Atx1. *Science (New York, N. Y.), 278*(5339), 853–856. https://doi.org/10.1126/science.278.5339.853.

Qiu, H., Chereji, R. V., Hu, C., Cole, H. A., Rawal, Y., Clark, D. J., & Hinnebusch, A. G. (2016). Genome-wide cooperation by HAT Gcn5, remodeler SWI/SNF, and chaperone Ydj1 in promoter nucleosome eviction and transcriptional activation. *Genome Research, 26*(2), 211–225. https://doi.org/10.1101/gr.196337.115.

Rawal, Y., Chereji, R. V., Qiu, H., Ananthakrishnan, S., Govind, C. K., Clark, D. J., & Hinnebusch, A. G. (2018). SWI/SNF and RSC cooperate to reposition and evict promoter nucleosomes at highly expressed genes in yeast. *Genes & Development, 32*(9–10), 695–710. https://doi.org/10.1101/gad.312850.118.

Rawal, Y., Qiu, H., & Hinnebusch, A. G. (2022). Distinct functions of three chromatin remodelers in activator binding and preinitiation complex assembly. *PLoS Genetics, 18*(7), e1010277. https://doi.org/10.1371/journal.pgen.1010277.

Sahu, R. K., Dhakshnamoorthy, J., Jain, S., Folco, H. D., Wheeler, D., & Grewal, S. I. S. (2024). Nucleosome remodeler exclusion by histone deacetylation enforces heterochromatic silencing and epigenetic inheritance. *Molecular Cell, 84*(17), 3175–3191.e8. https://doi.org/10.1016/j.molcel.2024.07.006.

Sahu, R. K., Singh, S., & Tomar, R. S. (2020). The mechanisms of action of chromatin remodelers and implications in development and disease. *Biochemical Pharmacology, 180*, 114200. https://doi.org/10.1016/j.bcp.2020.114200.

Sahu, R. K., Singh, S., & Tomar, R. S. (2021). The ATP-dependent SWI/SNF and RSC chromatin remodelers cooperatively induce unfolded protein response genes during endoplasmic reticulum stress. *Biochimica et Biophysica Acta. Gene Regulatory Mechanisms, 1864*(11–12), 194748. https://doi.org/10.1016/j.bbagrm.2021.194748.

Schmitt, M. E., Brown, T. A., & Trumpower, B. L. (1990). A rapid and simple method for preparation of RNA from *Saccharomyces cerevisiae. Nucleic Acids Research, 18*(10), 3091–3092. https://doi.org/10.1093/nar/18.10.3091.

Sharma, V. M., Li, B., & Reese, J. C. (2003). SWI/SNF-dependent chromatin remodeling of *RNR3* requires TAF(II)s and the general transcription machinery. *Genes & Development, 17*(4), 502–515. https://doi.org/10.1101/gad.1039503.

Shen, C. H., Leblanc, B. P., Alfieri, J. A., & Clark, D. J. (2001). Remodeling of yeast *CUP1* chromatin involves activator-dependent repositioning of nucleosomes over the entire gene and flanking sequences. *Molecular and Cellular Biology, 21*(2), 534–547. https://doi.org/10.1128/MCB.21.2.534-547.2001.

Shen, C.-H., Leblanc, B. P., Neal, C., Akhavan, R., & Clark, D. J. (2002). Targeted histone acetylation at the yeast *CUP1* promoter requires the transcriptional activator, the TATA boxes, and the putative histone acetylase encoded by *SPT10. Molecular and Cellular Biology, 22*(18), 6406–6416. https://doi.org/10.1128/MCB.22.18.6406-6416.2002.

Singh, S., & Hinnebusch, A. G. (2025). SAGA subunits Spt3 and Spt8 act directly and nonredundantly with TFIID in TBP recruitment in the Gcn4 transcriptome. *Nucleic Acids Research, 53*(13), https://doi.org/10.1093/nar/gkaf598.

Singh, S., Sahu, R. K., Sugathan, A., & Tomar, R. S. (2021). The H2A N-terminal tail is required to alleviate copper-induced stress in *Saccharomyces cerevisiae. FEMS Yeast Research, 21*(8), https://doi.org/10.1093/femsyr/foab061.

Singh, S., Sahu, R. K., & Tomar, R. S. (2021). The N-terminal tail of histone H3 regulates copper homeostasis in *Saccharomyces cerevisiae. Molecular and Cellular Biology, 41*(2), https://doi.org/10.1128/MCB.00210-20.

Szczypka, M. S., & Thiele, D. J. (1989). A cysteine-rich nuclear protein activates yeast metallothionein gene transcription. *Molecular and Cellular Biology, 9*(2), 421–429. https://doi.org/10.1128/mcb.9.2.421-429.1989.

Thakre, P. K., Golla, U., Biswas, A., & Tomar, R. S. (2020). Identification of histone H3 and H4 amino acid residues important for the regulation of arsenite stress signaling in *Saccharomyces cerevisiae. Chemical Research in Toxicology, 33*(3), 817–833. https://doi.org/10.1021/acs.chemrestox.9b00471.

Thakre, P. K., Sahu, R. K., & Tomar, R. S. (2021). Substitution of histone H3 arginine 72 to alanine leads to the deregulation of isoleucine biosynthesis in the budding yeast *Saccharomyces cerevisiae*. *Biochemistry and Cell Biology = Biochimie et Biologie Cellulaire, 99*(5), 636–644. https://doi.org/10.1139/bcb-2020-0651.

Thakre, P. K., Sv, A., Golla, U., Chauhan, S., & Tomar, R. S. (2019). Previously uncharacterized amino acid residues in histone H3 and H4 mutants with roles in DNA damage repair response and cellular aging. *The FEBS Journal, 286*(6), 1154–1173. https://doi.org/10.1111/febs.14723.

Thiele, D. J. (1988). ACE1 regulates expression of the *Saccharomyces cerevisiae* metallothionein gene. *Molecular and Cellular Biology, 8*(7), 2745–2752. https://doi.org/10.1128/mcb.8.7.2745-2752.1988.

Tomar, R. S., Psathas, J. N., Zhang, H., Zhang, Z., & Reese, J. C. (2009). A novel mechanism of antagonism between ATP-dependent chromatin remodeling complexes regulates *RNR3* expression. *Molecular and Cellular Biology, 29*(12), 3255–3265. https://doi.org/10.1128/MCB.01741-08.

Weiner, A., Chen, H. V., Liu, C. L., Rahat, A., Klien, A., Soares, L., ... Rando, O. J. (2012). Systematic dissection of roles for chromatin regulators in a yeast stress response. *PLoS Biology, 10*(7), e1001369. https://doi.org/10.1371/journal.pbio.1001369.

Welch, J., Fogel, S., Buchman, C., & Karin, M. (1989). The *CUP2* gene product regulates the expression of the *CUP1* gene, coding for yeast metallothionein. *The EMBO Journal, 8*(1), 255–260. https://doi.org/10.1002/j.1460-2075.1989.tb03371.x.

Wimalarathna, R. N., Pan, P. Y., & Shen, C.-H. (2012). Chromatin repositioning activity and transcription machinery are both recruited by Ace1p in yeast *CUP1* activation. *Biochemical and Biophysical Research Communications, 422*(4), 658–663. https://doi.org/10.1016/j.bbrc.2012.05.047.

Winge, D. R., Nielson, K. B., Gray, W. R., & Hamer, D. H. (1985). Yeast metallothionein. Sequence and metal-binding properties. *The Journal of Biological Chemistry, 260*(27), 14464–14470.

Yamaguchi-Iwai, Y., Serpe, M., Haile, D., Yang, W., Kosman, D. J., Klausner, R. D., & Dancis, A. (1997). Homeostatic regulation of copper uptake in yeast via direct binding of MAC1 protein to upstream regulatory sequences of *FRE1* and *CTR1*. *The Journal of Biological Chemistry, 272*(28), 17711–17718. https://doi.org/10.1074/jbc.272.28.17711.

Yudkovsky, N., Logie, C., Hahn, S., & Peterson, C. L. (1999). Recruitment of the SWI/SNF chromatin remodeling complex by transcriptional activators. *Genes & Development, 13*(18), 2369–2374. https://doi.org/10.1101/gad.13.18.2369.

Zhu, Z., Song, M., Ren, J., Liang, L., Mao, G., & Chen, M. (2024). Copper homeostasis and cuproptosis in central nervous system diseases. *Cell Death & Disease, 15*(11), 850. https://doi.org/10.1038/s41419-024-07206-3.

CHAPTER FIVE

Zinc-finger ubiquitin-binding domain of histone deacetylase 6 modulates Tau structure and function

Abhishek Ankur Balmik[a,*] and Subashchandrabose Chinnathambi[b,*]

[a]Department of Pharmacology and Chemical Biology, Emory University School of Medicine, Atlanta, GA, United States

[b]Department of Neurochemistry, National Institute of Mental Health and Neuro Sciences Hospital (NIMHANS), Institute of National Importance, Bangalore, Karnataka, India

*Corresponding authors. e-mail address: abalmik@emory.edu; subashneuro@nimhans.ac.in

Contents

1. Introduction 82
   1.1 Pathological hallmarks of AD 83
   1.2 Role of acetylation as a post-translational modification in AD 83
   1.3 Histone deacetylases – classification and functions 84
   1.4 HDACs in AD pathology 86
2. Rationale 89
3. Analysis of HDAC6 ZnF UBP function in modulation of Tau structure and function 92
   3.1 Plasmid preparation and protein expression 92
   3.2 Protein analysis 100
   3.3 Tau – HDAC6 interaction analysis 101
   3.4 Aggregation studies 104
   3.5 Structural and morphological analysis of Tau 105
4. Concluding remarks 107

References 108

Abstract

Histone deacetylase 6 (HDAC6), a predominantly cytoplasmic deacetylase, has emerged as a critical regulator of protein homeostasis, and plays an important role in neurodegenerative disorders characterized by Tau pathology. We investigated the role of zinc-finger ubiquitin-binding domain (ZnF UBP) of HDAC6 in modulating tau structure and function. HDAC6 is well known for its deacetylase activity and involvement in aggresome formation but the functional contribution of its ZnF UBP domain in tau biology remains not well explored. Using a combination of biochemical, biophysical, and molecular approaches, we demonstrate that the ZnF UBP

Methods in Enzymology, Volume 733
ISSN 0076-6879, https://doi.org/10.1016/bs.mie.2026.05.036

domain of HDAC6 directly interacts with tau, influencing its conformational dynamics and aggregation propensity. Our findings reveal that this interaction is independent of HDAC6's catalytic deacetylase activity, highlighting a non-canonical mechanism through which HDAC6 regulates tau. Structural analyses indicate that binding of ZnF UBP induces conformational rearrangements in tau, potentially altering its microtubule-binding capacity and aggregation behavior. Furthermore, we show that the HDAC6–tau interaction modulates tau aggregation, suggesting a protective or regulatory role in preventing aberrant tau aggregation. Functional assays support the notion that ZnF UBP-mediated modulation of tau contributes to maintaining cytoskeletal integrity and cellular proteostasis. These observations suggests that HDAC6 functions not only as a deacetylase but also performs regulatory functions through ZnF UBP domain that can influence the function and properties of intrinsically disordered proteins (IDPs) like tau. Overall, this chapter provides mechanistic insights into the non-enzymatic functions of HDAC6, specifically through its ZnF UBP domain, in regulating tau structure and function. Understanding this interaction expands the current perspective on HDAC6 biology and underscores its potential as a therapeutic target in tauopathies, including Alzheimer's disease and targeting domain-specific interactions of HDAC6 may offer a novel therapeutic strategy to modulate tau pathology.

1. Introduction

Alzheimer's disease (AD) is the most prevalent form of dementia which affects millions of people worldwide, reducing the life expectancy in ageing population (Knopman et al., 2021; Livingston et al., 2020; Long & Holtzman, 2019). Clinical manifestation of AD is marked with progressive decline in memory along with cognitive and behavioural impairment (Knopman et al., 2021; Long & Holtzman, 2019). The etiological aspect of AD is complex and involves multiple factors such as pathological accumulation and propagation of toxic proteinaceous aggregates, neuroinflammation, altered PTM profile of key proteins leading to their dysfunction, loss of synaptic plasticity, disruption in effective neuronal transmission and organellar functions leading to neurotoxicity and neurodegeneration (De Strooper & Karran, 2016; Guan & Wang, 2023; Long & Holtzman, 2019). In the current scenario, the therapeutics designed for the treatment of AD provides symptomatic relief without effectively checking or reversing the progression of disease (Cummings et al., 2023). Thus, it is important to have a deeper understanding of various pathological aspects of AD; especially the role of proteins involved in pathological events, their interconnections and an urgent need to develop therapeutics targeting these various aspects under different stages of disease progression.

1.1 Pathological hallmarks of AD

The major clinical hallmark of AD is the accumulation of extracellular amyloid- β (Aβ) plaques made up of aggregation prone peptides derived from amyloid precursor protein (APP) and intracellular neurofibrillary tangles (NFT) made by the aggregation of hyperphosphorylated microtubule-associated protein Tau (MAPT) (Balmik et al., 2022; Hardy & Selkoe, 2002; Long & Holtzman, 2019; Sonawane et al., 2022; Wang & Mandelkow, 2016). Protein aggregation in AD leads to the disruption of various neuronal functions involving synaptic transmission, cargo transport, cytoskeletal assembly, and protein homeostasis ultimately resulting in neuronal loss (De Strooper & Karran, 2016; Hardy & Selkoe, 2002). The events that lead to protein aggregation has been attributed to genetic predisposition as well as external factors which increases genetic susceptibility elevating the risk of developing AD. Autosomal dominant mutations in the gene encoding for APP the genes that are involved in APP processing such as presenilin 1 and presenilin 2 leads to the production of aggregation prone Aβ peptides, particularly Aβ-42, thereby causing Aβ plaque deposition in the extracellular region (Balmik et al., 2022; Hardy & Selkoe, 2002; Karran et al., 2011). On the other hand, mutations in Tau protein are not directly responsible for AD but typically cause several dominant tauopathies collectively termed as frontotemporal lobar degeneration with tau pathology (FTLD-tau) (Spillantini & Goedert, 2013; Wang & Mandelkow, 2016). Neurodegenerative disorders such as frontotemporal dementia with parkinsonism linked to chromosome-17 (FTDP-17) are closely related to AD and comprises of pathological hallmarks such as Tau hyperphosphorylation, misfolding and NFT formation (Spillantini & Goedert, 2013; Wang & Mandelkow, 2016).

1.2 Role of acetylation as a post-translational modification in AD

Another important aspect of AD which governs the etiology of disease is the altered post-translational modifications (PTMs) of key proteins involved in the neuronal functions (Guan & Wang, 2023; Narayan et al., 2015). In this context, lysine modification in histone and non-histone proteins is a critical yet less explored aspect which modulates the pathophysiological pathways in AD (Gräff & Tsai, 2013; Narayan et al., 2015). Lysine residues undergoes PTMs such as acetylation, methylation, ubiquitination, sumoylation and succinylation which serves in regulating the protein function in various ways (Narayan et al., 2015). Among these acetylation and methylation play major role in the regulation of gene expression by chromatin remodelling, epigenetic and transcriptional regulation. These regulatory mechanisms play a crucial role in neuronal health through modulation of synaptic function, neuronal plasticity and neuronal survival signifying the crucial role of PTMs in AD both at the level of gene regulation and protein (Gräff & Tsai, 2013; Guan & Wang, 2023).

1.3 Histone deacetylases – classification and functions

Histone deacetylases or more precisely and in broader perspective – protein deacetylases or lysine deacetylases (KDACs), are the family of enzymes which modulate the functions of substrate proteins through removal of acetyl groups from specific lysine residues (Choudhary et al., 2014; Haberland et al., 2009; Yang & Seto, 2008). HDACs play a central role in various biological functions either modifying the acetylation signature of histone proteins thereby modulating the chromatin structure during cellular development or under different signaling cues. Different HDAc isoforms are required under different conditions for removing acetyl groups from the histone tails and inducing chromatin condensation and gene repression (Gräff & Tsai, 2013; Haberland et al., 2009).

There are 18 isoforms in HDAC family which have been categorized into four distinct classes – I, II, III and IV, based on phylogenetic analysis and structural homology to yeast deacetylases (Haberland et al., 2009; Seto & Yoshida, 2014) (Fig. 1). The HDACs of Class I, II and IV are Zn^{2+} dependent enzymes while class III deacetylases require NAD+ as a cofactor for their activity (Finnin et al., 1999; Seto & Yoshida, 2014). HDACs functions in removing the acetyl groups from ε-acetyl lysine residues on their substrate proteins (Choudhary et al., 2014). The Zn^{2+} dependent HDACs requires a Zn^{2+} ion at the catalytic site to activate a water molecule into a nucleophile (Finnin et al., 1999). The nucleophilic attack on carbonyl carbon of acetyl group results in release of acetate and subsequent deacetylation of lysine residue. Class III deacetylases, sirtuins requires NAD^+ as a co-substrate for their function. The mechanism involves cleavage of NAD^+ into nicotinamide and ADP-ribose, followed by the transfer of acetyl group from lysine residue to ADP-ribose to form O-acetyl ADP-ribose, finally leading to deacetylated lysine (Imai & Guarente, 2014). Since NAD^+ is required for the action of sirtuins, their function is dependent on cellular metabolic status (Imai & Guarente, 2014).

Class I HDACs are the homologs of yeast RPD3 protein, comprises of HDAC1, HDAC2, HDAC3 and HDAC8 (Haberland et al., 2009; Seto &

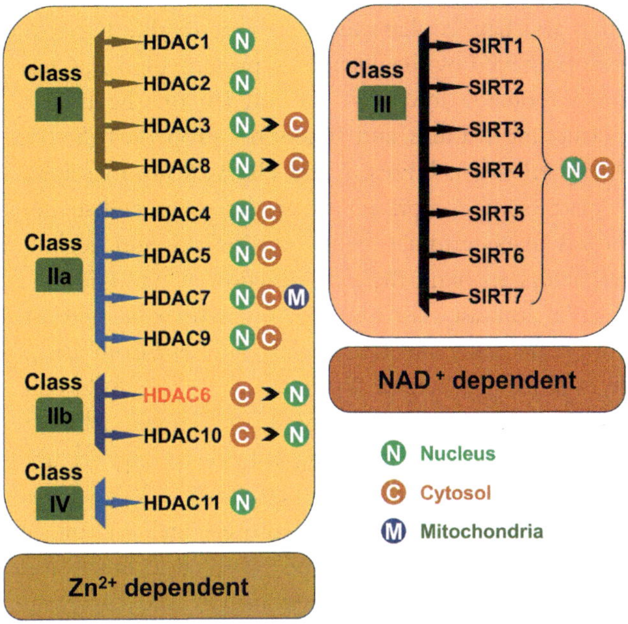

Fig. 1 Classification of histone deacetylases. Histones deacetylases (HDACs) are the class of enzymes well-studied for its role in transcriptional modulation and epigenetic regulation by carrying out deacetylation on histone protein. However, there are many non-histone substrates of HDACs present in cells and their activity is regulated by deacetylation. HDACs consist of family of 18 proteins which are grouped on the basis of their homology with yeast histone deacetylases. Functionally, they are classified on the basis of their enzymatic mechanism as Zn^{2+} dependent and NAD dependent HDACs. Most HDACs are present in both nucleus and cytosol, some of which can shuttle between the two compartments while others are exclusively found in nucleus. In cytosol, HDACs perform a wide variety of functions like cytoskeletal regulation, growth and development etc. by either functioning alone or as a part of multi-protein complexes. HDAC6 is majorly a cytosolic HDAC distinct from other HDACs by possessing two functional catalytic domains and a C-terminal ZnF UBP domain. HDAC6 is involved in the regulation of cytoskeletal dynamics by acting on α-tubulin and cortactin, transcriptional regulation, and chaperone function. The ZnF UBP domain of HDAC6 is involved in protein homeostasis by mediating the formation of aggresomes in order to degrade the protein aggregates by autophagy.

Yoshida, 2014). These enzymes are predominantly present in nucleus and serves their function as transcriptional co-repressor to regulate several processes related to cellular health such as DNA repair, cell differentiation, maintenance of genomic stability, chromatin remodelling and learning and memory formation during neuronal development (Gräff & Tsai, 2013; Haberland et al., 2009). Class II HDACs are the homologs of yeast protein

HDA1 protein and have a characteristic ability to shuttle between nucleus and cytoplasm facilitated by the chaperone protein 14–3-3. Class II HDACs are further differentiated into class IIa and IIb on the basis of sequence homology, structural features and functional properties (Seto & Yoshida, 2014). HDAc4, HDAC5, HDAC7 and HDAC9 are members of class IIa subfamily. Apart from comprising of catalytic deacetylase domain, these HDACs consists of N-terminal regulatory domains and docking motifs for transcription factors enabling them to function as transcriptional repressors (Parra, 2015). In contrast class IIb HDACs, which includes HDAC6 and HDAC10, are structurally and functionally different from class IIa HDACs (d'Ydewalle et al., 2012; Seto & Yoshida, 2014). HDAC6 consists of two functional catalytic domains and a ZnF UBP domain which can bind to poly-ubiquitinated substrates facilitating their recycling via ubiquitin proteasome system (UPS) (Balmik et al., 2020, 2021; d'Ydewalle et al., 2012). HDAC6 is primarily a cytoplasmic deacetylase which acts on substrates like α-tubulin, tau, cortactin and HSP90 and involved in functions like regulation of microtubule dynamics, axonal transport and aggresome-mediated degradation of protein aggregates (Balmik et al., 2020, 2021; d'Ydewalle et al., 2012). It also comprises NLS, NES1(Nuclear Export/Localization Signals) and domains unique to HDAC6 – dynein motor binding (DMB) and SE14 (Ser-Glu tetradecapeptide repeat domain) serving the function in intracellular transport via and cytoplasmic retention respectively (Fig. 2). HDAC10 is known to act on acetylated polyamines and primarily function towards autophagy regulation, lysosomal activity and stress response (Hai et al., 2017). It consists of a catalytically active and a functionally incomplete deacetylase domain along with a polyamine binding pocket and a ZnF UBP domain (Hai et al., 2017; Imai & Guarente, 2014). Class III protein deacetylases are NAD^+ dependent enzymes known as sirtuins, comprising of SIRT1-SIRT7 (RH, 2012). Sirtuins are also ubiquitously distributed across cytoplasm, nucleus and mitochondria where they are involved in various functions related to cellular survival, genomic and metabolic regulation (RH, 2012; Yeong et al., 2020). Class IV consists of HDAC11, which is structurally related to class I and class II HDACs, and known to be play regulatory role in cell growth and development, immune and metabolic functions (Chen et al., 2016).

1.4 HDACs in AD pathology

Acetylation, as a PTM is involved in different aspects of AD pathology in the modification of histone and various non-histone proteins (Gräff & Tsai, 2013;

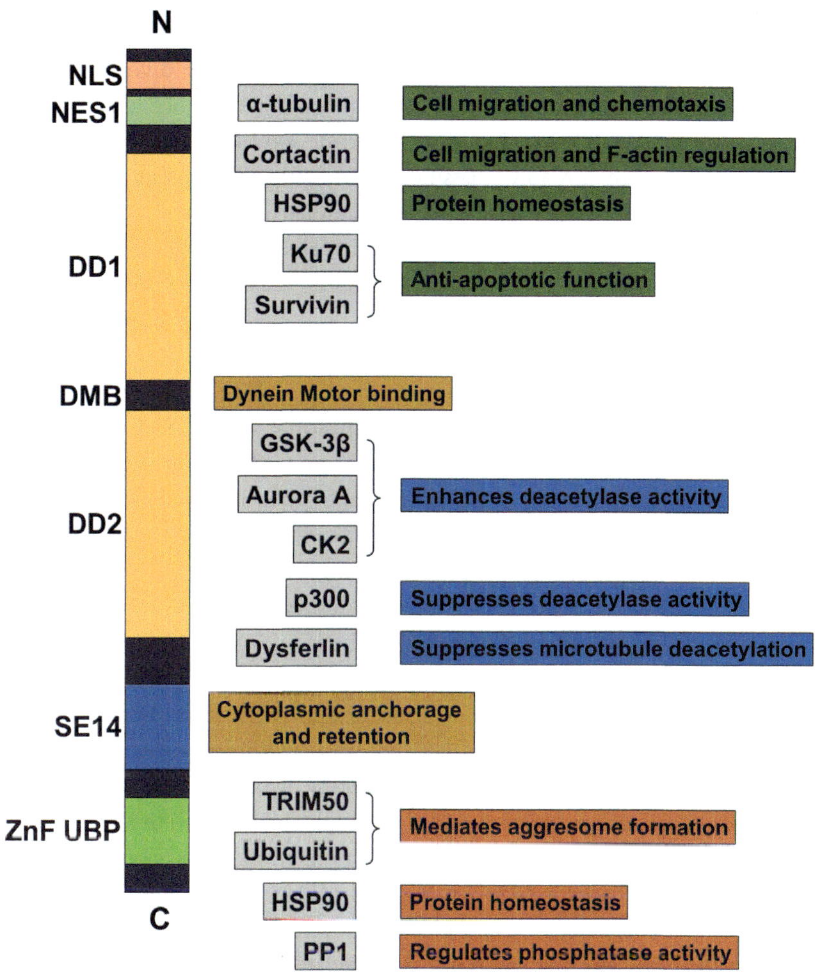

Fig. 2 Domain wise functions of HDAC6. HDAC6 consists of two catalytic deacetylase domains (DD1 and DD2) and a C-terminal Zinc-finger ubiquitin binding protein domain (ZnF UBP). It also consists of nuclear export sequence (NES1), nuclear localization sequence (NLS) and a cytoplasmic anchoring domain (SE14) in the form of a tetradecapeptide marked by serine and glutamine. HDAC6 performs a large number of cellular functions based on its deacetylase activity or direct interaction to regulate other proteins. There are some proteins which enhances HDAC6 activity through its phosphorylation e.g. GSK-3β or suppress its activity e.g. p300. HDAC6 regulates functions such as cell migration, apoptosis, protein homeostasis by deacetylation of its target proteins – tubulin, cortactin, HSP90 etc. HDAC6 ZnF UBP domain interacts with other proteins to mediate aggresome formation, protein homeostasis and regulation of phosphatase activity of protein phosphatase 1 (PP1).

Narayan et al., 2015; Yang et al., 2017). This chapter focuses on the role of acetylation in AD and regulation of protein acetylation by a diverse family of enzymes called histone deacetylases (HDACs) in different subcellular locations and with respect to different functions (Seto & Yoshida, 2014; Yang et al., 2017). HDACs are broadly classified into zinc dependent or NAD+ dependent enzymes, each with distinct subcellular localization and involved in various functions by acting on a wide array of substrates (Haberland et al., 2009; Seto & Yoshida, 2014). HDACs have been widely studied for their role in neurodegenerative diseases and found to be crucial in transcriptional regulation as well as in direct modulation of neurodegenerative disease associated proteins such as Tau and APP (Chan & Fletcher, 2021; Yang et al., 2017). HDACs are mainly involved in the transcriptional suppression of genes via chromatin remodelling and epigenetic regulation and their overexpression or increased activity is known to be associated with synaptic dysfunction and memory deficit in AD (Chan & Fletcher, 2021; Guan et al., 2009; Yang et al., 2017). Increased transcriptional levels of certain HDACs in AD post -mortem brain tissue has been linked to repressed levels of genes associated with neuroplasticity and cognitive function (Guan et al., 2009). On the other hand, activity of NAD+ dependent deacetylases classified as sirtuins, are associated with neuroprotective and anti-inflammatory pathways (Donmez & Outeiro, 2013; Yeong et al., 2020). Classically, almost all the members of HDAC family play an important role in various aspects of neuronal health. HDACs are the key epigenetic regulators responsible for chromatin condensation and transcriptional repression via removal of acetyl groups from histone proteins. The wide range of functions where various HDACs are known to be associated includes memory formation and development, neuronal survival, synaptic plasticity, intracellular trafficking and neuroinflammation (Chan & Fletcher, 2021; Yang et al., 2017).

Dysregulation of specific HDAC isoforms have been associated with neurodegenerative diseases at multiple levels, owing to its wide array of regulatory roles and function specific localization (Chan & Fletcher, 2021; Yang et al., 2017). Different isoforms of HDACs have distinct and sometimes opposing effects on neuronal functions (Chan & Fletcher, 2021; Guan et al., 2009). For example, HDAC1 has a protective role by mediating DNA damage repair and maintenance of genomic stability while HDAC2 has been found to be a negative regulator of synaptic plasticity and cognition (Guan et al., 2009). Class II HDACs, HDAC4 and HDAC5 are involved in activity-dependent neuronal signaling under normal conditions while their dysregulation leads to nuclear accumulation, which has been linked to apoptosis (Seto & Yoshida, 2014). HDAC6 is a unique deacetylase classified under class IIb, with having two catalytic deacetylase domains and a Zinc finger ubiquitin binding protein (ZnF UBP) domain (d'Ydewalle et al., 2012; LoPresti, 2020). It serves as a crucial mediator of aggresome formation under increased protein aggregate burden in cells by binding to poly-ubiquitinated protein aggregates through its ZnF UBP domain (d'Ydewalle et al., 2012). However, increased levels of HDAC6 have been reported under the conditions of progressive neurodegeneration (d'Ydewalle et al., 2012; LoPresti, 2020). Thus, it is plausible that HDAc6 has a dual role in neurodegeneration and neuroprotection through its catalytic domains and ZnF UBP domain respectively. The role of most members of Class III protein deacetylases called SIRT has not been well defined with respect to AD. SIRT1 exhibit neuroprotection by being involved in functions such as neuronal survival, anti-inflammatory pathways and tau/Aβ metabolism (AlHayani et al., 2025; Yeong et al., 2020). On the other hand, SIRT2 has been linked with oxidative stress, neuroinflammation and tau hyperphosphorylation in AD (de Oliveira et al., 2012; Donmez & Outeiro, 2013). Others members of class III like SIRT3 and SIRT6 play important role in mitochondrial function, DNA repair, and neuronal plasticity but their dysregulation under pathological conditions may lead to detrimental effects in AD (Donmez & Outeiro, 2013; Yeong et al., 2020).

2. Rationale

Tau is a microtubule-associated protein that plays a central role in maintaining neuronal structure and axonal transport (Cleveland et al., 1977; Weingarten et al., 1975). Under physiological conditions, tau exists as an intrinsically disordered protein which can adopt dynamic transient conformations, enabling it to interact with microtubules and stabilize their assembly (Jeganathan et al., 2006; Mukrasch et al., 2009). However, in neurodegenerative diseases such as AD, tau undergoes conformational that promote its detachment from microtubules, misfolding, and aggregation into toxic oligomers and neurofibrillary tangles (Goedert, 2015; Spillantini & Goedert, 2013; Wang & Mandelkow, 2016). Understanding the molecular determinants that regulate tau structure and its transition from a functional to a pathological state is a critical determinant in the pathophysiology of AD (Busche & Hyman, 2020).

HDAC6 has emerged as an important cytoplasmic regulator of protein quality control, owing to its unique domain organization and ability to interact with ubiquitinated substrates (Boyault, Zhang, et al., 2007; Deakin & Turner, 2014). Unlike other HDAC family members, HDAC6 contains a zinc-finger ubiquitin-binding protein (ZnF UBP) domain, which enables it to function in aggresome formation, autophagic clearance, and cellular stress responses (Bonnet et al., 2008; Boyault, Sadoul, et al., 2007; Kawaguchi et al., 2003; Seigneurin-Berny et al., 2001) (Fig. 3). While HDAC6 has been extensively studied for its deacetylase activity, particularly with respect to its action on α-tubulin regulating cytoskeletal dynamics, its non-catalytic domains ZnF UBP remain comparatively underexplored in the context of tau biology (Hubbert et al., 2002). Emerging evidence suggests that HDAC6 is closely associated with tau pathology, including its co-localization with tau aggregates and its involvement in proteostasis pathways that govern tau turnover (Cook et al., 2014; Li et al., 2022; Perez et al., 2009; Yan, 2014). However, most studies have focused on HDAC6-mediated deacetylation and its indirect effects on tau. This leaves a significant gap in understanding whether HDAC6 can directly modulate tau structure and function through ZnF UBP domain specific, non-enzymatic interactions. Tau is a natively disordered protein and has a large array of binding partners interacting transiently with Tau. It is plausible that interaction with HDAC6 ZnF UBP domain could influence Tau conformational and aggregation properties (Darling & Uversky, 2018; Mukrasch et al., 2009).

The rationale for this study is therefore based on the hypothesis that the HDAC6 ZnF UBP domain serves as a direct modulator of tau, independent of deacetylase function of HDAc6. Insights into this interaction, helps in elucidate the effect of HDAC6 ZnF UBP on Tau structural properties, aggregation properties and consequently influence on its functional role in microtubule stabilization. Such an approach addresses a critical gap in the current understanding of HDAC6 biology, shifting the focus from its enzymatic function to its role as a scaffolding and regulatory protein. Furthermore, targeting non-catalytic domains offers a promising therapeutic strategy, as it may allow selective modulation of disease-relevant interactions without broadly inhibiting HDAC6 activity and disrupting essential cellular processes (Fazal et al., 2021).

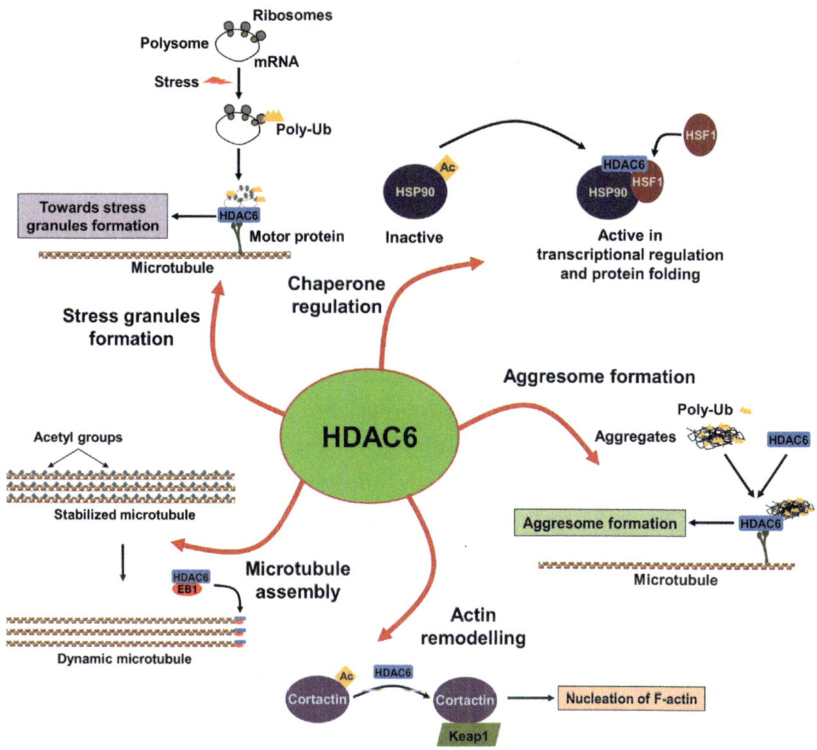

Fig. 3 Role of HDAC6 in neurodegenerative diseases/neuronal functions. HDAC6 plays a major role in various aspects of neurodegenerative diseases as it modulates many cellular functions. HDAC6 is known to regulate activation of chaperones either directly or indirectly. Heat shock protein 90 (HSP90) under basal conditions remains inactive in its acetylated form. Under stress conditions, HDAC6 binds to HSP90 and activates it via deacetylation. It also releases heat shock factor 1 (HSF1) which functions in transcriptional regulation and protein folding along with HSP90. Upon failure of chaperone system in protein refolding as well as proteasomal degradation of protein aggregates, HDAC6 mediates the formation of aggresomes which are cleared by autophagy. HDAC6 binds to poly-ubiquitinated (Poly-Ub) aggregates and help to carry them towards perinuclear region via dynein motor in order to form aggresomes. Actin dynamics during the process of cell migration, cell attachment, neurite extensions and formation of other membrane projections like filopodia, lamellipodia and podosomes require HDAC6 deacetylase activity. Cortactin is a nucleation factor for F-actin formation which is deacetylated by HDAC6. Cortactin initiates F-actin formation by associating with other nucleating factors such as Keap1 and Arp2/3. Acetylated α-tubulin in microtubules result in stabilized microtubules. Microtubules are maintained in a dynamic stage by HDAC6 through deacetylation and regulation of growing MT ends through association with end-binding protein 1 (EB1). Stress granules are the structures formed during cellular stress which contains mRNA, ribosomes and proteins associated with protein translation. Protein translation gets halted during stress conditions which triggers stress granules formation where mRNA and ribosomes are salvaged that can be reused later. HDAC6 helps in the transport of poly-ubiquitinated aggregates containing polysomes and other proteins in order to form stress granules.

3. Analysis of HDAC6 ZnF UBP function in modulation of Tau structure and function

Transition of Tau from a soluble, functional state to β-sheet-rich aggregates is governed by multiple cellular factors, including molecular chaperones, PTMs, and transiently interacting proteins (Avila et al., 2016; Goedert, 2015; Medina et al., 2016). The role of HDAC6 ZnF UBP domain, has been implicated in protein quality control pathways and tauopathies; however, its direct role in modulating Tau structure and aggregation remains insufficiently understood (Saha & Pahan, 2006; Singh et al., 2024).

To understand the role of HDAC6 ZnF UBP in regulating Tau structure and function, a combination of biochemical and biophysical approaches was employed (Fig. 4). Recombinant full-length Tau, repeat Tau, and HDAC6 ZnF UBP were expressed and purified, followed by systematic evaluation of their interactions and functional consequences. Aggregation kinetics, structural transitions, and stability of Tau were assessed using fluorescence-based assays, circular dichroism spectroscopy, transmission electron microscopy, and chromatography-based techniques (Balmik et al., 2020; Fitzpatrick et al., 2017; Greenfield, 2006; LeVine, 1999). Additionally, degradation and disaggregation assays were performed to evaluate the modulatory effects of HDAC6 ZnF UBP and small molecules on Tau dynamics (Balmik et al., 2020). These integrated approaches enable the characterization of catalytic-independent functions of HDAC6 ZnF UBP in influencing Tau conformational states and aggregation properties.

3.1 Plasmid preparation and protein expression

3.1.1 Plasmid isolation and purification

3.1.1.1 Reagents

1. Luria Broth (HiMedia)
2. Ampicillin (MP Biomedicals)
3. Kanamycin (Gibco)
4. Plasmid purification kit (Macherey-Nagel)
5. Ethanol (MP Biomedicals)
6. Plasmids – pET28a LIC HDAC6 ZnF UBP (Kindly gifted by Prof. Cheryl Aerosmith, University of Toronto, Canada), pT7C K18WT (Kindly gifted by Prof. Roland Brandt, University of Osnabruck, Germany), pT7C hTau40WT (Kindly gifted by Prof. Roland Brandt, University of Osnabruck, Germany)
7. DH5α competent *E. coli* cells (New England Biolabs)

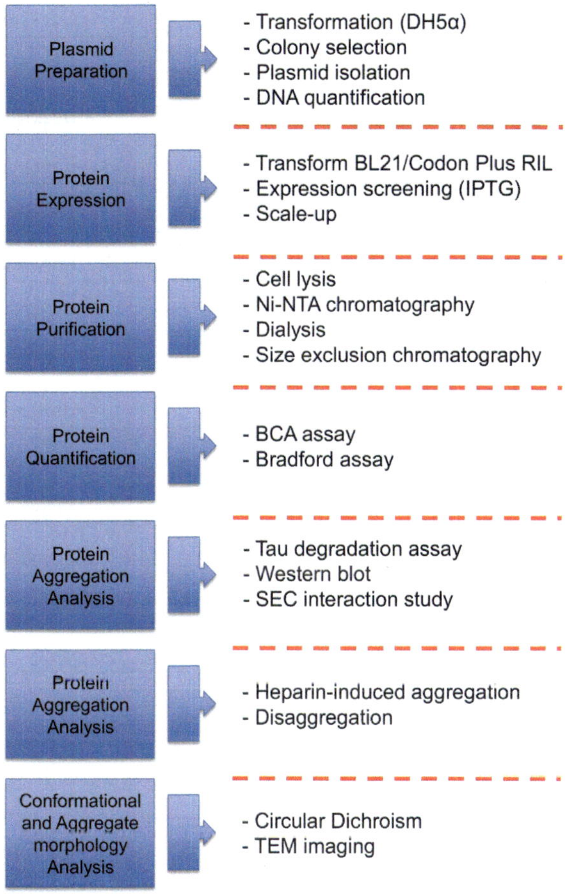

Fig. 4 Experimental workflow for analysis of Tau–HDAC6 interaction and aggregation. Plasmids encoding Tau and HDAC6 ZnF UBP were prepared and transformed into expression strains, followed by protein expression and large-scale production. Proteins were purified using affinity and size-exclusion chromatography and quantified using BCA/Bradford assays. Interaction between Tau and HDAC6 was assessed through degradation assays, Western blotting, and size-exclusion chromatography. Tau aggregation and disaggregation were monitored using heparin-induced assays and Thioflavin S fluorescence. Structural and morphological changes were analyzed using circular dichroism spectroscopy and transmission electron microscopy.

3.1.1.2 Equipment

Equipment	Company
AKTA Pure FPLC system	GE Healthcare
AKTA Start FPLC system	GE Healthcare

Amersham Imager 600	GE Healthcare
Amersham Semi Dry blotting apparatus	GE Healthcare
Avanti JXN26 High speed Centrifuge	Beckman Coulter
High Speed Centrifuge 5804 R	Eppendorf
Homogenizer	Constant Systems Ltd.
Microcentrifuge 5418 R	Eppendorf
MicroCal PEAQ-ITC	Malvern
Microplate reader Infinite 200 PRO	Tecan
Molecular Imager Gel Doc™ XR+	BioRad
Optima XPN10 Ultracentrifuge	Beckman Coulter
MicroCal PEAQ-ITC	Malvern
Shaker Incubator (H1010-MR)	Benchmark Scientific
Shaker Incubator Multitron Standard	Infors HT
UV-Visible spectrophotometer V-530	Jasco
Zeiss Axio observer 7 microscope with Apotome 2.0	Zeiss
Far-UV CD spectrometer J-815	Jasco
T20 Transmission Electron Microscope	Tecnai
Avance III HD 700 MHz NMR spectrometer equipped with TXI probe	Bruker

Miscellaneous equipments.

Equipment	Company
Analytical balance ME204	Mettler Toledo
Analytical balance ME4001E	Mettler Toledo
Autoclave	Spire
Dry bath	Genei
Forma 900 series −80 °C	Thermo Scientific
Gel rocker	Benchmark
Heratherm Hot Air Oven	Thermo Scientific

Heraeus Incubator	Thermo Scientific
Magnetic Stirrer	Genei
MiliQ Unit Direct 16	Millipore
Miniprotean Tetra vertical electrophoresis system	Biorad
MiniVE Vertical electrophoresis System	Amersham
MiniSpin Plus Table top centrifuge	Eppendorf
Nanodrop	Simplinano
pH meter Five Easyplus	Mettler Toledo
Vacuum Pump	Millipore
Vortexer mixer	Genei
Water bath	Genei

Softwares.

Software	Company/organization
ImageJ	NIH
Image Lab	BioRad
ImageQuant TL	GE
Magellan™ Data analysis software	Tecan
MicroCal PEAQ-ITC analysis software	Malvern
MS Excel	Microsoft
Sigma Plot 10.0	Systat softwares
Spectra Manager™	Jasco
Topspin 3.2	Bruker
Zen 2.3	Zeiss

3.1.1.3 Procedure

1. Transform plasmids encoding Tau or HDAC6 ZnF UBP into DH5α cells and plate on antibiotic-containing LB agar.
2. Select single isolated colonies and inoculate into 5 mL LB medium with appropriate antibiotic.

3. Incubate overnight at 37 °C with shaking.
4. Harvest cells by centrifugation at 11,000 *g* for 30 s at room temperature.
5. Resuspend pellets in 250 μL Buffer A1.
6. Add 250 μL Buffer A2, mix gently, and incubate for 5 min
7. Add 300 μL Buffer A3 and mix by inversion.
8. Clarify lysate by centrifugation at 11,000 *g* for 30 s
9. Load supernatant onto silica columns and centrifuge.
10. Wash columns with 600 μL Buffer A4 (with ethanol).
11. Dry columns by centrifugation for 2 min
12. Elute plasmid DNA with 50 μL Buffer AE.
13. Measure DNA concentration using NanoDrop.

3.1.1.4 Notes

1. DH5α cells are preferred due to *recA1* and *endA1* mutations, ensuring plasmid stability and quality.
2. Avoid vigorous mixing after lysis to prevent genomic DNA contamination.

3.1.2 Protein expression screening and scale-up

Buffers used for HDAC6 ZnF UBP preparation

Extraction/lysis buffer
50 mM Tris-Cl pH 8.0
500 mM NaCl
1 mM PMSF
20 mM Imidazole
Protease Inhibitor Cocktail

Ni-NTA affinity chromatography

Wash buffer	Elution buffer
50 mM Tris-Cl pH 8.0	50 mM Tris-Cl pH 8.0
500 mM NaCl	500 mM NaCl
20 mM Imidazole	1 M Imidazole
	5 % Glycerol

Dialysis buffer	Size-exclusion chromatography buffer
50 mM Tris-Cl pH 8.0	50 mM Tris-Cl pH 8.0
100 mM NaCl	100 mM NaCl
2 mM TCEP/DTT	1 mM TCEP/DTT
5 % Glycerol	

3.1.2.1 Reagents

1. LB medium (for Tau) (HiMedia)
2. Terrific Broth (TB) (for HDAC6 ZnF UBP) (Invitrogen)
3. Ampicillin (100 μg/mL) (MP Biomedicals)
4. Kanamycin (50 μg/mL) (Gibco)
5. IPTG (MP Biomedicals)
6. Glycerol (MP Biomedicals)
7. Plasmids – pET28a LIC HDAC6 ZnF UBP (Kindly gifted by Prof. Cheryl Aerosmith, University of Toronto, Canada), pT7C K18WT (Kindly gifted by Prof. Roland Brandt, University of Osnabruck, Germany), pT7C hTau40WT (Kindly gifted by Prof. Roland Brandt, University of Osnabruck, Germany)
8. BL21 Codon Plus (RIL) (New England Biolabs)

3.1.2.2 Equipment

1. Shaker incubator (37 °C) (Benchmark Scientific)
2. Shaker incubator (16 °C) (Infors HT)
3. Spectrophotometer (OD_{600} measurement)
4. Centrifuge (Eppendorf 5418 R)
5. High speed Centrifuge (Avanti JXN26 – Beckman Coulter)
6. SDS-PAGE apparatus (Bio-Rad)

3.1.2.3 Procedure

1. Transform plasmids into BL21* (for Tau) or BL21 Codon Plus (RIL) cells (for HDAC6 ZnF UBP) in LB and TB medium respectively.
2. Select single colonies and inoculate into 10 mL LB/TB medium with antibiotic.
3. Incubate at 37 °C until OD600 reaches ~0.6.
4. Divide cultures into induced and uninduced sets.
5. Induce Tau expression with 0.5 mM IPTG at 37 °C for 4–5 h.

6. For HDAC6 ZnF UBP, pre-incubate at 16 °C for 1 h, then induce with 0.2 mM IPTG and incubate overnight at 16 °C. For Tau,
7. Harvest cells by centrifugation.
8. Lyse cells in appropriate lysis buffer.
9. Analyze expression by SDS-PAGE (Tau ~55 kDa; HDAC6 ZnF UBP ~14 kDa).
10. Select colonies with highest expression.
11. Prepare glycerol stocks (20 % final glycerol) and store at −80 °C.

3.1.2.4 Scale-up procedure

1. Prepare primary inoculum (100 mL culture) from selected colony.
2. Inoculate secondary culture (1:100 ratio) in LB/TB medium.
3. Grow until OD_{600} ~0.6.
4. Induce protein expression as described above.
5. Harvest cells by centrifugation at 4 °C.
6. Store pellets at −80 °C or proceed for purification.

3.1.2.5 Notes

1. Lower temperature (16 °C) improves folding and solubility of HDAC6 ZnF UBP.
2. BL21 CodonPlus (RIL) cells help overcome codon bias.

3.1.3 Purification of HDAC6 ZnF UBP

3.1.3.1 Reagents

1. Lysis buffer (50 mM Tris-Cl pH 8.0, 500 mM NaCl, 20 mM imidazole, PMSF, protease inhibitors) (Fig. 5)
2. Wash buffer (50 mM Tris-Cl pH 8.0, 500 mM NaCl, 20 mM imidazole)
3. Elution buffer (50 mM Tris-Cl pH 8.0, 500 mM NaCl, 1 M imidazole, 5 % glycerol)
4. Dialysis buffer (50 mM Tris-Cl pH 8.0, 100 mM NaCl, 2 mM TCEP/DTT, 5 % glycerol)

3.1.3.2 Equipment

1. High-pressure cell disruptor (Constant Systems Ltd.)
2. Ultracentrifuge (Optima XPN10 Ultracentrifuge – Beckman Coulter)
3. Ni-NTA Agarose (Qiagen)
4. Polypropylene chromatography columns (Bio-Spin – Bio-Rad)
5. Size-exclusion chromatography system (Superdex 75 HiLoad 16/600)
6. Chromatography system (ÄKTA Pure – GE)

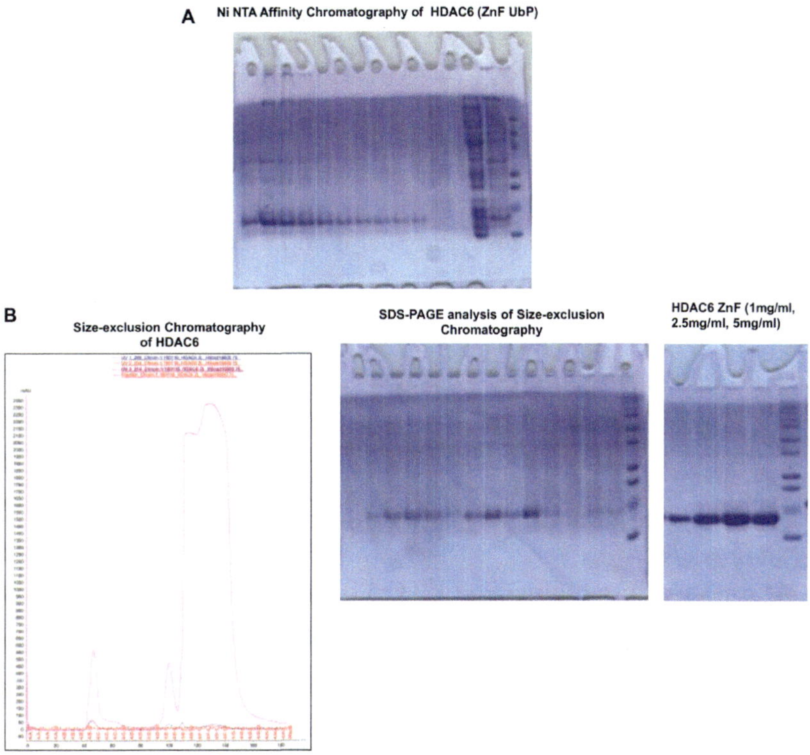

Fig. 5 Recombinant HDAC6 ZnF UBP preparation. Various biochemical and chromatography techniques have been to purify the protein. **A.** Preparation and purification of HDAC6 ZnF-UBP using Ni-NTA affinity chromatography. SDS-PAGE analysis clearly confirms the successful expression and purification of the target protein, demonstrating the presence of HDAC6 ZnF-UBP at the expected molecular weight. **B.** Size-exclusion chromatography (SEC) profile of HDAC6 ZnF-UBP. The collected elution fractions were analyzed by SDS-PAGE, which clearly confirmed the presence and purity of the target protein at the expected molecular weight.

3.1.3.3 Procedure

1. Resuspend cell pellet in lysis buffer.
2. Lyse cells using high-pressure disruption (~15,000 kpsi).
3. Centrifuge lysate at 20,000 rpm for 30 min
4. Incubate supernatant with Ni-NTA resin (1:30 ratio) at 4 °C for 1 h.
5. Wash with wash buffer to remove non-specific proteins.
6. Elute protein using high imidazole buffer.
7. Dialyze eluted fractions to remove imidazole.
8. Perform size-exclusion chromatography using Superdex 75 column.
9. Collect fractions corresponding to ~110 mL elution volume.

10. Pool, concentrate, and store protein at −80 °C.
11. Determine protein concentration using Bradford assay.

3.1.3.4 Notes

1. Maintain cold conditions (4 °C) throughout purification to prevent degradation.
2. Imidazole concentration in wash buffer is critical to reduce non-specific binding.

3.2 Protein analysis

3.2.1 Protein estimation (BCA assay)

3.2.1.1 Reagents/equipment

1. Bicinchoninic acid (BCA) reagent (Sigma)
2. Copper (II) sulfate ($CuSO_4$) solution (Sigma)
3. Bovine Serum Albumin (BSA) standards (0–250 μg/mL) (Sigma)
4. MiliQ water
5. Microplate reader Infinite 200 PRO (Tecan)

3.2.1.2 Procedure

1. Prepare working reagent (BCA: $CuSO_4$, 50:1).
2. Prepare BSA standards (0–250 μg/mL).
3. Mix 25 μL sample with 200 μL reagent.
4. Incubate at 37 °C for 1 h.
5. Measure absorbance at 562 nm.
6. Determine concentration using standard curve.

3.2.1.3 Notes

1. Ensure all samples and standards are prepared in the same buffer to avoid interference.
2. BCA assay is compatible with detergents but sensitive to strong reducing agents.
3. Perform measurements in duplicates or triplicates for accuracy.

3.2.2 Protein estimation (bradford assay)

3.2.2.1 Reagents/equipments

1. Bradford reagent (Bio-Rad)
2. MiliQ water
3. Bovine Serum Albumin (BSA) standards (Sigma)
4. Microplate reader Infinite 200 PRO (Tecan)

3.2.2.2 Procedure

1. Prepare working reagent (4:1 H_2O: Bradford reagent).
2. Mix 10 μL sample with 200 μL reagent.
3. Incubate for 5 min at room temperature.
4. Measure absorbance at 595 nm.
5. Determine concentration using standard curve.

3.2.2.3 Notes

1. Bradford assay is rapid but sensitive to detergents such as SDS.
2. Perform measurements quickly as the color complex is time-dependent.
3. Use consistent timing across samples for reliable comparisons.

3.3 Tau - HDAC6 interaction analysis

3.3.1 Tau degradation analysis

Buffers and reagents

SDS-PAGE running buffer
25 mM Tris-Cl
250 mM Glycine
10 % SDS

Tracking dye/sample buffer
300 mM Tris-Cl pH 6.8
600 mM DTT
6 % BPB
12 % SDS
60 % Glycerol

3.3.1.1 Reagents

1. 6X SDS loading buffer
2. Polyacrylamide gel (10–12 %)
3. Running buffer (Tris-Glycine-SDS) (MP Biomedicals)
4. Protein molecular weight marker (Bio-Rad)
5. Coomassie Brilliant Blue or staining reagent (BioRad)
6. PVDF membrane (Bio-Rad)

7. PBST (Phosphate Buffered Saline with Tween-20)
8. Skimmed milk (5 % blocking solution)
9. Primary antibodies: Tau: Pan Tau (K9JA) (A0024) – Dako, HDAC6 ZnF UBP: Anti-HDAC6 (C-terminal) (ab196351) – Abcam, β-tubulin (loading control): (BT7R) – Thermofisher
10. HRP-conjugated secondary antibodies: Donkey anti-goat IgG (ab6885) – Abcam, Goat anti-rabbit IgG HRP (31460) – Thermofisher
11. Enhanced chemiluminescence (ECL) reagent (Clarity™ Western ECL Substrate – Bio-Rad)
12. Recombinant Tau protein
13. HDAC6 ZnF UBP protein
14. BES buffer (pH 7.4)
15. Protease inhibitor cocktail (Roche)

3.3.1.2 Procedure

Buffers used.

SDS-PAGE running buffer	**Transfer buffer**
25 mM Tris-Cl	25 mM Tris-Cl
250 mM Glycine	250 mM Glycine
10 % SDS	20 % Methanol
Wash Buffer (1X PBST)	**Blocking Buffer**
1X PBS	5 % Skimmed Milk in 1X PBST
0.1 % Tween 20	

(A) Tau degradation assay setup.

1. Incubate recombinant Tau with HDAC6 ZnF UBP in BES buffer.
2. Maintain reactions at 37 °C.
3. Incubate both proteins in varying molar ratios (1:2, 1:1, 2:1).
4. Include appropriate controls (Tau alone, HDAC6 alone).
5. Collect samples at 0, 24, and 72 h time points.

(B) SDS-PAGE and western blotting analysis.

1. Mix collected samples with SDS loading buffer and heat prior to loading.
2. Separate proteins using SDS-PAGE.
3. Transfer proteins from gel to PVDF membrane.
4. Block membrane with 5% skimmed milk in PBST.

5. Incubate with respective primary antibodies.
6. Wash membrane with PBST.
7. Incubate with HRP-conjugated secondary antibodies.
8. Develop signals using ECL reagent.

3.3.1.3 Notes

1. Use appropriate gel percentage based on protein size. 12% SDS-PAGE is suitable to accommodate optimum migration for both HDAC6 ZnF UBP (~14 kDa and Tau ~55 kDa).
2. Maintain identical incubation and sampling conditions across all reactions for accurate comparison.
3. Protease inhibitors prevent non-specific degradation during incubation.
4. Confirm efficient protein transfer using Ponceau S staining.
5. Optimize antibody dilutions to minimize background and enhance specificity.
6. β-tubulin serves as a loading control to normalize protein levels.

3.3.2 Size-exclusion chromatography for HDAC6 ZnF UBP–Tau interaction

3.3.2.1 Reagents

1. Recombinant Tau protein
2. HDAC6 ZnF UBP protein
3. SEC running buffer (e.g., PBS or Tris-based buffer)

3.3.2.2 Equipment

1. Superdex Increase 10/300 GL column
2. Chromatography system (ÄKTA Pure – GE)

3.3.2.3 Procedure

1. Incubate Tau and HDAC6 ZnF UBP (1:1 ratio) at 37 °C.
2. Inject 100 μL sample onto Superdex Increase 10/300 GL column.
3. Perform runs at defined time points (0, 1, 2 h).
4. Monitor elution profiles to assess complex formation.

3.3.2.4 Notes

1. Filter and degas buffers before use to avoid column clogging.
2. Calibrate the column with molecular weight standards for accurate interpretation.
3. Shifts in elution volume indicate complex formation or conformational changes.

3.4 Aggregation studies

3.4.1 Aggregation and disaggregation assay setup

3.4.1.1 Reagents

1. Full-length Tau and repeat Tau
2. Heparin (MW ~17,500) (MP Biomedicals)
3. BES buffer (20 mM, pH 7.4) (Sigma)
4. DTT (1 μM) (Calbiochem)
5. NaCl (25 μM) (MP Biomedicals)
6. Sodium azide (0.01 %) (MP Biomedicals)
7. Protease inhibitor cocktail (Roche)

3.4.1.2 Procedure

1. Prepare aggregation reactions using Tau and heparin in a 4:1 (Tau: heparin) ratio.
2. Maintain Tau concentration at 20 μM for aggregation assays.
3. Prepare higher concentration stocks (100 μM Tau) for disaggregation experiments.
4. Dilute aggregates to 20 μM for disaggregation assays unless stated otherwise.
5. Perform all aggregation and disaggregation measurements in triplicates.

3.4.1.3 Notes

1. Heparin acts as a polyanionic inducer to promote Tau aggregation.
2. Maintain consistent buffer conditions across experiments for comparability.

3.4.2 Thioflavin S (ThS) fluorescence assay

3.4.2.1 Reagents

1. Thioflavin S (ThS) (Sigma)
2. Ammonium acetate buffer (50 mM, pH 7.0) (MP Biomedicals)

3.4.2.2 Procedure

1. Prepare ThS working solution in ammonium acetate buffer.
2. Mix Tau (2 μM) with ThS (8 μM) in a 1:4 ratio.
3. Incubate samples for 10 min in 384-well black plates.
4. Measure fluorescence using excitation at 440 nm and emission at 521 nm.
5. Record readings at multiple time points to monitor aggregation kinetics.

3.4.2.3 Notes

1. ThS selectively binds β-sheet-rich aggregates, enhancing fluorescence.

3.5 Structural and morphological analysis of Tau

3.5.1 Far-UV circular dichroism (CD) spectroscopy

3.5.1.1 Reagents

1. Sodium phosphate buffer (50 mM, pH 6.8) (MP Biomedicals)

3.5.1.2 Equipment

1. CD spectrometer (Jasco J-815)
2. Quartz cuvette (1 mm path length)

3.5.1.3 Procedure

1. Prepare samples of soluble Tau, aggregated Tau, and Tau with HDAC6 ZnF UBP in different ratios.
2. Maintain concentrations at 2 μM (full-length Tau) or 5 μM (repeat Tau).
3. Record spectra in the range of 190–250 nm.
4. Use scan speed of 100 nm/min and bandwidth of 1 nm.
5. Acquire 5 scans and average the data.
6. Perform measurements at 25 °C under nitrogen atmosphere.
7. Subtract buffer baseline from all spectra and plot final spectra using SigmaPlot software.

3.5.1.4 Notes

1. CD spectroscopy provides insights into secondary structural transitions during aggregation.
2. Data are represented as differential absorbance ($\Delta A = AL - AR$).

3.5.2 Transmission electron microscopy (TEM)

3.5.2.1 Reagents

1. Uranyl acetate (2 %, aqueous)
2. MiliQ water

3.5.2.2 Equipment

1. Carbon-coated copper grids (400 mesh)
2. Transmission Electron Microscope (Tecnai T20, 120 kV)

3.5.2.3 Procedure

1. Apply 2 μM Tau sample onto copper grids for 45–60 s
2. Wash grids twice with MiliQ water.
3. Stain with 2 % uranyl acetate for 1–2 min
4. Air-dry grids for 24 h.
5. Visualize samples using TEM at 120 kV.

3.5.3 ^{15}N labelled repeat Tau preparation

Repeat Tau labelling with ^{15}N was carried out by using ^{15}N containing minimal media. The labelled repeat Tau was prepared to be used in NMR studies (Balmik et al., 2022; Karran et al., 2011; Narayan et al., 2015; Sonawane et al., 2022; Spillantini & Goedert, 2013).

Procedure.

1. The pT7C repeat Tau transformed BL21*cells were initially grown in LB (6 L) with 100 μg/mL Ampicillin at 37 °C till the OD reaches 6 at 600 nm.
2. The cells were pelleted down and washed with 1X M9 salts.
3. The 1X M9 salts washed cells were re-suspended in minimal media containing 15 N labelled ammonium chloride.
4. In order to clear unlabelled metabolites, the cells were grown at 37 °C for 1 h prior to induction with 0.5 mM IPTG and harvesting after 4–5 h.
5. The labelled protein was prepared similarly as previously described method for unlabelled Tau protein.

Stock solutions.

5X M9 salts (autoclaved).

Salt	Weight (gm/100 mL)
KH_2PO_4	1.5
Na_2HPO_4	3
NaCl	0.25
NH_4Cl (^{15}N labelled)	0.5

The pH of the 5X M9 salts is adjusted to 7.2 before autoclaving.

1X M9 salts unlabelled (autoclaved).

Salt	Weight (gm/1000 mL)
KH_2PO_4	3
Na_2HPO_4	5
NaCl	0.5

The pH of the 1X M9 salts is adjusted to 7.2 before autoclaving.

Minimal Growth Media (1500 mL).

5X M9 Salts	300 mL
20 % D-Glucose (filter sterilized)	30 mL
Thiamine hydrochoride (5 mg/mL) (filter sterilized)	9 mL
1 M $MgSO_4$ (autoclaved)	3 mL
1 M $CaCl_2$ (autoclaved)	150 μl

The final volume is made up to 1500 mL using autoclaved MiliQ water.

4. Concluding remarks

The present study employs a combination of biochemical and biophysical approaches to explore the role of HDAC6 as a critical non-catalytic regulator of Tau structure and function as observed by its effect on Tau conformational dynamics, aggregation kinetics, and stability. These findings suggest a dual role of HDAC6, owing to its catalytic and non-catalytic functions and its direct involvement relevant to tauopathies (Cook et al., 2014; Fazal et al., 2021).

The in vitro analyses utilized in this study provide valuable mechanistic insights into the direct interaction between HDAC6 ZnF UBP and Tau, enabling to understand the domain-specific functions under controlled conditions. However, these systems have inherent limitations in consolidating the complexity of the intracellular environment, where additional cofactors, post-translational modifications, and cellular pathways may further influence Tau regulation (Avila et al., 2016; Medina et al., 2016). Therefore, integrating in vitro findings with cellular and in vivo studies will be crucial to validate the physiological and pathological relevance of HDAC6 ZnF UBP in Tau biology.

Overall, the complementary use of biochemical assays and advanced model systems provides a comprehensive framework to understand the mechanistic basis of HDAC6–Tau interactions and its importance in Tau function. This integrated strategy not only strengthens the biological significance of the findings but also provides an avenue for targeting specific domains of HDAC6 as a refined therapeutic approach in Tau-associated neurodegenerative disorders.

References

AlHayani, D. A., Kubaev, A., Uthirapathy, S., Mandaliya, V., Ballal, S., Kalia, R., ... Kadhim, A. J. (2025). Insights into the therapeutic potential of SIRT1-modifying compounds for Alzheimer's disease: A focus on molecular mechanisms. *Journal of Molecular Neuroscience, 75*(1), 29.

Avila, J., Pallas, N., Bolós, M., Sayas, C. L., & Hernandez, F. (2016). Intracellular and extracellular microtubule associated protein tau as a therapeutic target in Alzheimer disease and other tauopathies. *Expert Opinion on Therapeutic Targets, 20*(6), 653–661.

Balmik, A. A., Chidambaram, H., Dangi, A., Marelli, U. K., & Chinnathambi, S. (2020). HDAC6 ZnF UBP as the modifier of tau structure and function. *Biochemistry, 59*(48), 4546–4562.

Balmik, A. A., Das, R., Sonawane, S. K., & Chinnathambi, S. (2022). Aggregation species of amyloid-β and Tau oligomers in Alzheimer's disease: Role in therapeutics and diagnostics. 2022.

Balmik, A. A., Sonawane, S. K., & Chinnathambi, S. (2021). The extracellular HDAC6 ZnF UBP domain modulates the actin network and post-translational modifications of Tau. *Cell Communication and Signaling, 19*(1), 49.

Bonnet, J., Romier, C., Tora, L., & Devys, D. (2008). Zinc-finger UBPs: Regulators of deubiquitylation. *Trends in Biochemical Sciences, 33*(8), 369–375.

Boyault, C., Sadoul, K., Pabion, M., & Khochbin, S. (2007). HDAC6, at the crossroads between cytoskeleton and cell signaling by acetylation and ubiquitination. *Oncogene, 26*(37), 5468–5476.

Boyault, C., Zhang, Y., Fritah, S., Caron, C., Gilquin, B., Kwon, S. H., ... Matthias, P. (2007). HDAC6 controls major cell response pathways to cytotoxic accumulation of protein aggregates. *Genes & Development, 21*(17), 2172–2181.

Busche, M. A., & Hyman, B. T. (2020). Synergy between amyloid-β and tau in Alzheimer's disease. *Nature Neuroscience, 23*(10), 1183–1193.

Chan, A. M., & Fletcher, S. (2021). Shifting the paradigm in treating multi-factorial diseases: Polypharmacological co-inhibitors of HDAC6. *RSC Medicinal Chemistry, 12*(2), 178–196.

Chen, J., Sahakian, E., Powers, J., Lienlaf, M., Perez-Villarroel, P., Knox, T., & Villagra, A. (2016). *Histone deacetylases: Methods and protocols*. Springer 147–165.

Choudhary, C., Weinert, B. T., Nishida, Y., Verdin, E., & Mann, M. (2014). The growing landscape of lysine acetylation links metabolism and cell signalling. *Nature Reviews. Molecular Cell Biology, 15*(8), 536–550.

Cleveland, D. W., Hwo, S.-Y., & Kirschner, M. W. (1977). Purification of tau, a microtubule-associated protein that induces assembly of microtubules from purified tubulin. *Journal of Molecular Biology, 116*(2), 207–225.

Cook, C., Carlomagno, Y., Gendron, T. F., Dunmore, J., Scheffel, K., Stetler, C., ... DeTure, M. (2014). Acetylation of the KXGS motifs in tau is a critical determinant in modulation of tau aggregation and clearance. *Human Molecular Genetics, 23*(1), 104–116.

Cummings, J. L., Osse, A. M. L., & Kinney, J. W. (2023). Alzheimer's disease: Novel targets and investigational drugs for disease modification. *Drugs, 83*(15), 1387–1408.

Darling, A. L., & Uversky, V. N. (2018). Intrinsic disorder and posttranslational modifications: The darker side of the biological dark matter. *Frontiers in genetics, 9*, 158.

de Oliveira, R. M., Sarkander, J., Kazantsev, A. G., & Outeiro, T. F. (2012). SIRT2 as a therapeutic target for age-related disorders. *Frontiers in Pharmacology, 3*, 82.

De Strooper, B., & Karran, E. (2016). The cellular phase of Alzheimer's disease. *Cell, 164*(4), 603–615.

Deakin, N. O., & Turner, C. E. (2014). Paxillin inhibits HDAC6 to regulate microtubule acetylation, Golgi structure, and polarized migration. *Journal of Cell Biology, 206*(3), 395–413.

Donmez, G., & Outeiro, T. F. (2013). SIRT1 and SIRT2: Emerging targets in neurodegeneration. *EMBO Molecular Medicine, 5*(3), 344–352.

d'Ydewalle, C., Bogaert, E., & Van Den Bosch, L. (2012). HDAC 6 at the intersection of neuroprotection and neurodegeneration. *Traffic (Copenhagen, Denmark), 13*(6), 771–779.

Fazal, R., Boeynaems, S., Swijsen, A., De Decker, M., Fumagalli, L., Moisse, M., ... Vercruysse, T. (2021). HDAC6 inhibition restores TDP-43 pathology and axonal transport defects in human motor neurons with TARDBP mutations. *The EMBO Journal, 40*(7), EMBJ2020106177.

Finnin, M. S., Donigian, J. R., Cohen, A., Richon, V. M., Rifkind, R. A., Marks, P. A., ... Pavletich, N. P. (1999). Structures of a histone deacetylase homologue bound to the TSA and SAHA inhibitors. *Nature, 401*(6749), 188–193.

Fitzpatrick, A. W., Falcon, B., He, S., Murzin, A. G., Murshudov, G., Garringer, H. J., ... Scheres, S. H. (2017). Cryo-EM structures of tau filaments from Alzheimer's disease. *Nature, 547*(7662), 185–190.

Goedert, M. (2015). Alzheimer's and Parkinson's diseases: The prion concept in relation to assembled Aβ, tau, and α-synuclein. *Science (New York, N. Y.), 349*(6248), 1255555.

Gräff, J., & Tsai, L.-H. (2013). Histone acetylation: Molecular mnemonics on the chromatin. *Nature Reviews. Neuroscience, 14*(2), 97–111.

Greenfield, N. J. (2006). Using circular dichroism spectra to estimate protein secondary structure. *Nature Protocols, 1*(6), 2876–2890.

Guan, J.-S., Haggarty, S. J., Giacometti, E., Dannenberg, J.-H., Joseph, N., Gao, J., ... Mazitschek, R. (2009). HDAC2 negatively regulates memory formation and synaptic plasticity. *Nature, 459*(7243), 55–60.

Guan, P.-P., & Wang, P. (2023). The involvement of post-translational modifications in regulating the development and progression of Alzheimer's disease. *Molecular Neurobiology, 60*(7), 3617–3632.

Haberland, M., Montgomery, R. L., & Olson, E. N. (2009). The many roles of histone deacetylases in development and physiology: Implications for disease and therapy. *Nature Reviews. Genetics, 10*(1), 32–42.

Hai, Y., Shinsky, S. A., Porter, N. J., & Christianson, D. W. (2017). Histone deacetylase 10 structure and molecular function as a polyamine deacetylase. *Nature Communications, 8*(1), 15368.

Hardy, J., & Selkoe, D. J. (2002). The amyloid hypothesis of Alzheimer's disease: Progress and problems on the road to therapeutics. *Science (New York, N. Y.), 297*(5580), 353–356.

Hubbert, C., Guardiola, A., Shao, R., Kawaguchi, Y., Ito, A., Nixon, A., ... Yao, T.-P. (2002). HDAC6 is a microtubule-associated deacetylase. *Nature, 417*(6887), 455–458.

Imai, S.-i, & Guarente, L. (2014). NAD+ and sirtuins in aging and disease. *Trends in Cell Biology, 24*(8), 464–471.

Jeganathan, S., von Bergen, M., Brutlach, H., Steinhoff, H.-J., & Mandelkow, E. (2006). Global hairpin folding of tau in solution. *Biochemistry, 45*(7), 2283–2293.

Karran, E., Mercken, M., & Strooper, B. D. (2011). The amyloid cascade hypothesis for Alzheimer's disease: An appraisal for the development of therapeutics. *Nature Reviews. Drug Discovery, 10*(9), 698–712.

Kawaguchi, Y., Kovacs, J. J., McLaurin, A., Vance, J. M., Ito, A., & Yao, T.-P. (2003). The deacetylase HDAC6 regulates aggresome formation and cell viability in response to misfolded protein stress. *Cell, 115*(6), 727–738.

Knopman, D. S., Amieva, H., Petersen, R. C., Chételat, G., Holtzman, D. M., Hyman, B. T., ... Jones, D. T. (2021). Alzheimer disease. *Nature Reviews Disease Primers, 7*(1), 33.

LeVine, H., III (1999). *Methods in enzymology*. Elsevier 274–284.

Li, Y., Li, S., & Wu, H. (2022). Ubiquitination-proteasome system (UPS) and autophagy two main protein degradation machineries in response to cell stress. *Cells, 11*(5), 851.

Livingston, G., Huntley, J., Sommerlad, A., Ames, D., Ballard, C., Banerjee, S., ... Cooper, C. (2020). Dementia prevention, intervention, and care: 2020 report of the Lancet Commission. *The Lancet, 396*(10248), 413–446.
Long, J. M., & Holtzman, D. M. (2019). Alzheimer disease: An update on pathobiology and treatment strategies. *Cell, 179*(2), 312–339.
LoPresti, P. (2020). HDAC6 in diseases of cognition and of neurons. *Cells, 10*(1), 12.
Medina, M., Hernández, F., & Avila, J. (2016). New features about tau function and dysfunction. *Biomolecules, 6*(2), 21.
Mukrasch, M. D., Bibow, S., Korukottu, J., Jeganathan, S., Biernat, J., Griesinger, C., ... Zweckstetter, M. (2009). Structural polymorphism of 441-residue tau at single residue resolution. *PLoS Biology, 7*(2), e1000034.
Narayan, P. J., Lill, C., Faull, R., Curtis, M. A., & Dragunow, M. (2015). Increased acetyl and total histone levels in post-mortem Alzheimer's disease brain. *Neurobiology of Disease, 74*, 281–294.
Parra, M. (2015). Class II a HDAC s–new insights into their functions in physiology and pathology. *The FEBS Journal, 282*(9), 1736–1744.
Perez, M., Santa-Maria, I., De Barreda, E. G., Zhu, X., Cuadros, R., Cabrero, J. R., ... Perry, G. (2009). Tau–an inhibitor of deacetylase HDAC6 function. *Journal of Neurochemistry, 109*(6), 1756–1766.
RH, H. (2012). Sirtuins as regulators of metabolism and healthspan. *Nature Reviews. Molecular Cell Biology, 13*, 225–238.
Saha, R., & Pahan, K. (2006). HATs and HDACs in neurodegeneration: A tale of disconcerted acetylation homeostasis. *Cell Death & Differentiation, 13*(4), 539–550.
Seigneurin-Berny, D., Verdel, A., Curtet, S., Lemercier, C., Garin, J., Rousseaux, S., & Khochbin, S. (2001). Identification of components of the murine histone deacetylase 6 complex: Link between acetylation and ubiquitination signaling pathways. *Molecular and Cellular Biology, 21*(23), 8035–8044.
Seto, E., & Yoshida, M. (2014). Erasers of histone acetylation: The histone deacetylase enzymes. *Cold Spring Harbor Perspectives in Biology, 6*(4), a018713.
Singh, R., Rathore, A. S., Dilnashin, H., Keshri, P. K., Gupta, N. K., Prakash, S. A. S., ... Singh, S. P. (2024). HAT and HDAC: Enzyme with contradictory action in neurodegenerative diseases. *Molecular Neurobiology, 61*(11), 9110–9124.
Sonawane, S. K., Dubey, T., Balmik, A. A., Das, R., & Chinnathambi, S. (2022). Alzheimer's disease pathology. A Tau Perspective. 76-108.
Spillantini, M. G., & Goedert, M. (2013). Tau pathology and neurodegeneration. *The Lancet Neurology, 12*(6), 609–622.
Wang, Y., & Mandelkow, E. (2016). Tau in physiology and pathology. *Nature Reviews. Neuroscience, 17*(1), 22–35.
Weingarten, M. D., Lockwood, A. H., Hwo, S.-Y., & Kirschner, M. W. (1975). A protein factor essential for microtubule assembly. *Proceedings of the National Academy of Sciences, 72*(5), 1858–1862.
Yan, J. (2014). Interplay between HDAC6 and its interacting partners: Essential roles in the aggresome-autophagy pathway and neurodegenerative diseases. *DNA and Cell Biology, 33*(9), 567–580.
Yang, S.-s, Zhang, R., Wang, G., & Zhang, Y.-f (2017). The development prospection of HDAC inhibitors as a potential therapeutic direction in Alzheimer's disease. *Translational Neurodegeneration, 6*(1), 19.
Yang, X.-J., & Seto, E. (2008). The Rpd3/Hda1 family of lysine deacetylases: From bacteria and yeast to mice and men. *Nature Reviews. Molecular Cell Biology, 9*(3), 206–218.
Yeong, K. Y., Berdigaliyev, N., & Chang, Y. (2020). Sirtuins and their implications in neurodegenerative diseases from a drug discovery perspective. *ACS Chemical Neuroscience, 11*(24), 4073–4091.

CHAPTER SIX

Sirtuins in mitochondrial physiology and energy homeostasis

Arshdeep Kaur[a] and Ullas Kolthur Seetharam[a,b,c,*]
[a]Tata Institute of Fundamental Research Mumbai, Mumbai, Maharashtra, India
[b]Tata Institute of Fundamental Research Hyderabad, Hyderabad, Telangana, India
[c]Centre for DNA Fingerprinting and Diagnostics, Hyderabad, Telangana, India
*Corresponding author. e-mail address: ullas@tifr.res.in

Contents

1. Introduction 112
2. Rationale and experimental challenges in studying mitochondrial sirtuins 114
   2.1 Relative abundance of mitochondrial sirtuins 114
   2.2 NAD^+ availability and mitochondrial compartmentalization 115
   2.3 Substrate specificity overlapping 116
   2.4 Non-canonical localization and the importance of tissue specificity 116
   2.5 Post-translational regulation of mitochondrial sirtuins 117
3. Experimental models and design considerations 118
   3.1 Pharmacological vs genetic approaches 118
   3.2 Metabolic state as a critical experimental variable 119
4. Measurement of sirtuins enzymatic activity 120
   4.1 Measuring catalytic activity 120
   4.2 Assessing NAD^+ dependence 126
   4.3 Monitoring acylation dynamics using chemical biology approaches 129
5. Assessment of mitochondrial protein acylation 129
   5.1 Mitochondrial respiration and bioenergetics 130
   5.2 Mitochondrial redox balance and reactive oxygen species 132
   5.3 Mitochondrial dynamics and quality control 134
6. Cellular energy state and whole-body homeostasis 135
7. Conclusion 136
Acknowledgements 137
References 137

Abstract

Mitochondrial sirtuins integrate cellular metabolic and energetic state with post-translational regulation of mitochondrial proteins, thereby influencing bioenergetics, redox homeostasis, and metabolic flexibility. Despite extensive study of individual sirtuins, a comprehensive understanding of their function remains limited

Methods in Enzymology, Volume 733
ISSN 0076-6879, https://doi.org/10.1016/bs.mie.2026.05.029

by methodological and conceptual challenges. These arise from the complex metabolic environment in which mitochondrial sirtuins act, including compartmentalized NAD^+ pools, metabolite-driven non-enzymatic acylation, overlapping substrate specificities, and limited tools to accurately characterize enzyme-specific activity, particularly for SIRT4.

This review aims to highlight current experimental approaches used to study mitochondrial sirtuins, and the need to integrate enzymatic measurements with metabolic and physiological readouts. We discuss the importance of considering mitochondrial spatial heterogeneity within cells, tissue-specific metabolic context, and temporal dynamics of metabolic state when interpreting sirtuin activity. Recent advances in quantitative proteomics, metabolite profiling, and mitochondria-specific analyses provide new opportunities to resolve these complexities. Moving forward, integrating these approaches with a systems-level and quantitative framework will be critical to fully understand how mitochondrial sirtuins orchestrate metabolic regulation across cellular and organismal scales.

1. Introduction

Sirtuins constitute a conserved family of NAD^+-dependent enzymes that link cellular metabolic state to post-translational regulation of proteins and metabolic homeostasis. Although conserved from archaea to humans, the number and functions seem to have diversified with seven mammalian sirtuins, each displaying distinct subcellular localization, enzymatic activity and substrate specificity (Michan & Sinclair, 2007). Through their dependence on NAD^+, sirtuins uniquely function as metabolic sensors, integrating calorie/metabolic status, redox balance, and energetics to mediate physiological homeostasis at cellular and organismal levels (Li & Kazgan, 2011).

Among the seven mammalian sirtuins, SIRT3, SIRT4, and SIRT5 are the mitochondrial proteins that have emerged as key regulators of mitochondrial physiology and energy homeostasis (Zhong & Mostoslavsky, 2011). SIRT3 is the primary mitochondrial deacetylase and targets proteins involved in oxidative phosphorylation, β-oxidation, tricarboxylic acid (TCA) cycle, and antioxidant defense (Zhang et al., 2020, p. 3). The enzymatic activities of mitochondrial sirtuins extend beyond classical lysine deacetylation. SIRT4 has been implicated in ADP-ribosylation and deacylation of proteins involved in amino acid metabolism, insulin secretion, electron transport chain (ETC), ATP homeostasis and anaplerotic flux (Ahuja et al., 2007; Ho et al., 2013). SIRT5 exhibits robust desuccinylase, demalonylase, and deglutarylase activities, thereby controlling lysine acyl modifications that are particularly enriched on mitochondrial factors (Schlicker et al., 2008). Together, these mitochondrial sirtuins coordinate mitochondrial substrate utilization, respiratory capacity, and redox balance, underscoring their central role in cellular and organismal energy metabolism (Zhong & Mostoslavsky, 2011).

Although few, studies have illustrated evolutionary conservation of mitochondrial sirtuins (Chang et al., 2025; Vassilopoulos et al., 2011; Wood et al., 2018). Despite this, overlapping and non-overlapping functions of mitochondrial sirtuins are poorly understood. As detailed later, dissecting such redundant and non-redundant roles seem to be limited by studies which have investigated substrate specificities and metabolism state dependent expression and interactions. The diversity of sirtuins and their catalytic activities highlights the need for methodologically rigorous approaches to study enzymatic activity and to accurately assess specific acyl modifications under different metabolic transitions.

Owing to their NAD dependence, it is imperative to tease out differential engagement of mitochondrial sirtuins across physiological states, including feed-fast cycles and chronic dietary/lifestyle perturbations which are known to affect cellular NAD^+ levels (Høyland et al., 2024; Stein & Imai, 2012; Yang et al., 2007). Therefore, sirtuin activity cannot be inferred solely from protein expression levels but must be evaluated in the context of mitochondrial metabolic flux and cellular energy state. For instance, fasting can elevate mitochondrial NAD^+ levels and increase SIRT3-dependent deacetylation of enzymes involved in oxidative metabolism (Lu et al., 2015). In contrast, SIRT4 has been shown to hamper glutamine anaplerosis through inhibition of glutamate dehydrogenase, thereby limiting excessive substrate flux into the TCA cycle (Haigis et al., 2006).

While sirtuins have been largely considered as lysyl deacylases, studies have also indicated that they could catalyze deacylations on cysteine residues. Moreover, it is well known that protein acylations, both enzymatic and non-enzymatic and largely brought about by acyl CoA levels, are inherently coupled to cellular metabolism (Wagner & Hirschey, 2014; Weinert et al., 2015). High rates of TCA cycle activity and amino acid catabolism increase succinyl-CoA and related intermediates, promoting protein succinylation and thereby increasing the functional relevance of SIRT5-mediated desuccinylation (Rardin et al., 2013). Similarly, changes in carbohydrate and lipid utilization alter the availability of acetyl-CoA and mitochondrial acetylation patterns, directly impacting SIRT3 substrate engagement (Shimazu et al., 2010; Talwadekar et al., 2024; Weinert et al., 2015). Thus, mitochondrial sirtuins operate at the intersection of enzymatic and metabolic regulation, responding not only to NAD^+ availability but also to the abundance and distribution of acyl donors. Variations in metabolic state, cell type, tissue origin, and experimental timing can significantly influence measured outcomes.

Despite decades of work that have unequivocally demonstrated evolutionarily conserved pivotal roles of mitochondrial sirtuins, tissue and physiological context dependent significance is still limited due to methodological limitations in dissecting out interactions/substrate and catalytic specificities. Given this we have not only described methodological approaches that are currently utilized but also highlight technical gaps and opportunities necessary to provide comprehensive understanding of mitochondrial sirtuin biology. In addition, we posit that development of new assays and insights will be crucial to also consider therapeutic interventions to modulate SIRT3, SIRT4 and SIRT5 activities in various disease conditions.

2. Rationale and experimental challenges in studying mitochondrial sirtuins

As noted, mitochondrial sirtuin abundance does not reliably reflect catalytic activity, which is also governed by NAD^+ availability, substrate context, and competing non-enzymatic acylation driven by mitochondrial acyl-CoA levels. Some of the key factors that potentially limit physiological deductions/interpretations vis-à-vis methodological approaches are discussed in the following sections.

2.1 Relative abundance of mitochondrial sirtuins

A fundamental but often overlooked aspect of mitochondrial sirtuin biology is the relative abundance of different sirtuins within the mitochondrion. Since these enzymes coexist in the same cellular compartment and utilize a common co-substrate (NAD^+), their effective contribution to protein deacylation is likely dictated by their relative concentrations in conjunction with their catalytic efficiencies and substrate preferences. However, most studies have examined mitochondrial sirtuins individually, and quantitative comparisons of their abundance across tissues or metabolic states remain limited.

In principle, the steady-state level of a given acyl modification reflects the balance between non-enzymatic acylation and the combined activity of multiple deacylases. Therefore, the relative abundance of mitochondrial sirtuins, together with their kinetic parameters (e.g., Km for NAD^+ and acyl substrates), will influence which enzyme predominates under specific metabolic conditions. This lack of comparative information oversimplifies interpretation of mitochondrial acylation phenotypes.

Another emerging consideration is the extent of intracellular heterogeneity in their distribution and function. Mitochondria are highly dynamic organelles that undergo continuous fission, fusion, and spatial redistribution, leading to functional heterogeneity within a single cell. Increasingly, cellular metabolism is understood in a spatially resolved context, where distinct mitochondrial subpopulations can differ in membrane potential, substrate utilization, and redox state. It is unlikely that all mitochondria within a cell maintain identical sirtuin composition or activity. Instead, local variations in NAD^+ availability, metabolite gradients, and protein interactions may give rise to heterogeneous activation of SIRT3, SIRT4, and SIRT5 across mitochondrial subdomains.

2.2 NAD^+ availability and mitochondrial compartmentalization

Mitochondrial sirtuins operate within a metabolically dynamic organelle whose function is influenced by both cell-autonomous and systemic signals. Nutrient availability, hormonal cues, circadian oscillations, and tissue-specific metabolic demands all modulate mitochondrial redox state and substrate flux, thereby influencing sirtuin activity (Pirinen et al., 2012).

Another layer of complexity arises from compartmentalization of NAD^+ pools. Mitochondrial NAD^+ concentrations are not necessarily equivalent to cytosolic or nuclear pools and may vary across tissues depending on metabolic demand (Høyland et al., 2024). Therefore, systemic metabolic changes do not uniformly translate into mitochondrial sirtuin activation Table 1.

Table 1 Approximate NAD^+ concentrations across subcellular compartments (Cambronne et al., 2016; Stein & Imai, 2012).

Compartment	NAD^+ concentration	NAD^+/NADH ratio
Mitochondrial matrix	~250–500 μM	Low (7−10)
Cytosol	~50–100 μM	High (60−100)
Nucleus	Similar to cytosol (~80–100 μM)	Similar to cytosol

Although several studies have reported changes in mitochondrial NAD^+ levels in response to physiological perturbations such as fasting, exercise, and/or aging, tissue wide comparisons of NAD^+ pools are limited. A further question is how the fluctuations in mitochondrial NAD^+ pools might differentially regulate individual sirtuins, as SIRT3, SIRT4, and SIRT5 possess different catalytic efficiencies and Km values for NAD^+ utilization, as indicated in Table 3. Future experiments will be required to establish whether these differential sensitivities in NAD^+ utilization play a role in tissue-specific regulation of mitochondrial sirtuins.

2.3 Substrate specificity overlapping

A major experimental challenge in the study of mitochondrial sirtuins is the extensive overlap in substrate specificity among family members. Although SIRT3, SIRT4, and SIRT5 exhibit preferential catalytic activities, many mitochondrial proteins harbor multiple lysine acyl modifications that can potentially be regulated by more than one sirtuin. Consequently, assigning substrate-enzyme relationships based solely on global acylation changes frequently leads to ambiguous interpretation.

Substrate specificity in vivo is influenced not only by intrinsic catalytic preference but also by local metabolite availability, protein-protein interactions, and sub-mitochondrial localization. Local concentrations of acyl donors, NAD^+ availability, and spatial redox states influence which substrates become modified and therefore accessible to deacylation. As a result, substrate selection reflects the integration of enzyme chemistry with mitochondrial metabolic state rather than strict sequence recognition.

2.4 Non-canonical localization and the importance of tissue specificity

Although initially classified predominantly based on subcellular localization, recent studies suggests that their spatial distribution is more dynamic and context dependent than originally appreciated. SIRT3, SIRT4, and SIRT5 are enriched within mitochondria through N-terminal targeting sequences; however, multiple studies demonstrate that localization can vary across tissues, developmental states, and metabolic conditions.

SIRT3 is widely regarded as the principal mitochondrial deacetylase, yet both full-length and processed forms have been detected outside mitochondria under specific conditions, suggesting regulated import and maturation (Iwahara et al., 2012; Scher et al., 2007). Initial studies demonstrated that SIRT3, once inside the mitochondria, is proteolytically cleaved to an enzymatically active shorter isoform within the matrix (Onyango et al., 2002; Schwer et al., 2002). Subsequent work indicates that precursor localization and import efficiency may vary with cellular stress and metabolic demand, implying that substrate selectivity may partly be a function of accessibility.

SIRT4 and SIRT5 similarly exhibit context-dependent cellular localization. While primarily mitochondrial, SIRT5 has been detected in cytosolic and peroxisomal fractions in certain tissues, consistent with its role in regulating diverse acyl modifications derived from intermediary metabolism (Nishida et al., 2015). These observations highlight that enzyme localization and metabolite availability jointly determine effective substrate engagement.

Tissue specificity is a critical determinant of mitochondrial sirtuin function because mitochondrial metabolism varies substantially across organs. Liver, skeletal muscle, adipose tissue, brain, and heart have differential substrate preference, respiratory demand, and mitochondrial density, resulting in distinct acylation landscapes. For example, SIRT3-mediated regulation of fatty acid oxidation is significant in liver and oxidative muscle, whereas SIRT4 regulated amino acid metabolism and insulin secretion pathways in pancreatic and hepatic contexts underscoring that functional outputs depend strongly on tissue metabolic specialization (Haigis et al., 2006). Quantitative proteomic studies have shown that the stoichiometry of mitochondrial protein acylation varies significantly in oxidative versus glycolytic tissues. Therefore the experimental design should take into consideration tissue-specific models and metabolic conditions.

2.5 Post-translational regulation of mitochondrial sirtuins

Owing to the nature of the control that sirtuins in general and mitochondrial sirtuins in particular exert over metabolic fluxes, mitochondrial energetics, and stress-responsive signaling pathways, it is imperative to expect PTM based regulation. In fact, several studies in the field have clearly established such control of SIRT1, SIRT6 and SIRT2. Relatively little is known about PTM based modulation of mitochondrial sirtuins. This is important because it is well known that a plethora of proteins in the mitochondrial matrix are subject to a diverse set of PTMs. Few studies have indicated phosphorylation, and other regulatory modifications on SIRT3, SIRT4, and SIRT5, suggesting that upstream signaling pathways may directly influence their catalytic function, although these mechanisms remain less well characterized (Liu et al., 2015). Consequently, assessing mitochondrial sirtuin function would require consideration of their post-translational modifications in addition to measurements of enzyme abundance and cellular metabolic parameters.

3. Experimental models and design considerations

Studies on sirtuin mediated regulation of mitochondrial physiology generally rely along two independent dimensions: the method of manipulation (pharmacological, genetic, or physiological) and the duration of intervention. Metabolic inputs are often simplified and tightly controlled in cell culture, enabling mechanistic analysis of enzyme function. However, these systems may not fully recapitulate the complexity of in vivo metabolic regulation. In contrast, whole-animal models incorporate systemic regulation but introduce additional variables, including endocrine signaling, inter-organ crosstalk, and behavioral factors such as feeding patterns and physical activity.

3.1 Pharmacological vs genetic approaches

Pharmacological approaches provide the temporal resolution required to investigate direct enzymatic regulation. Acute manipulation of sirtuin activity, either using small-molecule activators and inhibitors or through interventions that alter mitochondrial NAD^+ availability, enables rapid assessment of changes in protein acylation, respiration, and redox balance prior to transcriptional or adaptive remodeling. For instance, the widely used SIRT3 activator, honokiol, has been shown to maintain mitochondrial ROS homeostasis and improve respiratory capacity in acute conditions. However, these effects are observed within a relatively narrow concentration window (typically ~1–10 μM), beyond which off-target antioxidant and membrane-modulatory effects become prominent (Pillai et al., 2017). Conversely, pharmacological inhibitors such as 3-TYP (3-(1H-1,2,3-triazol-4-yl) pyridine), frequently used to inhibit SIRT3 activity, lead to a reduction in mitochondrial protein deacetylation within a few hours at concentrations of approximately 50–100 μM. It is widely utilized in experimental models investigating oxidative stress and mitochondrial dysfunction (Shi et al., 2022). However, interpretation of pharmacological experiments requires caution, as many reported modulators exhibit limited specificity and frequently induce secondary metabolic alterations.

In addition, genetic models are particularly useful for understanding long-term metabolic adaptation and systemic energy homeostasis. The use of knockout and overexpression model systems have shown the importance of mitochondrial sirtuins in fatty acid oxidation, amino acid metabolism, and oxidative stress regulation (Ahuja et al., 2007; Chen et al., 2011; Hirschey et al., 2010) . A well-characterized example of genetic manipulation is the global deletion of SIRT3. SIRT3 knockout mice exhibit hyperacetylation of mitochondrial proteins, leading to impaired fatty acid oxidation and reduced activity of key metabolic enzymes such as long-chain acyl-CoA dehydrogenase. At the mitochondrial level, this manifests as reduced respiratory efficiency and increased levels of reactive oxygen species due to reduced activation of antioxidant systems such as MnSOD (Someya et al., 2010; Tao et al., 2010). Importantly, these mitochondrial changes translate to whole-body physiological phenotypes, including increased susceptibility to metabolic syndrome, cold intolerance, and impaired adaptation to fasting, highlighting a critical role for SIRT3 in maintaining metabolic flexibility (Hirschey et al., 2010; Someya et al., 2010).

Nevertheless, chronic genetic manipulation often induces compensatory changes in mitochondrial dynamics, transcriptional reprogramming, and metabolic pathway rewiring, complicating attribution of phenotypes to direct enzymatic effects (Dittenhafer-Reed et al., 2015). Conditional or inducible models improve temporal and tissue specificity, reducing developmental compensation and enabling investigation of metabolically specialized tissues.

Importantly, neither pharmacological nor genetic strategies alone are sufficient to fully resolve sirtuin function. Integration of acute and chronic perturbations provides stronger causal inference by distinguishing primary enzymatic regulation from secondary physiological adaptation.

3.2 Metabolic state as a critical experimental variable

Physiological manipulations such as fasting, calorie restriction, exercise, or nutrient switching provide valuable approaches to modulate endogenous sirtuin activity in a metabolically relevant manner.

Comparing sirtuin activity between fed and starved states enables delineation of their functions across distinct metabolic conditions. In the fed state, increased reducing equivalents lower the NAD^+/NADH ratio, generally limiting sirtuin activity. In contrast, fasting or increased energetic demand elevates NAD^+ availability through enhanced oxidative metabolism and activation of NAD^+ salvage pathways, facilitating sirtuin-dependent metabolic adaptation (Chalkiadaki & Guarente, 2012).

Differences in NAD^+ affinity may also create hierarchical activation across physiological states. Some sirtuins may respond rapidly to acute energetic stress, while others might become fully engaged only during prolonged metabolic adaptation, such as sustained fasting or endurance training. This temporal resolution is particularly important when interpreting experimental outcomes, as acute pharmacological perturbations and chronic physiological interventions may engage distinct regulatory pathways.

The circadian regulation introduces another layer of control. Oscillations in NAD^+ biosynthesis generate daily rhythms in sirtuin activity, aligning mitochondrial function with feeding-fasting cycles (Peek et al., 2013; Zhuang et al., 2024). Consequently, experiments performed at different times of day can have different outcomes.

4. Measurement of sirtuins enzymatic activity

In addition to the limitations mentioned above, our understanding of catalytic activities is restricted because of non-availability of methods of choice.

4.1 Measuring catalytic activity

Mitochondrial sirtuins catalyze NAD^+-dependent lysine deacylation reactions in which cleavage of NAD^+ generates nicotinamide and O-acyl-ADP-ribose alongside the deacylated substrate. Direct quantification of these reaction products provides the most definitive measurement of enzymatic activity.

Immunoblotting using pan-acetyl-lysine, succinyl-lysine, or malonyl-lysine antibodies is frequently used as an initial readout of activity changes in cells or tissues. While technically straightforward, these approaches measure steady-state modification levels rather than catalytic rates.

4.1.1 Fluorometric assay

The most direct assessment of sirtuin activity relies on enzymatic assays that quantify removal of acyl modifications from lysine residues. In vitro fluorometric assays using synthetic acylated peptides coupled to fluorescent reporters allow rapid measurement of deacylation kinetics. Kinetic parameters such as Km and Kcat can be estimated by measurement of reaction velocity across substrate concentrations, providing quantitative comparison of catalytic efficiencies among SIRT1, SIRT6, and mitochondrial isoforms SIRT3–5 (Chattopadhyay et al., 2020).

Key reagents- Recombinant or immunoprecipitated sirtuin protein, fluorogenic acylated peptide substrate, NAD^+, developer reagent, assay buffer, nicotinamide, positive inhibitor control, black 96-well plates.

Typical reaction components for fluorometric assay of sirtuin activity

Component	Final concentration
Assay buffer	1×
Fluorogenic substrate	10–50 μM
NAD^+	0.5–2 nM
Recombinant sirtuin	1–5 U

Reaction components-

Assay procedure-

1. Prepare all reagents on ice and thaw enzyme rapidly to preserve activity.
2. Add recombinant or immunoprecipitated sirtuin protein to the wells of a black 96-well plate.
3. Prepare the substrate mixture containing fluorogenic substrate and NAD^+ in assay buffer.
4. Prepare inhibitor stocks at 5× final concentration.
5. To each well, assemble the reaction mixture in a total volume of 50 μL.
6. Initiate the reaction by adding the substrate/NAD^+ mixture.
7. Incubate the plate at 37 °C for 30–40 min. The incubation time should be optimized empirically to maintain measurements within the linear phase of substrate turnover.
8. Terminate the reaction by adding 50uL developer solution II containing protease and 2 mM nicotinamide.
9. Incubate the plate at 37 °C for 30–45 min in the dark for fluorophore development.
10. Measure fluorescence using a microplate reader at excitation 360 nm and emission 460 nm.

Controls- Include the following controls:

1. No enzyme (background)
2. No NAD^+ (reaction specificity)
3. Time zero (developer added before substrate)
4. With nicotinamide or with sirtuin specific inhibitor
5. Heat inactivated enzyme

Enzyme activity may be expressed as relative fluorescence units (RFU) or fold activity over control. For kinetic analysis, vary substrate concentration while maintaining constant enzyme concentration and fit the data to Michaelis menten curve to calculate Km and Vmax (Fig. 1).

Radiometric assays using labeled acyl substrates provide higher sensitivity and can be used for low-activity enzymes such as SIRT4 (Tanny & Moazed, 2001). In contrast to SIRT3 and SIRT5, the lack of well-defined substrates and dominant catalytic activity has limited the development of robust assays for SIRT4, with most studies relying on indirect functional readouts rather than direct measurements of enzymatic activity.

However, simplified peptide assays do not mimic cellular mitochondrial organization or substrate accessibility. To improve specificity, activity should be assessed on validated endogenous substrates, such as mitochondrial metabolic enzymes, rather than relying solely on global acylation signals (Table 2).

High-performance liquid chromatography (HPLC) provides a direct quantitative approach for measuring sirtuin enzymatic activity by separating and quantifying reaction products (Hong et al., 2018). In our previous work, immunoprecipitated or recombinant SIRT1 protein was incubated with defined or saturating concentrations of an acetylated H3K9 peptide substrate in the presence of NAD^+. Reactions were terminated at defined time points to maintain measurements within the linear range of product formation and subsequently analyzed by reverse-phase HPLC (RP-HPLC). Quantification of product allowed calculation of reaction velocities, and kinetic parameters were determined by fitting the data to Michaelis-Menten models (Deota et al., 2017). This chromatographic approach provides an antibody-independent and highly reproducible method for assessing catalytic activity and is particularly useful for validating substrate specificity and comparing enzymatic efficiencies under controlled conditions.

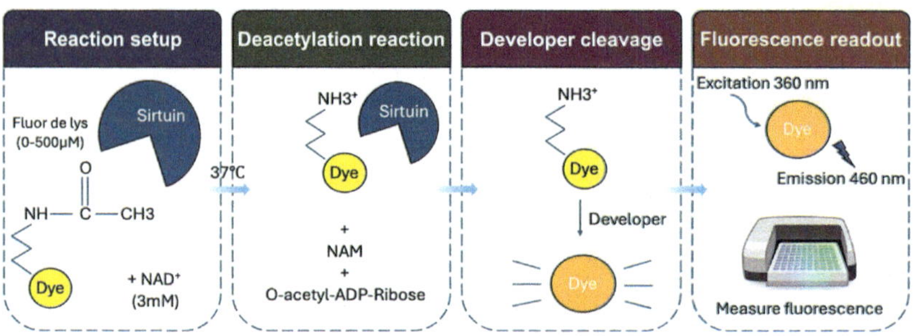

Fig. 1 Fluorometric assay for measuring sirtuin deacylase activity.

Table 2 Enzymatic activities of mitochondrial sirtuins and current assay limitations.

Sirtuin	proposed/known activity	Available kits	Scope for development
SIRT3	Robust NAD^+-dependent deacetylase; regulates oxidative phosphorylation, TCA cycle, fatty acid β-oxidation, and antioxidant enzymes	Widely available fluorometric deacetylase activity kits using acetyl-lysine substrates	Development of assays that better reflect mitochondrial context (e.g., native substrates, physiological NAD^+ levels)
SIRT4	Weak deacetylase activity, ADP-ribosyltransferase, lipoamidase; regulates amino acid metabolism (e.g., GDH), insulin secretion, and anaplerotic flux	No robust or standardized direct activity assay kits; only ELISA-based protein quantification kits available	Need for substrate-specific high-throughput assays; identification of physiological substrates; development of reliable readouts for ADP-ribosylation
SIRT5	NAD^+-dependent desuccinylase, demalonylase, and deglutarylase; regulates TCA cycle, urea cycle, and amino acid metabolism	Fluorometric assay kits available using succinylated peptide substrates; suitable for in vitro enzymatic assays	Multiplex detection of multiple acyl modifications; physiologically relevant substrate assays, Assays for endogenous activity in tissues

4.1.2 Mass spectrometry-based profiling of mitochondrial acylation

The most comprehensive method for determining sirtuin dependent changes in acylation in vivo is through mass spectrometry-based acyl-proteomics. This method allows identification of site-specific acylation changes which enables estimation of stoichiometric modification levels across metabolic states. Importantly, comparisons between fed and fasted conditions help distinguish enzymatic deacylation from metabolite-driven acyl group accumulation (Jayasena et al., 2016; Yang et al., 2016).

This workflow involves mitochondrial enrichment, protein extraction, proteolytic digestion, enrichment of acylated peptides, and liquid chromatography-tandem mass spectrometry (LC-MS/MS) analysis (Hebert et al., 2013; Schilling et al., 2019; Still et al., 2013).

Materials and reagents- Fresh cells or tissue samples, mitochondrial isolation buffer, protease inhibitor cocktail, nicotinamide, urea (8 M), trichostatin

A (TSA), dithiothreitol (DTT), iodoacetamide (IAA), trypsin, anti-acetyllysine antibody beads (for acetylome profiling), anti-succinyllysine antibody beads (for succinylome profiling), C18 desalting columns, LC-MS-grade water, acetonitrile, formic acid, high-resolution LC-MS/MS system (e.g., Orbitrap Mass Spectrometer).

Procedure.

1. Workflow for LC/MS based analysis

Mitochondrial isolation	**1.** Collect cells or freshly dissected tissues and wash with ice-cold PBS. **2.** Resuspend samples in mitochondrial isolation buffer containing 250 mM sucrose, 10 mM HEPES (pH 7.4), 0.5 mM EGTA, protease inhibitors, and 5 mM nicotinamide. Mince tissue into small pieces. **3.** Homogenize using a Dounce homogenizer on ice (number of strokes optimized depending on tissue type). **4.** Centrifuge at 800g for 10 min at 4 °C to remove nuclei and debris. **5.** Transfer supernatant and centrifuge at 12,000 g for 15 min at 4°C to pellet mitochondria. **6.** Wash mitochondrial pellet once in isolation buffer. **7.** Store mitochondrial pellets at −80 °C or proceed immediately.
Protein extraction	**1.** Lyse mitochondrial pellets in 8 M urea lysis buffer containing 100 mM TEAB (pH 8.5), 8 M urea, protease inhibitors, 5 mM nicotinamide, 75 mM NaCl, and 5 μM Trichostatin A. **2.** Incubate on ice for 30 min **3.** Clarify lysates by centrifugation at 14,000 *g* for 15 min at 4 °C. **4.** Determine protein concentration using BCA Protein Assay. Normalize all samples to equal protein input (typically 0.5−1 mg total protein) using lysis buffer prior to downstream processing.
Reduction and alkylation	**1.** Reduce normalized protein samples with 10 mM DTT at 37 °C for 45 min **2.** Cool samples to room temperature and alkylate cysteines with 20 mM Iodoacetamide for 30 min at room temperature in the dark. Add 5 mM DTT to quench excess iodoacetamide. **3.** Dilute reduced and alkylated proteins tenfold with 50 mM TEAB to reduce urea concentration below 2 M prior to trypsin digestion.

Trypsin digestion	**1.** Add sequencing-grade trypsin at an enzyme to protein ratio of 1:50 (w/w). **2.** Digest overnight at 37 °C. **3.** Stop digestion by acidifying with formic acid to a final concentration of 1 %. **4.** Centrifuge digested peptides at 14,000 g for 10 min to remove insoluble debris.
Peptide desalting	**1.** Desalt peptides using C18 columns according to the manufacturer's instructions. **2.** Dry peptides using vacuum centrifugation.
Enrichment of acylated peptides	**1.** Resuspend peptides in immunoaffinity enrichment buffer. **2.** Incubate with anti-acetyl lysine or anti-succinyl lysine antibody-conjugated agarose overnight at 4 °C with gentle rotation. **3.** Wash beads sequentially with immunoprecipitation buffer and water. **4.** Elute bound peptides using 0.1 % trifluoroacetic acid. **5.** Dry the eluted peptides in a vacuum centrifuge.
LC-MS/MS analysis	**1.** Resuspend peptides in 0.1 % formic acid. **2.** Load peptides onto a reverse-phase C18 nano-LC column coupled to an Orbitrap Mass Spectrometer. **3.** Separate peptides using an appropriate mobile phase, typically a gradient of 5−35 % acetonitrile over 60−120 min **4.** Analyze on a high-resolution mass spectrometer operating in data-dependent acquisition mode (MS1 resolution 70,000, MS/MS resolution 15,000).
Database search and quantification	**1.** Analyze raw files using either open source or instrumentation specific software like MaxQuant or Proteome Discoverer. **2.** Set database search parameters according to the acyl modification under investigation. Variable modifications should include the relevant lysine acylation events such as lysine acetylation (+42.0106 Da), lysine succinylation (+100.0160 Da), lysine malonylation (+86.0004 Da), or lysine fatty acylation (e.g., myristoylation, +210.1984 Da; palmitoylation, +238.2297 Da) depending on the experimental design. In addition, carbamidomethylation of cysteine (+57.0215 Da) should be set as a fixed modification to account for alkylation during sample preparation. **3.** Apply a false discovery rate (FDR) threshold of 1–5 % at peptide and protein levels.
Data analysis	**1.** Quantify modification site intensities and normalize to total protein abundance. **2.** Compare acylation occupancy across experimental groups. **3.** Perform pathway enrichment analysis to identify regulated mitochondrial metabolic pathways.

Controls.

1. Wild type vs sirtuin knockout
2. Input proteome (Total digested peptides prior to immuno enrichment)
3. Bead only

Candidate substrates may be validated by immunoblotting or targeted enzymatic assays. Acetylome profiling is most informative for SIRT3, whereas succinylome and malonylome profiling are more informative for SIRT5. For SIRT4, substrate identification remains challenging and may require targeted approaches beyond standard acyl-proteomics.

To assess substrate specificity, traditional approaches relied on immunoprecipitation of candidate substrates followed by assessment of acylation status in wild-type versus sirtuin-deficient systems. While informative, these methods are limited by antibody specificity and bias toward known targets. Quantitative mass spectrometry-based acyl-proteomics has significantly expanded substrate identification, enabling unbiased mapping of acetylation, succinylation, and malonylation sites. However, proteomic datasets alone cannot establish direct enzyme-substrate relationships.

To strengthen causal inference, candidate substrates are typically validated through complementary strategies, such as genetic rescue with catalytically inactive variants, targeted analysis of endogenous protein complexes, and site-directed mutagenesis of modified lysine residues to evaluate functional consequences.

4.2 Assessing NAD^+ dependence

A defining feature of sirtuin catalysis is strict dependence on NAD^+, making cofactor availability a critical experimental variable (Avalos et al., 2005).

Table 3 Kinetic parameters and substrate specificity of mitochondrial sirtuins.

Sirtuin	Preferred substrate	Km (NAD^+)	Km (acyl substrate)	Key references
SIRT3	Acetyl-lysine	880 μM	150 mM	Hirschey et al. (2011)
SIRT5	Succinyl-, malonyl-, glutaryl-lysine	980 μM	450 μM (Succinyl-lysine)	Fischer et al. (2012), Rardin et al. (2013)
SIRT4	GDH, PDH complex (lipoylated proteins)	35 μM	Not well defined	Laurent et al. (2013)

Table 4 Comparative analysis of analytical methods for sirtuin activity.

Method	Principle	Readout	Advantages	Limitations	References
Mass spectrometry (LC-MS/MS)	Detects mass shift after deacetylation or other acyl removal	Product/substrate ratios, site-specific activity	High specificity, detects multiple PTMs simultaneously	Expensive instrumentation, labor intensive, complex analysis	Du et al. (2011)
RP-HPLC (reverse-phase HPLC)	Separates acetylated substrate and deacetylated product after enzymatic reaction	Direct quantification of deacetylated peptide, kinetic parameters (Km, Vmax)	Gold-standard, quantitative, label-free, suitable for enzyme kinetics	Low throughput, expensive instrumentation	Borra et al. (2005), Hong et al. (2018)
Fluorogenic peptide assays (Fluor-de-Lys type)	Deacetylation exposes fluorophore after developer cleavage	Fluorescence intensity	Simple, Moderate throughput, screening compatible	Artificial substrates artifacts, two step reaction	Marcotte et al. (2004), Wegener et al. (2003)
Radiometric assays	Uses ^{3}H- or ^{14}C-acetyl-lysine substrates	Radioactive signal loss or transfer	Extremely sensitive, direct quantitative measurement	Hazardous radioactivity handling	Tanny and Moazed (2001)
Coupled NAD^{+} consumption assays	Deacetylation consumes NAD^{+}, coupled enzymatic reactions generate fluorescence/color	NAD^{+} or NADH levels	Continuous monitoring, adaptable to kinetics	Indirect measurement	Smith et al. (2009)

(continued)

Table 4 Comparative analysis of analytical methods for sirtuin activity. *(cont'd)*

Method	Principle	Readout	Advantages	Limitations	References
Nicotinamide (NAM) release assays	Quantifies NAM produced during reaction	NAM concentration	Direct enzymatic product measurement	Multi-step detection, moderate throughput	Hubbard and Sinclair (2013)
Western blot-based deacetylation assays	Measures loss of acetylation on protein substrates	Change in acetyl-lysine signal	Physiologically relevant substrates	Semi-quantitative, antibody dependent	Vaziri et al. (2001)
Immuno precipitation + activity assay	Endogenous sirtuin isolated then assayed ex vivo	Enzymatic activity from native complexes	Reflects physiological regulation	Co-purifying factors may confound	Rodgers et al. (2005)
Cell-based reporter assays	Sirtuin regulates acetylation-dependent transcription reporters	Luciferase/ fluorescent reporter output	Measures functional activity in cells	Indirect, pathway dependent	Amano et al. (2019), Satterstrom and Haigis, (2014)
Metabolite-based assays (OAADPR detection)	Detects formation of O-acetyl-ADP-ribose co-product	OAADPR levels	Robust, mechanistically specific readout	Technically demanding	Jackson and Denu (2002)
Thermal shift/binding assays	Measures ligand binding affecting stability	ΔTm shifts	Useful for inhibitor studies	Not catalytic activity per se	Hubbard et al. (2013)

Enzymatic assays should therefore be performed across a range of NAD^+ concentrations rather than under saturating conditions (Schultz et al., 2018).

In cellular systems, NAD^+ dependence can be verified by experimentally altering NAD^+ metabolism. Supplementation with NAD^+ precursors (nicotinamide, niacin, nicotinamide mononucleotide), inhibition of NAD^+ consumption pathways (PARP inhibition), or induction of energetic stress by substrate limitation or mitochondrial perturbation allows assessment of how changes in NAD^+/NADH ratio influence deacylation activity. Parallel quantification of intracellular and mitochondrial NAD^+ pools using enzymatic cycling assays or targeted metabolomics is essential to link enzymatic output with cofactor availability (Kanamori et al., 2018; Trammell & Brenner, 2013).

In summary, measurement of mitochondrial sirtuin activity requires consideration of catalytic specificity, substrate availability, and NAD^+ dynamics within defined metabolic states. Relative efficiencies toward distinct acyl modifications and differential sensitivity to NAD^+ position sirtuins as context-dependent regulators rather than constitutively active enzymes (Table 3).

4.3 Monitoring acylation dynamics using chemical biology approaches

Chemical biology approaches, particularly click chemistry-based labeling strategies, provide a complementary means to examine acylation dynamics in living systems. In these methods, bioorthogonal acyl analogs containing alkyne or azide functional groups are incorporated into protein modifications, enabling selective tagging through click reactions for subsequent enrichment and identification. These approaches are useful for detecting newly formed acylation events and monitoring modification turnover during metabolic transitions (Talwadekar et al., 2024). Pulse-chase labeling paradigms can be employed to temporally monitor acylation kinetics (Table 4).

5. Assessment of mitochondrial protein acylation

Mitochondrial protein acylation represents a key functional outcome of mitochondrial sirtuin activity and provides an experimentally accessible link between enzymatic regulation and mitochondrial physiology. The steady-state level of protein acylation in mitochondria reflects the balance between metabolite-driven acylation and sirtuin-mediated deacylation, largely through

SIRT3 and, to a lesser extent, SIRT4 and SIRT5. Understanding the significance of acetylation changes therefore requires evaluating their downstream effects on mitochondrial function, particularly respiration, redox homeostasis, and organelle dynamics (Fig. 2).

5.1 Mitochondrial respiration and bioenergetics

One of the most direct consequences of changes in mitochondrial acetylation is the modulation of oxidative metabolism, as SIRT3 deacetylates key enzymes of the TCA cycle, fatty acid oxidation, and the electron transport chain. Measurement of mitochondrial respiration provides a functional readout of mitochondrial bioenergetics and can reveal changes in oxidative phosphorylation resulting from altered acetylation status.

Mitochondrial oxygen consumption rate (OCR) can be measured using an extracellular flux analyzer such as the Seahorse XFe24 analyzer (Agilent).

Materials and reagents- Seahorse XFe24 or XFe96 flux analyzer, Seahorse XF cell culture microplates, XF sensor cartridge, XF calibrant, XF base medium (pH 7.4), D-glucose, sodium pyruvate, L-Glutamine.

Mitochondrial stress test compounds (Final concentration)- Oligomycin (2uM), FCCP (2uM), Rotenone (0.5uM) Antimycin A (4uM).

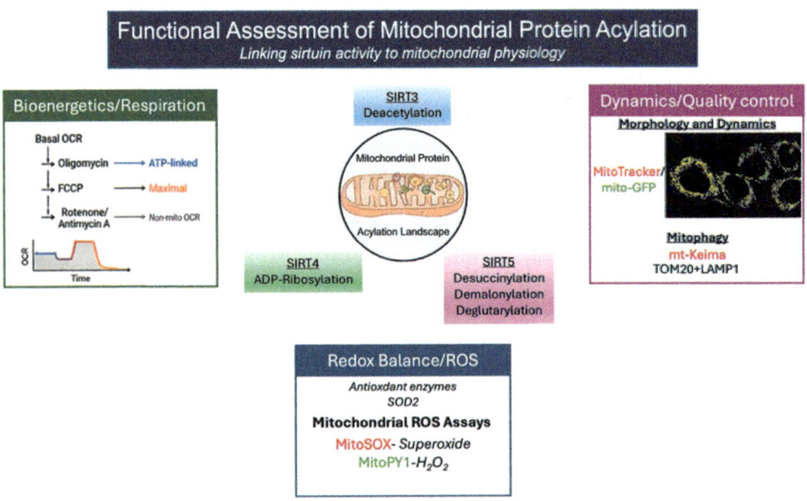

Fig. 2 Functional assessment of mitochondrial protein acylation.

Cell Preparation- Seed cells in a seahorse XFe24 cell culture plate, typically $2*10^4$ cells/well and allow cells to adhere overnight under standard culture conditions.

Prior to the day of assay, hydrate the sensor cartridge in XF calibrant for at least 12 h at 37 °C in a non-CO_2 incubator.

Assay medium preparation- Prepare Seahorse XF base medium by supplementing with 25 mM glucose, 1 mM sodium pyruvate and 2 mM L-glutamine. Adjust the pH to 7.4 at 37 °C with NaOH.

Assay procedure-

1. Remove culture medium and gently wash the cells twice with assay medium.
2. Add 500uL assay medium per well.
3. Incubate the cells in a non-CO_2 incubator for 45 min.
4. Prepare mitochondrial stress test inhibitors at 10X working concentration and load into the sensor cartridge ports in the order (A) Oligomycin (B) FCCP (C) Rotenone + Antimycin A.
5. Insert the hydrated sensor cartridge into the flux analyzer for calibration.
6. Following calibration, eject and replace the utility plate with the cell plate and initiate the assay.
7. Post assay, aspirate the assay medium, wash cells once with PBS.
8. Lyse cells directly in the seahorse plate using RIPA buffer.
9. Determine protein concentration using the BCA assay and normalize OCR values to total protein content.

Basal respiration is measured first, and then there is sequential injections of mitochondrial inhibitors. Oligomycin is used to estimate ATP-linked respiration, FCCP to determine maximal respiratory capacity, and rotenone combined with antimycin A to quantify non-mitochondrial oxygen consumption.

These parameters provide insight into how changes in mitochondrial acetylation influence oxidative metabolism. The experiments can be done in intact cells, permeabilized cells, or isolated mitochondria depending on whether the objective is to assess integrated cellular metabolism or intrinsic mitochondrial function. In isolated mitochondria, respiratory measurements can also be interpreted according to the classical respiratory states described by Britton Chance and G.R. Williams as shown in Fig. 3.

P/O ratio calculation- The phosphate-to-oxygen (P/O) ratio gives an estimate of mitochondrial coupling efficiency and reflects the amount of

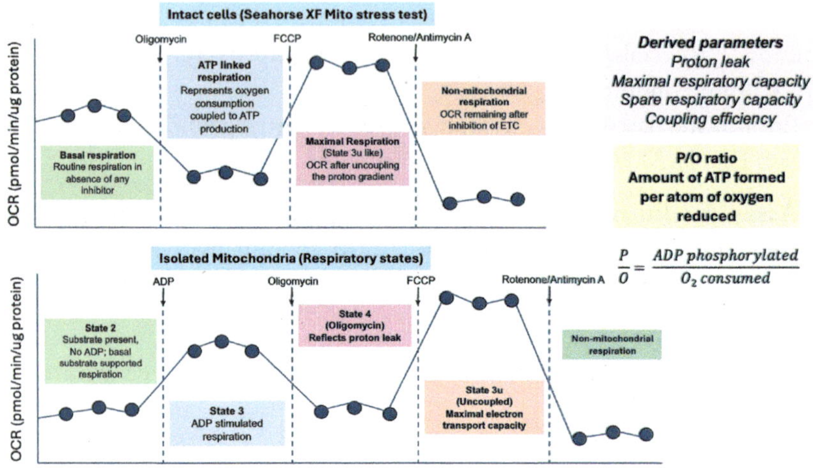

Fig. 3 Mitochondrial states of respiration: intact cells vs isolated mitochondria.

ATP generated per oxygen atom reduced during oxidative phosphorylation. Changes in P/O ratio may indicate altered coupling efficiency, proton leak, or defects in oxidative phosphorylation.

Further, specific assays with substrates such as pyruvate/malate or succinate can help in understanding complex-specific effects of acetylation-dependent regulation, which may vary across tissues and metabolic states.

Additional measurements of ATP production rates, mitochondrial membrane potential, and coupling efficiency strengthen interpretation and help define whether acetylation primarily affects electron transport, substrate oxidation, or ATP synthesis.

5.2 Mitochondrial redox balance and reactive oxygen species

Mitochondrial acetylation also influences redox homeostasis and oxidative stress responses. This is because several antioxidant enzymes are subject to regulation by acetylation. Alterations in sirtuin activity frequently affect mitochondrial ROS production indirectly by impacting the efficiency of the electron transport chain. An example is the deacetylation activity of the sirtuin SIRT3 on the antioxidant enzyme superoxide dismutase 2 (SOD2), which increases the antioxidant activity and limits the production of mitochondrial-derived superoxide (Qiu et al., 2010).

Experimentally, mitochondrial ROS can be measured using fluorescent probes targeted to mitochondria, alongside complementary biochemical measurements such as hydrogen peroxide production assays or redox-sensitive reporters.

MitoSOX Red is a commonly used fluorogenic dye which upon oxidation by superoxide, generates red fluorescence, enabling quantitative assessment of mitochondrial reactive oxygen species (ROS) generation in live cells.

Materials and reagents- MitoSOX red (stock 5 mM in DMSO), Hank's Balanced salt solution (HBSS), glass bottom imaging dishes, confocal microscope.

Cell preparation- Seed cells onto glass bottom dishes at a density to achieve 60–70% confluency on the day of imaging. Allow cells to adhere overnight under standard culture conditions.

Apply experimental treatments (nutrient deprivation, sirtuin modulation, metabolic perturbation) prior to ROS measurement.

MitoSOX staining-
1. Prepare a fresh working solution of MitoSOX Red at 0.5−1uM concentration in prewarmed HBSS immediately before use and protect from light.
2. Remove the culture medium and gently wash cells once with HBSS.
3. Add mitoSOX working solution and incubate at 37 °C for 10 min.
4. Following incubation, wash cells three times with HBSS to remove excess probe.

Measure fluorescence using a confocal microscope with excitation at 510 nm and emission range of 570–620 nm. Use identical imaging parameters (laser intensity, gain, exposure time) across all experimental groups to allow quantitative comparison.

Experimental controls-The following controls should be included:
1. No dye control for background autofluorescence
2. Vehicle-treated control for baseline ROS
3. Positive control using Antimycin A to induce mitochondrial superoxide production
4. Antioxidant-treated control (e.g., N-acetylcysteine) to assess signal specificity

Another example is the measurement of mitochondrial-derived hydrogen peroxide using the fluorescent dye MitoPY1, where isolated mitochondria or intact cells are incubated with probe-containing buffer and the fluorescence is monitored kinetically to measure the production of hydrogen peroxide. Because ROS measurements are highly sensitive to experimental conditions, careful calibration, normalization to mitochondrial content e.g., citrate synthase activity, mitochondrial protein levels, or mitochondrial dyes such as MitoTracker Green, and inclusion of antioxidant or electron transport inhibitors as controls are essential.

The interpretation of these results should take into consideration that increased levels of ROS can result from impaired respiratory efficiency or from adaptive changes in cellular signaling pathways. Thus, by combining the analysis of ROS levels with respiratory control analysis and acetylation profiles, it is possible to distinguish between pathological oxidative stress and adaptive changes in cellular metabolism due to altered sirtuin activity.

5.3 Mitochondrial dynamics and quality control

Beyond regulation of metabolic enzymes, mitochondrial sirtuins also influence structural and quality-control processes. Key regulators of mitochondrial fusion, fission, and mitophagy are responsive to metabolic signaling and can be modulated, directly or indirectly, by sirtuin-dependent deacetylation. For example, SIRT3-mediated deacetylation of OPA1 has been implicated in maintaining mitochondrial fusion and cristae organization (Samant et al., 2014).

Mitochondrial dynamics are typically assessed using imaging-based approaches. Fluorescence microscopy using mitochondrial markers like MitoTracker dyes or genetically encoded reporters such as mitochondria-targeted GFP (mito-GFP) can be used for the quantification of mitochondrial morphology, and spatial distribution within cells. Live-cell imaging provides additional information regarding fusion-fission kinetics and mitochondrial motility. In typical experiments, cells expressing mito-GFP are imaged at short acquisition intervals under controlled temperature and CO_2 conditions, and fusion or fission events are quantified using time-lapse analysis software. These measurements are particularly informative when combined with metabolic perturbations known to alter acetylation levels.

Quality control pathways can be evaluated by quantifying mitophagy, mitochondrial turnover, and proteostatic stress. Mitophagy is commonly assessed using mitochondria-targeted Keima (mt-Keima), a pH-sensitive fluorescent reporter that shifts excitation spectra upon lysosomal delivery. Alternatively, mitophagic flux can be estimated by co-localization analysis between mitochondrial markers like TOM20 and lysosomal proteins such as LAMP1.

Additional indicators of mitochondrial integrity include assessment of mitochondrial DNA (mtDNA) maintenance and proteostatic stress. mtDNA copy number is typically quantified by qPCR and normalized to nuclear DNA. Accumulation of unfolded or damaged mitochondrial proteins can be assessed using markers of the mitochondrial unfolded protein response (UPRmt). Sustained hyperacetylation has been associated with impaired mitochondrial proteostasis and turnover, potentially leading to accumulation of dysfunctional mitochondria and secondary metabolic defects. The combination of imaging-based measurements with biochemical assessment of acetylation therefore provides a mechanistic link for post-translational regulation to mitochondrial quality control.

6. Cellular energy state and whole-body homeostasis

The cellular energy state is most accurately represented by the relative abundance of adenine nucleotides ATP, ADP, and AMP rather than ATP levels alone. Measurement of these metabolites can be achieved using luciferase-based assays, high-performance liquid chromatography (HPLC), or targeted liquid chromatography mass spectrometry (LC-MS) approaches. Calculation of energy charge or ATP/ADP and AMP/ATP ratios provides a more physiologically meaningful readout of energetic stress than absolute metabolite concentrations.

Alterations in cellular energy status are tightly coupled to activation of AMP-activated protein kinase (AMPK). Importantly, AMPK and sirtuins function within an interconnected regulatory network linking cellular energy charge to mitochondrial adaptation. Increases in the AMP/ATP ratio during energetic stress activates AMPK, which promotes catabolic pathways and enhances NAD^+ biosynthesis through upregulation of nicotinamide phosphoribosyltransferase (NAMPT), thereby indirectly stimulating NAD^+-dependent sirtuin activity. Conversely, mitochondrial sirtuins, particularly SIRT3, improve oxidative metabolism and electron transport efficiency, contributing to restoration of cellular ATP levels and feedback regulation of AMPK signaling.

Experimentally, AMPK activation is most commonly assessed by immunoblotting for phosphorylation at Thr172 within the catalytic α-subunit, along with phosphorylation of downstream substrates such as acetyl-CoA carboxylase (ACC). Parallel measurement of intracellular NAD^+/NADH ratios provides insight into the metabolic conditions that favor coordinated AMPK-sirtuin engagement.

Metabolic flux analysis provides an additional layer of insight by quantifying pathway activity rather than metabolite abundance. The use of labeled substrates such as glucose, fatty acids, or amino acids in the medium for metabolic flux analysis helps to determine the flux through the glycolytic pathway, the tricarboxylic acid cycle, as well as biosynthetic pathways. Isotope incorporation patterns measured by mass spectrometry can be used to reconstruct the pathways utilized as well as the shifts in the preference for the substrates due to altered regulation of the mitochondria. These approaches are particularly relevant when assessing how changes in mitochondrial sirtuin activity influence anaplerosis, cataplerosis, or substrate competition.

While cellular and mitochondrial measurements define mechanistic regulation, understanding the physiological relevance of sirtuins-mediated metabolic control requires assessment at the organismal level. Sirtuins function as metabolic sensors that coordinate mitochondrial metabolism to cellular and systemic energy demands. Therefore, alterations in sirtuin activity at the mitochondrial level can influence whole body energy homeostasis. Indirect calorimetry represents the gold-standard approach for assessing these parameters in vivo.

By measuring oxygen consumption (VO_2) and carbon dioxide production (VCO_2), indirect calorimetry enables calculation of energy expenditure and the respiratory exchange ratio (RER), a key indicator of relative carbohydrate versus lipid utilization. Continuous monitoring across light-dark cycles enables assessment of circadian variation in metabolism, which is essential because metabolic rate and substrate utilization, and therefore sirtuin activity fluctuate substantially with behavioral state, feeding patterns, and hormonal signaling. Integration with physiological tests like glucose and insulin tolerance tests gives additional insight into how mitochondrial sirtuin regulation affects systemic glucose homeostasis and nutrient handling. When combined with cellular, biochemical, and molecular analyses of mitochondrial function, these organismal measurements enable a multiscale framework linking sirtuin-dependent mitochondrial regulation to whole-body metabolic homeostasis.

7. Conclusion

Mitochondrial sirtuins lie at the interface of metabolism, energetics and post-translational regulation within mitochondria. While substantial progress has been made in identifying their substrates and physiological roles, important gaps remain in understanding how their enzymatic activities are quantitatively regulated in vivo.

Apparent redundancy among mitochondrial sirtuins often reflects differences in spatial distribution and tissue-specific metabolic environments rather than true overlap in enzymatic function. The accessibility of candidate substrates is constrained by sub-mitochondrial localization, interaction networks, and tissue-dependent metabolic flux, all of which shape the effective substrate landscape encountered by individual sirtuins in vivo. Furthermore, most studies have examined individual sirtuins in isolation, limiting a systems-level understanding of how SIRT3, SIRT4, and SIRT5 collectively shape mitochondrial acylation landscapes across tissues and metabolic states.

Addressing these challenges will require integrated experimental approaches that combine quantitative proteomics, metabolite profiling, and robust enzymatic assays with spatially and temporally resolved analyses. This will be essential to move toward a more mechanistic understanding of how mitochondrial sirtuins coordinate metabolic adaptation and mitochondrial function under physiological and pathological conditions.

Acknowledgements

We acknowledge the financial support to A.K. and U.K-S. towards their salary and fellowship by TIFR and CDFD.

References

Ahuja, N., Schwer, B., Carobbio, S., Waltregny, D., North, B. J., Castronovo, V., ... Verdin, E. (2007). Regulation of insulin secretion by SIRT4, a mitochondrial ADP-ribosyltransferase*. *Journal of Biological Chemistry, 282*(46), 33583–33592. https://doi.org/10.1074/jbc.M705488200.

Amano, H., Chaudhury, A., Rodriguez Aguayo, C., Lu, L., Akhanov, V., Catic, A., ... Sahin, E. (2019). Telomere dysfunction induces sirtuin repression that drives telomere-dependent disease. *Cell Metabolism, 29*(6), 1274–1290.e9. https://doi.org/10.1016/j.cmet.2019.03.001.

Avalos, J. L., Bever, K. M., & Wolberger, C. (2005). Mechanism of sirtuin inhibition by nicotinamide: Altering the NAD+ cosubstrate specificity of a Sir2 enzyme. *Molecular Cell, 17*(6), 855–868. https://doi.org/10.1016/j.molcel.2005.02.022.

Borra, M. T., Smith, B. C., & Denu, J. M. (2005). Mechanism of human SIRT1 activation by resveratrol*. *Journal of Biological Chemistry, 280*(17), 17187–17195. https://doi.org/10.1074/jbc.M501250200.

Cambronne, X. A., Stewart, M. L., Kim, D., Jones-Brunette, A. M., Morgan, R. K., Farrens, D. L., ... Goodman, R. H. (2016). NAD+ biosensor reveals multiple sources for mitochondrial NAD+. *Science (New York, N. Y.), 352*(6292), 1474–1477. https://doi.org/10.1126/science.aad5168.

Chalkiadaki, A., & Guarente, L. (2012). Sirtuins mediate mammalian metabolic responses to nutrient availability. *Nature Reviews. Endocrinology, 8*(5), 287–296. https://doi.org/10.1038/nrendo.2011.225.

Chang, S. M., Franklin, L. P., Sattar, S., Moro, C. A., DeGennaro, M. V., LaGanke, N. G., & Hanna-Rose, W. (2025). Mitochondrial sirtuins sir-2.2 and sir-2.3 regulate lifespan in C. elegans. *Genetics, 231*(3), iyaf185. https://doi.org/10.1093/genetics/iyaf185.

Chattopadhyay, T., Maniyadath, B., Bagul, H. P., Chakraborty, A., Shukla, N., Budnar, S., ... Kolthur-Seetharam, U. (2020). Spatiotemporal gating of SIRT1 functions by O-GlcNAcylation is essential for liver metabolic switching and prevents hyperglycemia. *Proceedings of the National Academy of Sciences, 117*(12), 6890–6900. https://doi.org/10.1073/pnas.1909943117.

Chen, Y., Zhang, J., Lin, Y., Lei, Q., Guan, K.-L., Zhao, S., & Xiong, Y. (2011). Tumour suppressor SIRT3 deacetylates and activates manganese superoxide dismutase to scavenge ROS. *EMBO Reports, 12*(6), 534–541. https://doi.org/10.1038/embor.2011.65.

Deota, S., Chattopadhyay, T., Ramachandran, D., Armstrong, E., Camacho, B., Maniyadath, B., ... Kolthur-Seetharam, U. (2017). Identification of a tissue-restricted isoform of SIRT1 defines a regulatory domain that encodes specificity. *Cell Reports, 18*(13), 3069–3077. https://doi.org/10.1016/j.celrep.2017.03.012.

Dittenhafer-Reed, K. E., Richards, A. L., Fan, J., Smallegan, M. J., Fotuhi Siahpirani, A., Kemmerer, Z. A., ... Denu, J. M. (2015). SIRT3 Mediates Multi-Tissue Coupling for Metabolic Fuel Switching. *Cell Metabolism, 21*(4), 637–646. https://doi.org/10.1016/j.cmet.2015.03.007.

Du, J., Zhou, Y., Su, X., Yu, J. J., Khan, S., Jiang, H., ... Lin, H. (2011). Sirt5 Is a NAD-dependent protein lysine demalonylase and desuccinylase. *Science (New York, N. Y.), 334*(6057), 806–809. https://doi.org/10.1126/science.1207861.

Fischer, F., Gertz, M., Suenkel, B., Lakshminarasimhan, M., Schutkowski, M., & Steegborn, C. (2012). Sirt5 deacylation activities show differential sensitivities to nicotinamide inhibition. *PLoS One,* 7(9), e45098. https://doi.org/10.1371/journal.pone.0045098.

Haigis, M. C., Mostoslavsky, R., Haigis, K. M., Fahie, K., Christodoulou, D. C., Murphy, A. J., ... Guarente, L. (2006). SIRT4 inhibits glutamate dehydrogenase and opposes the effects of calorie restriction in pancreatic beta cells. *Cell, 126*(5), 941–954. https://doi.org/10.1016/j.cell.2006.06.057.

Hebert, A. S., Dittenhafer-Reed, K. E., Yu, W., Bailey, D. J., Selen, E. S., Boersma, M. D., ... Coon, J. J. (2013). Calorie restriction and SIRT3 trigger global reprogramming of the mitochondrial protein acetylome. *Molecular Cell, 49*(1), 186–199. https://doi.org/10.1016/j.molcel.2012.10.024.

Hirschey, M. D., Shimazu, T., Goetzman, E., Jing, E., Schwer, B., Lombard, D. B., ... Verdin, E. (2010). SIRT3 regulates mitochondrial fatty-acid oxidation by reversible enzyme deacetylation. *Nature, 464*(7285), 121–125. https://doi.org/10.1038/nature08778.

Hirschey, M. D., Shimazu, T., Jing, E., Grueter, C. A., Collins, A. M., Aouizerat, B., ... Verdin, E. (2011). SIRT3 deficiency and mitochondrial protein hyperacetylation accelerate the development of the metabolic syndrome. *Molecular Cell, 44*(2), 177–190. https://doi.org/10.1016/j.molcel.2011.07.019.

Ho, L., Titus, A. S., Banerjee, K. K., George, S., Lin, W., Deota, S., ... Kolthur-Seetharam, U. (2013). SIRT4 regulates ATP homeostasis and mediates a retrograde signaling via AMPK. *Aging (Albany NY), 5*(11), 835–849. https://doi.org/10.18632/aging.100616.

Hong, J. Y., Zhang, X., & Lin, H. (2018). HPLC-based enzyme assays for sirtuins. *Methods in Molecular Biology (Clifton, N. J.), 1813*, 225–234. https://doi.org/10.1007/978-1-4939-8588-3_15.

Høyland, L. E., VanLinden, M. R., Niere, M., Strømland, Ø., Sharma, S., Dietze, J., ... Ziegler, M. (2024). Subcellular NAD+ pools are interconnected and buffered by mitochondrial NAD+. *Nature Metabolism, 6*(12), 2319–2337. https://doi.org/10.1038/s42255-024-01174-w.

Hubbard, B. P., Gomes, A. P., Dai, H., Li, J., Case, A. W., Considine, T., ... Sinclair, D. A. (2013). Evidence for a common mechanism of SIRT1 regulation by allosteric activators. *Science (New York, N. Y.), 339*(6124), 1216–1219. https://doi.org/10.1126/science.1231097.
Hubbard, B. P., & Sinclair, D. A. (2013). Measurement of sirtuin enzyme activity using a substrate-agnostic fluorometric nicotinamide assay. *Methods in Molecular Biology (Clifton, N. J.), 1077*, 167–177. https://doi.org/10.1007/978-1-62703-637-5_11.
Iwahara, T., Bonasio, R., Narendra, V., & Reinberg, D. (2012). SIRT3 functions in the nucleus in the control of stress-related gene expression. *Molecular and Cellular Biology, 32*(24), 5022–5034. https://doi.org/10.1128/MCB.00822-12.
Jackson, M. D., & Denu, J. M. (2002). Structural identification of 2′- and 3′-O-acetyl-ADP-ribose as novel metabolites derived from the Sir2 family of beta -NAD +-dependent histone/protein deacetylases. *The Journal of Biological Chemistry, 277*(21), 18535–18544. https://doi.org/10.1074/jbc.M200671200.
Jayasena, T., Poljak, A., Braidy, N., Zhong, L., Rowlands, B., Muenchhoff, J., ... Sachdev, P. (2016). Application of targeted mass spectrometry for the quantification of sirtuins in the central nervous system. *Scientific Reports, 6*(1), 35391. https://doi.org/10.1038/srep35391.
Kanamori, K. S., de Oliveira, G. C., Auxiliadora-Martins, M., Schoon, R. A., Reid, J. M., & Chini, E. N. (2018). Two different methods of quantification of oxidized nicotinamide adenine dinucleotide (NAD+) and reduced nicotinamide adenine dinucleotide (NADH) intracellular levels: Enzymatic coupled cycling assay and ultra-performance liquidchromatography (UPLC)-mass spectrometry. *Bio-Protocol, 8*(14), e2937. https://doi.org/10.21769/BioProtoc.2937.
Laurent, G., German, N. J., Saha, A. K., de Boer, V. C. J., Davies, M., Koves, T. R., ... Haigis, M. C. (2013). SIRT4 coordinates the balance between lipid synthesis and catabolism by repressing malonyl CoA decarboxylase. *Molecular Cell, 50*(5), 686–698. https://doi.org/10.1016/j.molcel.2013.05.012.
Li, X., & Kazgan, N. (2011). Mammalian sirtuins and energy metabolism. *International Journal of Biological Sciences,* 7(5), 575–587. https://doi.org/10.7150/ijbs.7.575.
Liu, R., Fan, M., Candas, D., Qin, L., Zhang, X., Eldridge, A., ... Li, J. J. (2015). CDK1-mediated SIRT3 activation enhances mitochondrial function and tumor radioresistance. *Molecular Cancer Therapeutics, 14*(9), 2090–2102. https://doi.org/10.1158/1535-7163.MCT-15-0017.
Lu, Z., Chen, Y., Aponte, A. M., Battaglia, V., Gucek, M., & Sack, M. N. (2015). Prolonged fasting identifies heat shock protein 10 as a sirtuin 3 substrate: Elucidating a new mechanism linking mitochondrial protein acetylation to fatty acid oxidation enzyme folding and function*. *Journal of Biological Chemistry, 290*(4), 2466–2476. https://doi.org/10.1074/jbc.M114.606228.
Marcotte, P. A., Richardson, P. R., Guo, J., Barrett, L. W., Xu, N., Gunasekera, A., & Glaser, K. B. (2004). Fluorescence assay of SIRT protein deacetylases using an acetylated peptide substrate and a secondary trypsin reaction. *Analytical Biochemistry, 332*(1), 90–99. https://doi.org/10.1016/j.ab.2004.05.039.
Michan, S., & Sinclair, D. (2007). Sirtuins in mammals: Insights into their biological function. *Biochemical Journal, 404*(1), 1–13. https://doi.org/10.1042/BJ20070140.
Nishida, Y., Rardin, M. J., Carrico, C., He, W., Sahu, A. K., Gut, P., ... Verdin, E. (2015). SIRT5 regulates both cytosolic and mitochondrial protein malonylation with glycolysis as a major target. *Molecular Cell, 59*(2), 321–332. https://doi.org/10.1016/j.molcel.2015.05.022.
Onyango, P., Celic, I., McCaffery, J. M., Boeke, J. D., & Feinberg, A. P. (2002). SIRT3, a human SIR2 homologue, is an NAD-dependent deacetylase localized to mitochondria. *Proceedings of the National Academy of Sciences of the United States of America, 99*(21), 13653–13658. https://doi.org/10.1073/pnas.222538099.

Peek, C. B., Affinati, A. H., Ramsey, K. M., Kuo, H.-Y., Yu, W., Sena, L. A., ... Bass, J. (2013). Circadian clock NAD+ cycle drives mitochondrial oxidative metabolism in mice. *Science, 342*(6158), 1243417. https://doi.org/10.1126/science.1243417.

Pillai, V. B., Kanwal, A., Fang, Y. H., Sharp, W. W., Samant, S., Arbiser, J., & Gupta, M. P. (2017). Honokiol, an activator of Sirtuin-3 (SIRT3) preserves mitochondria and protects the heart from doxorubicin-induced cardiomyopathy in mice. *Oncotarget, 8*(21), 34082–34098. https://doi.org/10.18632/oncotarget.16133.

Pirinen, E., Lo Sasso, G., & Auwerx, J. (2012). Mitochondrial sirtuins and metabolic homeostasis. *Best Practice & Research Clinical Endocrinology & Metabolism, Mitochondria in Endocrinology, 26*(6), 759–770. https://doi.org/10.1016/j.beem.2012.05.001.

Qiu, X., Brown, K., Hirschey, M. D., Verdin, E., & Chen, D. (2010). Calorie restriction reduces oxidative stress by SIRT3-mediated SOD2 activation. *Cell Metabolism, 12*(6), 662–667. https://doi.org/10.1016/j.cmet.2010.11.015.

Rardin, M. J., He, W., Nishida, Y., Newman, J. C., Carrico, C., Danielson, S. R., ... Verdin, E. (2013). SIRT5 regulates the mitochondrial lysine succinylome and metabolic networks. *Cell Metabolism, 18*(6), 920–933. https://doi.org/10.1016/j.cmet.2013.11.013.

Rodgers, J. T., Lerin, C., Haas, W., Gygi, S. P., Spiegelman, B. M., & Puigserver, P. (2005). Nutrient control of glucose homeostasis through a complex of PGC-1α and SIRT1. *Nature, 434*(7029), 113–118. https://doi.org/10.1038/nature03354.

Samant, S. A., Zhang, H. J., Hong, Z., Pillai, V. B., Sundaresan, N. R., Wolfgeher, D., ... Gupta, M. P. (2014). SIRT3 deacetylates and activates OPA1 to regulate mitochondrial dynamics during stress. *Molecular and Cellular Biology, 34*(5), 807–819. https://doi.org/10.1128/MCB.01483-13.

Satterstrom, F. K., & Haigis, M. C. (2014). Luciferase-based reporter to monitor the transcriptional activity of the SIRT3 promoter. *Methods in Enzymology, 543*, 141–163. https://doi.org/10.1016/B978-0-12-801329-8.00007-6.

Scher, M. B., Vaquero, A., & Reinberg, D. (2007). SirT3 is a nuclear NAD+-dependent histone deacetylase that translocates to the mitochondria upon cellular stress. *Genes & Development, 21*(8), 920–928. https://doi.org/10.1101/gad.1527307.

Schlicker, C., Gertz, M., Papatheodorou, P., Kachholz, B., Becker, C. F. W., & Steegborn, C. (2008). Substrates and regulation mechanisms for the human mitochondrial sirtuins Sirt3 and Sirt5. *Journal of Molecular Biology, 382*(3), 790–801. https://doi.org/10.1016/j.jmb.2008.07.048.

Schilling, B., Meyer, J. G., Wei, L., Ott, M., & Verdin, E. (2019). High-resolution mass spectrometry to identify and quantify acetylation protein targets. *Methods in Molecular Biology (Clifton, N. J.), 1983*, 3–16. https://doi.org/10.1007/978-1-4939-9434-2_1.

Schultz, M. B., Lu, Y., Braidy, N., & Sinclair, D. A. (2018). Assays for NAD+-dependent reactions and NAD+ metabolites. *Methods in Molecular Biology (Clifton, N. J.), 1813*, 77–90. https://doi.org/10.1007/978-1-4939-8588-3_6.

Schwer, B., North, B. J., Frye, R. A., Ott, M., & Verdin, E. (2002). The human silent information regulator (Sir)2 homologue hSIRT3 is a mitochondrial nicotinamide adenine dinucleotide-dependent deacetylase. *The Journal of Cell Biology, 158*(4), 647–657. https://doi.org/10.1083/jcb.200205057.

Shi, C., Jiao, F., Wang, Y., Chen, Q., Wang, L., & Gong, Z. (2022). SIRT3 inhibitor 3-TYP exacerbates thioacetamide-induced hepatic injury in mice. *Frontiers in Physiology, 13*, 915193. https://doi.org/10.3389/fphys.2022.915193.

Shimazu, T., Hirschey, M. D., Hua, L., Dittenhafer-Reed, K. E., Schwer, B., Lombard, D. B., ... Verdin, E. (2010). SIRT3 deacetylates mitochondrial 3-hydroxy-3-methylglutaryl CoA synthase 2 and regulates ketone body production. *Cell Metabolism, 12*(6), 654–661. https://doi.org/10.1016/j.cmet.2010.11.003.

Smith, B. C., Hallows, W. C., & Denu, J. M. (2009). A continuous microplate assay for sirtuins and nicotinamide producing enzymes. *Analytical Biochemistry, 394*(1), 101–109. https://doi.org/10.1016/j.ab.2009.07.019.

Someya, S., Yu, W., Hallows, W. C., Xu, J., Vann, J. M., Leeuwenburgh, C., ... Prolla, T. A. (2010). Sirt3 mediates reduction of oxidative damage and prevention of age-related hearing loss under caloric restriction. *Cell, 143*(5), 802–812. https://doi.org/10.1016/j.cell.2010.10.002.

Stein, L. R., & Imai, S. (2012a). The dynamic regulation of NAD metabolism in mitochondria. *Trends in Endocrinology and Metabolism: TEM, 23*(9), 420–428. https://doi.org/10.1016/j.tem.2012.06.005.

Still, A. J., Floyd, B. J., Hebert, A. S., Bingman, C. A., Carson, J. J., Gunderson, D. R., ... Pagliarini, D. J. (2013). Quantification of mitochondrial acetylation dynamics highlights prominent sites of metabolic regulation. *The Journal of Biological Chemistry, 288*(36), 26209–26219. https://doi.org/10.1074/jbc.M113.483396.

Talwadekar, M., Khatri, S., Balaji, C., Chakraborty, A., Basak, N.-P., Kamat, S. S., & Kolthur-Seetharam, U. (2024). Metabolic transitions regulate global protein fatty acylation. *Journal of Biological Chemistry, 300*(1), https://doi.org/10.1016/j.jbc.2023.105563.

Tanny, J. C., & Moazed, D. (2001). Coupling of histone deacetylation to NAD breakdown by the yeast silencing protein Sir2: Evidence for acetyl transfer from substrate to an NAD breakdown product. *Proceedings of the National Academy of Sciences, 98*(2), 415–420. https://doi.org/10.1073/pnas.98.2.415.

Tao, R., Coleman, M. C., Pennington, J. D., Ozden, O., Park, S.-H., Jiang, H., ... Gius, D. (2010). Sirt3-mediated deacetylation of evolutionarily conserved lysine 122 regulates MnSOD activity in response to stress. *Molecular Cell, 40*(6), 893–904. https://doi.org/10.1016/j.molcel.2010.12.013.

Trammell, S. A., & Brenner, C. (2013). Targeted, LCMS-based metabolomics for quantitative measurement of NAD(+) metabolites. *Computational and Structural Biotechnology Journal, 4*, e201301012. https://doi.org/10.5936/csbj.201301012.

Vassilopoulos, A., Fritz, K. S., Petersen, D. R., & Gius, D. (2011). The human sirtuin family: Evolutionary divergences and functions. *Human Genomics, 5*(5), 485–496. https://doi.org/10.1186/1479-7364-5-5-485.

Vaziri, H., Dessain, S. K., Eaton, E. N., Imai, S.-I., Frye, R. A., Pandita, T. K., ... Weinberg, R. A. (2001). *hSIR2SIRT1* functions as an NAD-dependent p53 deacetylase. *Cell, 107*(2), 149–159. https://doi.org/10.1016/S0092-8674(01)00527-X.

Wagner, G. R., & Hirschey, M. D. (2014). Nonenzymatic protein acylation as a carbon stress regulated by sirtuin deacylases. *Molecular Cell, 54*(1), 5–16. https://doi.org/10.1016/j.molcel.2014.03.027.

Wegener, D., Wirsching, F., Riester, D., & Schwienhorst, A. (2003). A fluorogenic histone deacetylase assay well suited for high-throughput activity screening. *Chemistry & Biology, 10*(1), 61–68. https://doi.org/10.1016/s1074-5521(02)00305-8.

Weinert, B. T., Moustafa, T., Iesmantavicius, V., Zechner, R., & Choudhary, C. (2015). Analysis of acetylation stoichiometry suggests that SIRT3 repairs nonenzymatic acetylation lesions. *The EMBO Journal, 34*(21), 2620–2632. https://doi.org/10.15252/embj.201591271.

Wood, J. G., Schwer, B., Wickremesinghe, P. C., Hartnett, D. A., Burhenn, L., Garcia, M., ... Helfand, S. L. (2018). Sirt4 is a mitochondrial regulator of metabolism and lifespan in Drosophila melanogaster. *Proceedings of the National Academy of Sciences of the United States of America, 115*(7), 1564–1569. https://doi.org/10.1073/pnas.1720673115.

Yang, H., Yang, T., Baur, J. A., Perez, E., Matsui, T., Carmona, J. J., ... Sinclair, D. A. (2007). Nutrient-sensitive mitochondrial NAD+ levels dictate cell survival. *Cell, 130*(6), 1095–1107. https://doi.org/10.1016/j.cell.2007.07.035.

Yang, W., Nagasawa, K., Münch, C., Xu, Y., Satterstrom, K., Jeong, S., ... Haigis, M. C. (2016). Mitochondrial sirtuin network reveals dynamic SIRT3-dependent deacetylation in response to membrane depolarization. *Cell, 167*(4), 985–1000.e21. https://doi.org/10.1016/j.cell.2016.10.016.

Zhang, J., Xiang, H., Liu, J., Chen, Y., He, R.-R., & Liu, B. (2020). Mitochondrial sirtuin 3: New emerging biological function and therapeutic target. *Theranostics, 10*(18), 8315–8342. https://doi.org/10.7150/thno.45922.

Zhong, L., & Mostoslavsky, R. (2011). Fine tuning our cellular factories: Sirtuins in mitochondrial biology. *Cell Metabolism, 13*(6), 621–626. https://doi.org/10.1016/j.cmet.2011.05.004.

Zhuang, Y., Zhang, Y., Liu, C., & Zhong, Y. (2024). Interplay between the circadian clock and sirtuins. *International Journal of Molecular Sciences, 25*(21), https://doi.org/10.3390/ijms252111469.

CHAPTER SEVEN

DNA methylation and histone modifications in tumor hypoxia☆

Shruti Ganesh Dhamdhere[1], Parik Kakani[1], and Sanjeev Shukla*
Department of Biological Sciences, Indian Institute of Science Education and Research, Bhopal, Madhya Pradesh, India
*Corresponding author. e-mail address: sanjeevs@iiserb.ac.in

Contents

1. Introduction 144
2. Rationale 147
3. Analyses of differential DNA methylation and histone modifications in hypoxic cancers as compared to their normoxic counterparts 148
4. Isolation and immunoprecipitation of genomic DNA to assess methylation patterns 149
   4.1 Equipments 149
   4.2 Reagents 149
   4.3 Procedure 150
   4.4 Notes 151
5. Chromatin immunoprecipitation (ChIP) to determine occupancy of histone marks at specific locus 151
   5.1 Equipments 151
   5.2 Reagents 151
   5.3 Procedure 155
   5.4 Notes 157
6. Effect of presence of epigenetic modifications on transcription factor binding and gene regulation 157
7. De novo maintenance of methylation via dCas9-DNMT3A CRISPR systems under hypoxic condition 158
   7.1 Equipments 158
   7.2 Reagents 158
   7.3 Procedure 158
   7.4 Notes 159
8. De novo DNA methylation under hypoxia abolishes recruitment of hypoxic transcription factors HIF1α and CTCF to binding sites 159
9. De novo histone dissymmetric methylation under hypoxia by dCas9-PRMT5 regulates splicing outcome 159

☆ Epigenetics regulation in hypoxic cancer
[1] Contributed equally

Methods in Enzymology, Volume 733
ISSN 0076-6879, https://doi.org/10.1016/bs.mie.2026.05.035

10. Challenges and limitations 160
11. Concluding remarks 160
References 161

Abstract

Hypoxia, an inherent condition to tumors, contributes to cancer development, metastasis, and therapy resistance. Cancer cells employ epigenetic remodelling for survival under hypoxia. DNA methylation and various histone modifications are two key mechanisms of epigenetic regulation of gene expression. Histone modifications in collaboration with DNA methylation play a vital role in controlling gene expression and modulating chromatin structure. Recently it has become evident that hypoxia plays key roles in the EMT, angiogenesis, autophagy, and ECM remodeling via (i) modulating DNA methylation at the promoter of the gene or at loci undergoing alternative splicing; and (ii) locus-specific histone modifications such as H3K27Ac, H3K9Ac, H3K4me3, H4R3me2s, and H3R8me2s. Thus, it is important to mechanistically assess how hypoxia-mediated DNA methylation changes and histone modification affect tumorigenesis using suitable methods. In this chapter, we describe the methods used in our laboratory to assess the locus-specific DNA methylation and histone modifications profiling. The goals of this chapter are twofold. First, we discuss immunoprecipitation-based approaches applied to assess DNA methylation or hydroxymethylation, histone modification, and transcription factor binding profile at specific loci. Second, we discuss methods used to modify the epigenetic landscape, such as DNA methylation, binding of transcription factors such as HIF1α and CTCF, and histone marks on a given locus using dCas9 fused with epigenetic modifiers such as DNMT3A, TET1, PRMT5, etc. Collectively, these approaches provide a template that can be applied to evaluate roles of any epigenetic mark involved in development and/or progression of cancer.

1. Introduction

Epigenetics is a fine-tuning regulator of gene expression that introduces reversible modifications in the genome without altering the underlying DNA sequence. DNA methylation and histone modifications constitute the two principal epigenetic mechanisms widely studied in the context of transcriptomic modalities in cancers (Esteller et al., 2024). During DNA methylation, the 5′ carbon of cytosine residues within CpG dinucleotides is covalently added with a methyl group by DNA methyltransferases (DNMTs). In the mammalian genomes, DNA methylation at regions rich in CpG dinucleotides, often located at gene promoters, is associated with the transcriptional activity of the gene. Actively transcribing genes typically have unmethylated promoters, whereas silenced genes usually harbor hypermethylated promoters (Attwood et al., 2002; Moore et al., 2013). DNA methylation silences transcription by blocking the binding of transcription factors (TFs) at gene promoters, since several TFs require unmethylated CpG within their recognition sequences to bind efficiently to the DNA (for example, SP1, HIF1α, CTCF, KLF4, YY1, etc) (D'Anna et al., 2020; Hashimoto et al., 2017; Kim & Kim, 2009; Monteagudo-Sánchez et al., 2024; Tian et al., 2015). Methylated DNA also recruits methylated-CpG binding proteins and co-repressive complexes that make DNA inaccessible to the transcriptional machinery (Ng et al., 1999; Zou et al., 2012). In parallel, the chromatin also undergoes post-translational modifications on the N-terminal tails of histone proteins, such as methylation, acetylation, phosphorylation, ubiquitination, SUMOylation, crotonylation, and more. These histone modifications are added, removed, and interpreted by enzymes referred to as writers, erasers, and readers, respectively (Berger, 2002; Zou et al., 2012). Histone modifications specifically at the gene promoters also regulate the gene transcriptional activities, either independently or in a coordinated fashion, mutually dependent on the DNA methylation status of the particular gene promoter (Miller & Grant, 2013; Zhang et al., 2005). The histone acetyltransferases (HATs) add the acetyl groups on the lysine residues, thereby neutralizing the positive charges and weakening the DNA-histone interactions, which facilitates transcription by making the chromatin open and accessible to the transcriptional machinery. On the other hand, histone deacetylases (HDACs), as their name suggests, remove the acetylation marks from histone tails, leading to chromatin compaction and transcriptional repression. However, histone methylation marks function in context-dependent manner, such as H3K4me3, which is a mark associated with active promoters, whereas H3K9me3 and H3K27me3 are linked with a silenced promoter (Bannister & Kouzarides, 2011). In addition to promoter-associated modifications, the epigenomic modifications within the gene bodies also contribute to the transcriptomic diversity. DNA methylation patterns and histone modifications at the splice sites near exon-intron junctions, or at distant regulatory sites, can influence the alternative splicing of genes (Shukla et al., 2011). These epigenetic marks can modulate the recruitment of splicing factors by altering the chromatin structure or by regulating the rate of RNA polymerase II elongation during transcription (Dhamdhere et al., 2025; Kakani et al., 2023; Mutnuru et al., 2025; Pandey et al., 2024; Yadav et al., 2023). The generation of multiple transcript variants of a single gene, as a consequence of epigenetic landscape-dependent inclusion or exclusion of specific exons, adds an additional layer of complexity to transcriptional regulation in response to internal or external stimuli (Fig. 1).

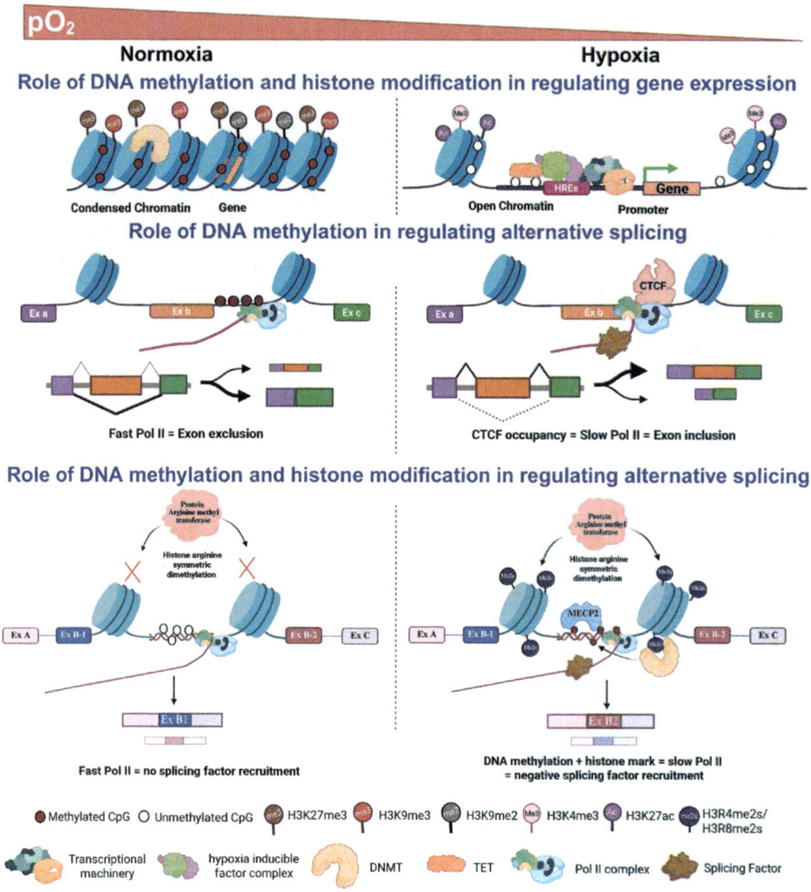

Fig. 1 Hypoxia-induced mechanisms that link differential DNA methylation, chromatin modification to gene expression regulation and switch in alternative splicing. A promoter level change in DNA methylation status and histone marks from repressive to active state under hypoxia allows binding of hypoxia inducible factor binding, thus transcriptional upregulation of the gene. Under hypoxia, in one mechanism, the change in DNA methylation allow binding of CTCF that cause RNA polymerase II pause and hence inclusion of exon while in an another mechanism histone modification by PRMT allows DNMT recruitment and causes selective DNA methylation of specific-loci. This allows binding of negative splicing regulator which causes switch in alternative splicing outcome.

In his recent review on hallmarks of cancer, Hanahan has introduced non-mutational epigenetic reprogramming as an emerging hallmark. Aberrant epigenetic reprogramming in cancer has been studied to influence the overall functioning in cancers, underscoring the need for detailed investigation of its underlying mechanisms and biological consequences (Hanahan, 2022). Within a solid tumor, the highly proliferative cancer cells outgrow their vasculature, and the incompensating irregular and tortuous vasculature leads to drastic differences in oxygen concentrations. These regions with diminished oxygen diffusion are referred to as the hypoxic niches. Hypoxia in cancers has been widely studied to have an immense impact on their phenotypes and has been linked with decreased disease-free survival. The hypoxic microenvironment drives extensive changes in the cellular transcriptome, ultimately altering various cancer cellular pathways. These transcriptional changes modulate key cancer hallmark features such as cell proliferation, angiogenesis, migration, invasion, metabolism, and therapy resistance (Chen et al., 2023; Muz et al., 2015). Under hypoxic condition the cancer cells also experience an extensive alteration in their epigenome, as the hypoxia-mediated signaling pathways, particularly those regulated by hypoxia-inducible factors (HIFs), influence the activity of several epigenetic regulators (Camuzi et al., 2019; Ma et al., 2022). These enzymes modify chromatin structure through DNA methylation and histone modifications, thereby altering chromatin accessibility and regulating gene expression in a context-dependent manner. Consequently, epigenetic remodeling under hypoxia contributes to transitioning into a transcriptomic state permissible to survival in the harsh microenvironmental conditions. Furthermore, such epigenetic changes can also influence alternative splicing, enhancer activity, and long-range chromatin interactions, collectively expanding the regulatory complexity of gene expression in cancer cells. Dysregulation of these epigenetic mechanisms under hypoxic stress has been implicated in tumor progression, metastasis, and therapeutic resistance (Camuzi et al., 2019; Verdikt & Thienpont, 2024; Watt et al., 2025).

2. Rationale

Epigenetic dysregulation manifested as global hypomethylation, promoter-specific hypermethylation, or misregulated histone marks is a defining feature of numerous diseases, particularly cancer, where it contributes to inappropriate gene silencing, activation of oncogenic pathways, aberrant alternative splicing and genomic instability. Additionally, understanding the interplay between epigenetic features such as DNA methylation and histone modifications is central to elucidating mechanisms of gene regulation in normal physiology and disease. Rapid epigenomic changes have been studied to be the major driving forces indispensable for the cancer progression and aggression. In case of hypoxia, there are reports which suggest that various genomic foci undergo alteration in terms of their methylation status as well as histone modifications. These changes ultimately lead to dysregulation of the transcriptional state and/or the splicing of various genes. Owing to the reversible nature of these epigenetic alterations, they have become lucrative targets for therapeutic interventions. Therefore, assessing the real time granularity of the cancer epigenome is essential to comprehend the extent of the impact that the altered epigenome imparts on the tumor progression. The neoteric development in cancer research pertaining to asssessing, understanding and targeting the cancer epigenome, emergence of sophisticated technologies and advancements in next-generation sequencing (NGS) will not only provide better understanding of the cancer itself but also aid in understanding the environmental influence on the cancer such as the external exposures and lifestyle factors. Epigenetic biomarkers would prove to be better over the traditional genetic and protein biomarkers as they would provide real time state of gene expression and the underlying mechanisms, which can be beneficial in diagonostics, prognostics and also in designing the therapeutic regimen of that patient.

3. Analyses of differential DNA methylation and histone modifications in hypoxic cancers as compared to their normoxic counterparts

Epigenomic status of a particular genetic locus, either at the gene promoter or within the gene body, influences the pre-transcription and post-transcription fate of a gene. To understand the change in methylation patterns or histone modifications at a given locus in cancer cells exposed to hypoxic conditions, immunoprecipitation-based techniques are commonly employed. The adherent cancer cells were seeded and cultured in cell culture plates for at least 12–16 h before exposure to the hypoxic conditions to ensure proper attachment and recovery. Hypoxic conditions were maintained using a hypoxic chamber set at 1% oxygen concentration. Following exposure, chromatin or DNA is immunoprecipitated using antibodies specific to the epigenetic modification of interest. The immunoprecipitated DNA or chromatin is quantified via qPCR to assess the prevalence of a specific genomic region.

4. Isolation and immunoprecipitation of genomic DNA to assess methylation patterns

4.1 Equipments

1. Hypoxia Chamber (Ruskinn INVIVO2 400)
2. CO2 incubator
3. Bioruptor sonicator (Diagenode)
4. qTOWER3-G (Analytik Jena)

4.2 Reagents

1. gDNA isolation kit (Sigma, G1N70)
2. 1X PBS
3. 5-methylCytosine (D3s2z) (Cell Signaling Technology Cat# 28692S)
4. Normal Rabbit IgG (Cell Signaling TechnologyCat# 2729S)
5. MeDIP buffer
6. Protein G Dynabeads(Invitrogen Cat# 10004D)
7. Elution buffer
8. QIAquick PCR purification Kit (QiagenCat# 28106)
9. KAPA SYBR FAST QPCR master mix (Sigma, KK4618)

Composition of MeDIP buffer

Component	Working concentration	Volume
100 mM Sodium phosphate pH 8	10 mM	5 mL
5 M NaCl	140 mM	1.4 mL
10% Triton X-100	2.5%	250 μL
Total Volume (make up with milliQ water)		10 mL

Composition of elution buffer

Component	Working concentration	Volume
1 M Tris HCl pH 8	50 mM	50 μL
0.5 M EDTA	10 mM	20 μL
10 % SDS	1%	100 μL
500 mM Sodium bicarbonate	50 mM	100 μL
Total Volume (make up with milliQ water)		1 mL

4.3 Procedure

1. Trypsinize the cells and pellet down by centrifuging at 1200 rpm for 2 mins at 4 °C.
2. Give 1X PBS wash to this pellet and centrifuge at 2000 rpm for 5 mins at 4 °C. Repeat this process twice.
3. The gDNA was isolated using gDNA isolation kit (Sigma, G1N70) according to the manufacturer's instructions and eluted in TE buffer (~150 μL). After quality check and quantification on nanodrop the gDNA was subjected to sonication.
4. Approximately 6.5 μg of gDNA per condition (i.e. 3 μg for immuno-precipitated with 5mC antibody, 3 μg for control IgG and 500 ng for running on agarose gel) from the isolated gDNA was used for sonication. For sonication the 1.5 mL Bioruptor Microtubes (Diagenode, cat no.# C30010016) was used and the volume was made upto 250 μL using TE buffer.
5. The samples were sonicated for 9 cycles with 30 s on and 30 s off to get fragment size ~250–100 bp. 400 ng of the sonicated samples was run on 2 % gel to check the fragment size along with 50 bp DNA ladder.
6. The samples with appropriate fragment size were transferred at 97 °C for 12 mins followed by immediately shifting on ice for 10 mins.
7. The remaining sonicated samples were divided into two halves (120ul) in 1.5 mL MCTs, (1) for adding 5mc antibody and (2) for control IgG. MeDIP buffer was added to the respective samples to make up the volume till 500 μL and 25 μL (5 %) was taken out as input from each tube before adding antibody and kept aside at 4 °C.
8. To one 1.5 mL MCT 5mC antibody was added and to the other MCT IgG rabbit (equal to the amount of 5mC antibody) was added. The MCTs were incubated overnight at 4 °C on rotation.
9. On the next day, 15 μL of Dynabeads Protein G was added to each sample (after removal of the storage buffer) and kept on rotation for another 1 h at 4 °C.
10. After the incubation the magnetic beads were washed using MeDIP buffer by keeping the samples on a magnetic rack, removing the buffer and adding 500 μL of the MeDIP buffer. The samples were incubated by keeping on rotation at 4 °C for 5 min. This process was repeated thrice.
11. Finally, the buffer was removed and 150 μL of elution buffer was added such that the beads were resuspended from the walls into the solution

and given quick spun. To this 1 μL of proteinase K solution was added. The 1.5 mL MCTs were then wrapped with parafilm and keep at 65 °C at 950 rpm overnight.

12. On the next day, all the samples were PCR purified along with input using QIAGEN PCR purification kit. The samples were stored at −20 °C until further use.
13. The immunoprecipitated samples along with input were quantitatively analysed by qPCR using KAPA SYBR FAST (Sigma, KK4618) in qTOWER3 Touch Real Time PCR (analytic jena). The PCR reaction was put as mentioned before in 2.5.3. The enrichment was calculated using input as normalization factor and the following formula: 2^(Ct_input − Ct_Immuno-precipitation). Additionally, the resultant values were normalized by the enrichment values of normal rabbit IgG. The final values are represented as mean ± SEM or mean ± SD as per requirement of triplicates.

4.4 Notes

1. Depending upon the transcriptomic kinetics of the gene of interest (GOI) the MeDIP-qPCR is carried out at varied time intervals of hypoxic exposure.
2. The normoxic controls are seeded and cultured at the same time along with the hypoxia exposed cells. Before starting a time kinetic experiment both the normoxic and hypoxic cells are supplemented with fresh cell culture media.
3. To avoid loss of hypoxic effect on the cells, trypsinize the hypoxic cells inside the hypoxic chamber and immediately shift on ice.

5. Chromatin immunoprecipitation (ChIP) to determine occupancy of histone marks at specific locus

5.1 Equipments

1. Hypoxia Chamber (Ruskinn INVIVO2 400)
2. CO2 incubator
3. Digenode Sonicator
4. qTOWER3 Touch Real Time PCR (analytic jena)

5.2 Reagents

1. Formaldehyde
2. Glycine

3. 1X PBS
4. Protease inhibitor cocktail
5. Anti-histone marks antibody H3K4trimethylation (H3K4me3), AbcamCat# ab8580; H3K27acetylation (H3K27ac), AbcamCat#ab4729; Anti-Symmetric Dimethyl-Histone H4 (Arg3), PTM biolabs, PTM-639; Anti-Symmetric Dimethyl-Histone H3 (Arg8) PTM biolabs, PTM-672
6. Normal Rabbit IgG(Cell Signaling TechnologyCat# 2729S)
7. Lysis buffers (LB) 1, 2, 3
8. Dynabeads Protein G
9. Elution buffer
10. Wash buffers (low salt immune complex wash buffer, high salt immune complex wash buffer, LiCl immune complex wash buffer, TE wash buffer)
11. QIAGEN PCR purification kit
12. KAPA SYBR FAST (Sigma, KK4618)

Composition of LB1 buffer

Component	Working concentration	Volume
1 M HEPES KOH pH 7.5	50 mM	500 μL
5 M NaCl	140 mM	280 μL
0.5 M EDTA	1 mM	20 μL
50 % glycerol	10 %	2 mL
10 % NP-40	0.50 %	500 μL
10% Triton X-100	0.25 %	250 μL
Total Volume (make up with milliQ water)		10 mL

Composition of LB2 buffer

Component	Working concentration	Volume
1 M Tris HCl pH 8	10 mM	100 μL
5 M NaCl	200 mM	400 μL
0.5 M EDTA	1 mM	20 μL
0.5 M EGTA	0.5 mM	10 μL
Total Volume (make up with milliQ water)		10 mL

Composition of LB3 buffer

Component	Working concentration	Volume
1 M Tris HCl pH 8	10 mM	100 μL
5 M NaCl	100 mM	200 μL
0.5 M EDTA	1 mM	20 μL
0.5 M EGTA	0.5 mM	10 μL
10 % Sodium deoxycholate	0.10 %	100 μL
20 % N-Laurylsarcosin	0.50 %	250 μL
Total Volume (make up with milliQ water)		10 mL

Composition of Elution buffer

Component	Working concentration	Volume
1 M Tris HCL pH 8	50 mM	50 μL
0.5 M EDTA	10 mM	20 μL
10 % SDS	1 %	100 μL
500 mM Sodium bicarbonate	50 mM	100 μL
Total Volume (make up with milliQ water)		1 mL

Composition of low salt immune complex wash buffer

Component	Working concentration	Volume
10 % SDS	0.10 %	1 mL
10 % Triton X-100	1.00 %	1 mL
0.5 M EDTA	2 mM	40 μL
1 M Tris HCl pH 8.1	20 mM	200 μL
5 M NaCl	150 mM	300 μL
Total Volume (make up with milliQ water)		10 mL

Composition of high salt immune complex wash buffer

Component	Working concentration	Volume
10 % SDS	0.10 %	1 mL
10 % Triton X-100	1.00 %	1 mL
0.5 M EDTA	2 mM	40 µL
1 M Tris HCl pH 8.1	20 mM	200 µL
5 M NaCl	500 mM	1 mL
Total Volume (make up with milliQ water)		10 mL

Composition of LiCl immune complex wash buffer

Component	Working concentration	Volume
8 M LiCl	0.15 mM	262 µL
10 % NP-40	1.00 %	1 mL
10 % Sodium deoxycholate	1.00 %	1 mL
0.5 M EDTA	20 mM	20 µL
1 M Tris HCl pH 8.1	10 mM	100 µL
Total Volume (make up with milliQ water)		10 mL

Composition of TE wash buffer

Component	Working concentration	Volume
1 M Tris HCl pH 8.1	10 mM	100 µL
0.5 M EDTA	20 mM	20 µL
Total Volume (make up with milliQ water)		10 mL

Preparation of 4 % formaldehyde solution: 1 mL of molecular biology grade formaldehyde (Sigma, F8775) was added to 36 mL 1X PBS.

Preparation of 0.125 M glycine solution: 0.93 gm glycine (Sigma, G8898,) was mixed in 10 mL autoclaved milliQ to prepare 1.25 M stock solution. 5 mL of stock solution was added to 45 mL 1X PBS to obtain 0.125 M working solution.

5.3 Procedure

1. Approximately 10 million cells were used per experiment. After the incubations, media was discarded, cells were washed twice with ice-cold 1X PBS and incubated with 1 % formaldehyde solution for 10 min at RT on rocker.
2. The process of fixation was stopped by discarding the formaldehyde solution, giving a quick 1X PBS (room tempertature) wash and adding 0.125 M glycine for 7 min at RT on rocker.
3. After discarding glycine solution, the cells were washed twice with cold 1X PBS and scraped in 1X PBS. The cell suspension was collected in 50 mL tube kept in ice and subjected to centrifugation at 2000 rpm for 5 min at 4 °C.
4. After decanting PBS, the cells were resuspended thoroughly in ice-cold 1 mL LB1 (supplemented with 1X PIC) and incubated for 10 min on ice. For proper lysis, after every 3 min, the cells were gently mixed 5 times using 1 mL pipette to ensure homogenous cell suspension. The cells were then centrifuged at 2000 rpm for 5 min at 4 °C.
5. The supernatant was discarded, and the pellet was then thoroughly resuspended in ice-cold 1 mL LB2 (supplemented with 1X PIC) and incubated for 10 min on ice. For proper lysis, after every 3 min, the cells were gently mixed 5 times using 1 mL pipette. The sample was then centrifuged at 2000 rpm for 5 min at 4 °C.
6. Finally, the supernatant was discarded, and the nuclear pellet was resuspended in ice-cold 500 µL LB3 (supplemented with 1X PIC) and incubated for 5 min on ice.
7. The nuclear lysate was resuspended 50 times using a 200 µL pipette and the suspension was passed thrice through 25-gauge needle before proceeding for sonication.
8. Sample was divided into two sonication tubes (each containing ~250 µL chromatin solution) and subjected to 5 sonication cycles (30 s ON followed by 30 s OFF).
9. Subsequently, all the sonicated chromatin of one sample was pooled and 1/20th volume of 20 % Triton X-100 was added. The 1.5 mL MCTs were then inverted few times and centrifuged for 15,000 g for 10 min at 4 °C.
10. The debris was discarded, and the supernatant was transferred to a fresh tube. A 30 µL aliquot of chromatin was reverse crosslinked by adding 1 µL RNase A (Invitrogen, 12091021, 1 µg/µL) and incubating at 37 °C for 30 min followed by addition of 100 µL elution buffer and 1 µL

proteinase K solution (Invitrogen, 25530049, 20 μg/μL). The tubes were parafilm sealed and incubated at 65 °C for 2 hrs with shaking.

11. The reverse cross-linked DNA was then purified using Qiagen PCR purification kit and eluted in 30 μL nuclease free water. The DNA concentration was measured using nanodrop and 500 ng DNA was run on 2 % agarose gel to confirm if the sonication DNA is within 150–300 bp size range.
12. From the sonicated chromatin, for histone marks 10 ug and for transcription factor 30 μg chromatin was taken for immunoprecipitation. The chromatin was diluted in LB3 (supplemented with 1X PIC) containing 1/20th volume of 20 % Triton X-100 to make final volume to 500 μL and 25 μL of this was kept as input before adding antibody.
13. The primary antibody and corresponding control IgG were added individually for pulldown. The tubes were incubated overnight at 4 °C on rocker.
14. On the next day, 15 μL of Dynabeads Protein G was added to each sample (after removing the storage buffer) and kept on rotation for another 2 h at 4 °C.
15. The tubes were then kept on a magnetic stand, and the supernatant was discarded. The pellet of magnetic beads was sequentially washed with 500 μL low salt, high salt, LiCl and TE wash buffers. Each wash was for 5 min at 4 °C on rocker. After completion of the washing steps, 150 μL elution buffer was added to the beads, sealed using parafilm and incubated at 65 °C with shaking (900 rpm) for 2 h.
16. The tubes were then kept on a magnetic stand, and the supernatant was transferred to a new tube. To the input tubes from the previous day 100 μL of elution buffer was added and 1 μL RNase A was added to each tube and incubated at 37 °C for 30 min
17. After the incubation 1 μL proteinase K solution was then added to each tube, parafilm sealed and incubated overnight at 65 °C with shaking.
18. On the next day, the samples were purified using Qiagen PCR purification kit and the DNA was eluted in 30 μL of nuclease free water.
19. The immunoprecipitated samples along with input were quantitatively analysed by qPCR using KAPA SYBR FAST (Sigma, KK4618) in qTOWER3 (Analytic Jena). The PCR reaction was put as mentioned before in MeDIP protocol. The enrichment was calculated using input as normalization factor and the following formula: 2^(Ct_input − Ct_Immunoprecipitation). Additionally, the resultant values were normalized by the enrichment values of IgG. The final values are represented as mean ± SEM or mean ± SD of triplicates as per requirnment.

5.4 Notes

1. Perform the formaldehyde fixation inside the hypoxic chamber for the hypoxic cells to preserve the hypoxic epigenomic state of cells.
2. Keep the plates on ice during the scrapping step and always keep the cells covered in PBS to avoid drying and loss.
3. Avoid clumping and frothing during the lysis steps.

6. Effect of presence of epigenetic modifications on transcription factor binding and gene regulation

Epigenetic modifications modulate the local biochemical environment, nucleosome positioning and chromatin accessibility of genomic loci, thereby regulating the binding of transcription factors to the DNA. DNA methylation and histone modifications can either promote or inhibit the binding of transcription factors depending on their binding preferences. Histone acetylation, such as H3K27ac, promotes an open chromatin conformation that facilitates transcription factor access, whereas histone deacetylation and repressive marks like H3K27me3 lead to chromatin compaction and reduced binding. DNA methylation at CpG sites can directly influence the binding affinity of transcription factors, whose binding is inhibited by methylation within their recognition motifs, thereby also altering chromatin looping and higher-order genome organization. Examples of such transcription factors are CTCF, HIF1α, ATF3, SP1, and so on. On the other hand, methyl-CpG-binding proteins such as MeCP2 preferentially bind to methylated DNA and recruit co-repressor complexes, including histone deacetylases, leading to chromatin condensation and transcriptional repression, which can further restrict access of other transcription factors. Under hypoxic conditions, oxygen sensitive enzymes such as TET DNA demethylases add an additional layer of regulation reshaping the transcription factor binding landscapes.

The use of ectopic epigenome editing tools provides a powerful approach to test whether a specific epigenetic mark is required for transcription factor binding. CRISPR-based tools with catalytically inactive Cas9 (dCas9) fused to epigenetic modifiers, such as DNMTs, TETs, p300, or KDM5A, can be targeted to precise genomic loci using guide RNAs. Such systems can be exploited for addition or removal or specific epigenomic marks without altering the underlying DNA sequence. Furthermore, the artificial alteration of the epigenetic landscape at a defined promoter, enhancer or sequences within the gene body provides a mechanistic insight into how epigenetic modifications enhances or impairs occupancy of DNA binding proteins. Subsequent ChIP assays for these DNA binding proteins and transcriptional assays can also help determine the direct relationship between epigenetic modifications and transcriptional networks. In case of hypoxia, factors such as HIF1α and CTCF show differential binding depending on the methylation status of the binding sites, which consequently reflects in the transcriptional changes during adaptation to hypoxic stress.

7. De novo maintenance of methylation via dCas9-DNMT3A CRISPR systems under hypoxic condition

7.1 Equipments

1. Hypoxia Chamber (Ruskinn INVIVO2 400)
2. CO2 incubator

7.2 Reagents

1. dCas9-DNMT3A plasmid (Addgene #74407)
2. Transfection media (Gibco Opti-MEM)
3. Transfection reagent (Thermo Scientific TurboFect)

7.3 Procedure

1. The genomic region of interest such as promoter, enhancer or transcription factor binding site or alternative splicing locus is identified and single guide RNAs (sgRNAs) targeting CpG-rich regions proximal to the locus are designed. Multiple sgRNAs were used to improve targeting efficiency and coverage across the region.
2. sgRNAs were cloned into the dCas9-DNMT3A plasmid using the cloning protocol provided by the depositor and sequence confirmed.
3. The cells were seeded at least 16 h before transfection to ensure attachment and optimal cell health. Transfection was performed when the cell confluency was around 50−60 %.
4. Cloned plasmids were transfected using transfection media, transfection reagent. Approximately 12 h after transfection, the media is changed, and the cells were transferred into the hypoxia chamber.
5. For checking the transfection efficiency, an extra control plate can be used to isolate protein and determine the presence of dCas9-DNMT3A fusion protein using Cas9 antibody via western blotting.

6. After the hypoxic exposure, the cells were subjected to genomic DNA isolation, followed by methylated DNA immunoprecipitation to confirm the maintenance of methylation at the regions of interest under hypoxia.

7.4 Notes

1. sgRNAs can be designed using online design tools such as CRISPick, CHOPCHOP, CRISPOR, Benchling, etc.
2. A dCas9-DNMT3A plasmid cloned with scrambled sgRNA, and a no-sgRNA plasmid are used as controls to rule out plasmid-based biases.
3. Off-target effects can be checked by assessing methylation changes at off-target sites predicted by tools such as Cas-OFFinder.
4. Similar strategy can be performed for all the other dCas9-epigenetic modifier plasmids such as dCas9-TET1, dCas9-PRMT5, dCas9-VP64, and dCas9-p300 etc.

8. De novo DNA methylation under hypoxia abolishes recruitment of hypoxic transcription factors HIF1α and CTCF to binding sites

1. Following confirmation of maintenance of methylation, the sgRNA with optimal increase in methylation as indicated by MeDIP-qPCR (using primers targeting the region of interest and compared to scrambled sgRNA/no transfection controls), was selected for subsequent experiments.
2. The entire procedure for ectopic maintenance of methylation was repeated using the selected sgRNA plasmids and control plasmids.
3. After transfection and exposure to hypoxic conditions, the cells are used for ChIP assays. Here, antibodies against the transcription factor of interest (HIF1α or CTCF) were used during the ChIP assay.
4. Further, the same primers used for MedIP-qPCR and ChIP-qPCR to assess methylation status and the binding of transcription factor of interest, respectively.

9. De novo histone dissymmetric methylation under hypoxia by dCas9-PRMT5 regulates splicing outcome

Following the similar strategies described above for dCas9_DNMT3A, the dCas9-PRMT5 was used to modify histone to add H3R4me2s and

H3R8me2 marks at the splicing locus to study the mechanism of alternative splicing under hypoxia. The change in histone marks was confirmed via ChIP-qPCR assay as described above.

10. Challenges and limitations

MeDIP and ChIP are principal tools for understanding DNA methylation status, transcriptional cascades, and histone modifications, respectively, for decoding information encoded in chromatin. The main limitation of the Methylated DNA immunoprecipitation (MeDIP) method is the genomic resolution of 150 bp compared to single-base resolution offered by bisulfite sequencing. The varying quality of 5mC antibodies, non-specific interactions by the antibody used in the procedure, and the requirement of single-stranded (denatured) DNA for analysis. The methylated DNA recovery by the antibody is affected by mC/mCG density, making regions of very low density that may be underrepresented or even interpreted as unmethylated. The antibody-based selection is biased towards hypermethylated regions. Chromatin Immunoprecipitation (ChIP) is a powerful but demanding technique with significant challenges, primarily rooted in the requirement for high-quality, specific antibodies, high sample input, and being time-consuming. ChIP is not always suitable for detecting low-abundance or transient protein-DNA interactions. Key challenges in dCas9-based epigenetic modification include off-target activity (unintended site modification), low editing efficiency, and the transient nature of modifications. While offering precise locus targeting, limitations include constraints from PAM sequences, delivery challenges, and the potential for temporary effects that fade over time.

11. Concluding remarks

The MedIP, ChIP, and dCas9-mediated epigenetic modification are invaluable tools to study hypoxia-mediated changes in DNA methylation status, transcription factor binding, and histone mark profiling on chromatin that drive cancer progression. To explore the CTCF-HIF1α axis as a potential therapeutic target for hypoxic breast cancer cells with involvement of many downstream pathways and genes, this model, the dCas9-DNMT3A epigenetic modifier system at the CTCF promoter, was used to determine the contribution of this axis in the invasive phenotype of hypoxic breast cancer. MeDIP and ChIP-qPCR assays examine the effects of DNA methylation and histone modifications at specific loci and are key tools to define the mechanism by which identified targets and effectors contribute to the invasiveness of cancer. However, MedIP and ChIP-qPCR have limitations in assessing global changes in DNA methylation changes and histone marks. On the other hand, next-generation sequencing techniques provide important platforms to study global aspects of these epigenetic changes. Therefore, both approaches are complementary, and utilizing both approaches ensures a comprehensive mechanistic understanding of the epigenetic remodeling.

References

Attwood, J. T., Yung, R. L., & Richardson, B. C. (2002). DNA methylation and the regulation of gene transcription. *Cellular and Molecular Life Sciences*. https://doi.org/10.1007/s00018-002-8420-z.

Bannister, A. J., & Kouzarides, T. (2011). Regulation of chromatin by histone modifications. *Cell Research*. https://doi.org/10.1038/cr.2011.22.

Berger, S. L. (2002). Histone modifications in transcriptional regulation. *Current Opinion in Genetics and Development*. https://doi.org/10.1016/S0959-437X(02)00279-4.

Camuzi, D., de Amorim, Í., Ribeiro Pinto, L., Oliveira Trivilin, L., Mencalha, A., & Soares Lima, S. (2019). Regulation Is in the air: The relationship between hypoxia and epigenetics in cancer. *Cells*. https://doi.org/10.3390/cells8040300.

Chen, Z., Han, F., Du, Y., Shi, H., & Zhou, W. (2023). Hypoxic microenvironment in cancer: Molecular mechanisms and therapeutic interventions. *Signal Transduction and Targeted Therapy, 8*(1), 70. https://doi.org/10.1038/s41392-023-01332-8.

D'Anna, F., Van Dyck, L., Xiong, J., Zhao, H., Berrens, R. V., Qian, J., ... Lambrechts, D. (2020). DNA methylation repels binding of hypoxia-inducible transcription factors to maintain tumor immunotolerance. *Genome Biology*. https://doi.org/10.1186/s13059-020-02087-z.

Dhamdhere, S. G., Bansal, A., Singh, P., Kakani, P., Agrawal, S., Samaiya, A., & Shukla, S. (2025). Hypoxia-induced ATF3 escalates breast cancer invasion by increasing collagen deposition via P4HA1. *Cell Death & Disease, 16*(1), 142. https://doi.org/10.1038/s41419-025-07461-y.

Esteller, M., Dawson, M. A., Kadoch, C., Rassool, F. V., Jones, P. A., & Baylin, S. B. (2024). The epigenetic hallmarks of cancer. *Cancer Discovery*. https://doi.org/10.1158/2159-8290.CD-24-0296.

Hanahan, D. (2022). Hallmarks of cancer: New dimensions. *Cancer Discovery*. https://doi.org/10.1158/2159-8290.CD-21-1059.

Hashimoto, H., Wang, D., Horton, J. R., Zhang, X., Corces, V. G., & Cheng, X. (2017). Structural basis for the versatile and methylation-dependent binding of CTCF to DNA. *Molecular Cell*. https://doi.org/10.1016/j.molcel.2017.05.004.

Kakani, P., Dhamdhere, S., Pant, D., Mishra, J., Samaiya, A., & Shukla, S. (2023). Hypoxia-induced CTCF mediates alternative splicing via coupling chromatin looping and RNA Pol II pause to promote EMT in breast cancer. *BioRxiv*, 2023.05.06.539689. https://doi.org/10.1101/2023.05.06.539689.

Kim, J. do, & Kim, J. (2009). YY1's longer DNA-binding motifs. *Genomics*. https://doi.org/10.1016/j.ygeno.2008.09.013.

Ma, S., Zhao, Y., Lee, W. C., Ong, L. T., Lee, P. L., Jiang, Z., ... Yu, Q. (2022). Hypoxia induces HIF1α-dependent epigenetic vulnerability in triple negative breast cancer to confer immune effector dysfunction and resistance to anti-PD-1 immunotherapy. *Nature Communications*. https://doi.org/10.1038/s41467-022-31764-9.

Miller, J. L., & Grant, P. A. (2013). The role of DNA methylation and histone modifications in transcriptional regulation in humans. *Subcellular Biochemistry*. https://doi.org/10.1007/978-94-007-4525-4_13.

Monteagudo-Sánchez, A., Richard Albert, J., Scarpa, M., Noordermeer, D., & Greenberg, M. V. C. (2024). The impact of the embryonic DNA methylation program on CTCF-mediated genome regulation. *Nucleic Acids Research*. https://doi.org/10.1093/nar/gkae724.

Moore, L. D., Le, T., & Fan, G. (2013). DNA methylation and its basic function. *Neuropsychopharmacology*. https://doi.org/10.1038/npp.2012.112.

Mutnuru, S. A., Yadav, P., Kakani, P., Dhamdhere, S. G., Kumari, P., Agrawal, S., ... Shukla, S. (2025). PRMT5 regulates alternative splicing of TCF3 under hypoxia to promote EMT and invasion in breast cancer. *PLOS Biology*. https://doi.org/10.1371/journal.pbio.3003444.

Muz, B., de la Puente, P., Azab, F., & Azab, A. K. (2015). The role of hypoxia in cancer progression, angiogenesis, metastasis, and resistance to therapy. *Hypoxia*. https://doi.org/10.2147/hp.s93413.

Ng, H. H., Zhang, Y., Hendrich, B., Johnson, C. A., Turner, B. M., Erdjument-Bromage, H., ... Bird, A. (1999). MBD2 is a transcriptional repressor belonging to the MeCP1 histone deacetylase complex. *Nature Genetics*. https://doi.org/10.1038/12659.

Pandey, A., Kakani, P., & Shukla, S. (2024). CTCF and BORIS-mediated autophagy regulation via alternative splicing of BNIP3L in breast cancer. *The Journal of Biological Chemistry, 300*(7), 107416. https://doi.org/10.1016/j.jbc.2024.107416.

Shukla, S., Kavak, E., Gregory, M., Imashimizu, M., Shutinoski, B., & Kashlev, M. (2011). CTCF-promoted RNA polymerase II. *Nature, V*. https://doi.org/10.1038/nature10442.

Tian, H. P., Lun, S. M., Huang, H. J., He, R., Kong, P. Z., Wang, Q. S., ... Feng, Y. M. (2015). DNA methylation affects the SP1-regulated transcription of FOXF2 in breast cancer cells. *Journal of Biological Chemistry*. https://doi.org/10.1074/jbc.M114.636126.

Verdikt, R., & Thienpont, B. (2024). Epigenetic remodelling under hypoxia. *Seminars in Cancer Biology*. https://doi.org/10.1016/j.semcancer.2023.10.005.

Watt, K., Dauber, B., Szkop, K. J., Lee, L., Jovanovic, P., Chen, S., ... Postovit, L. M. (2025). Epigenetic alterations facilitate transcriptional and translational programs in hypoxia. *Nature Cell Biology*. https://doi.org/10.1038/s41556-025-01786-8.

Yadav, P., Pandey, A., Kakani, P., Mutnuru, S. A., Samaiya, A., Mishra, J., & Shukla, S. (2023). Hypoxia-induced loss of SRSF2-dependent DNA methylation promotes CTCF-mediated alternative splicing of VEGFA in breast cancer. *IScience*. https://doi.org/10.1016/j.isci.2023.106804.

Zhang, Y., Fatima, N., & Dufau, M. L. (2005). Coordinated changes in DNA methylation and histone modifications regulate silencing/derepression of luteinizing hormone receptor gene transcription. *Molecular and Cellular Biology*. https://doi.org/10.1128/mcb.25.18.7929-7939.2005.

Zou, X., Ma, W., Solov'Yov, I. A., Chipot, C., & Schulten, K. (2012). Recognition of methylated DNA through methyl-CpG binding domain proteins. *Nucleic Acids Research*. https://doi.org/10.1093/nar/gkr1057.

CHAPTER EIGHT

Epigenetic modulators as therapeutics for neurodegenerative disorders/ Alzheimer's disease

Debojyoti Halder[a], Denish Prajapati[a], Tonmoy Banerjee[b], Swati Biswas[b], and Balaram Ghosh[a,*,1]

[a]Epigenetic Research Laboratory, Department of Pharmacy, Birla Institute of Technology and Science-Pilani, Hyderabad Campus, Shamirpet, Hyderabad, India

[b]Nanomedicine Research Laboratory, Department of Pharmacy, Birla Institute of Technology and Science-Pilani Hyderabad Campus, Shamirpet, Hyderabad, India

*Corresponding author. e-mail address: balaram@hyderabad.bits-pilani.ac.in

Contents

1. Introduction 164
2. Epigenetic regulation of the nervous system 167
   2.1 DNA methylation in neural regulation 167
   2.2 Histone modifications and transcriptional regulation 168
   2.3 Chromatin remodeling in neuronal function 168
   2.4 Non-coding RNA-mediated regulation 169
   2.5 Integration of epigenetic mechanisms in synaptic plasticity 169
   2.6 Epigenetic dysregulation in neurodegenerative disorders 171
3. Epigenetic dysfunction in AD 171
   3.1 DNA methylation in AD 172
   3.2 Histone modifications in AD 173
   3.3 Non-coding RNAs in AD 174
   3.4 Epigenetic regulation of peripheral immune responses in AD 174
   3.5 Experimental assays and analytical methods used in epigenetic studies of AD 176
4. Epigenetic modulators for neurodegenerative disorders 196
   4.1 Classification of epigenetic modulators 196
5. Prospects and emerging therapeutics 212
Acknowledgements 214
Declaration of generative AI and AI-assisted technologies in the writing process 214
Conflict of interest 214
References 214

[1] Associate Investigator, Non-Resident, Molecular Medicine Institute [MMI], Boston, USA

Methods in Enzymology, Volume 733
ISSN 0076-6879, https://doi.org/10.1016/bs.mie.2026.06.010

Abstract

Epigenetic dysregulation is recognized as a primary contributor to the pathogenesis of neurodegenerative disorders, especially Alzheimer`s disease (AD). Epigenetic mechanisms, including DNA methylation, histone modifications, and non-coding RNA regulation, govern gene expression without altering the DNA sequence, which plays a crucial role in neuronal development, synaptic plasticity, and memory formation. In AD, abnormal epigenetic alterations disrupt neuronal homeostasis, promote Aβ aggregation, tau hyperphosphorylation, and neuroinflammation, which leads to cognitive impairment. This chapter explores the assays performed and their processes using epigenetic modulators, along with their therapeutic potential in neurodegenerative diseases, especially AD. HDAC inhibitors, DNMT inhibitors, and emerging approaches, such as PROTACs for selective degradation of epigenetic enzymes, were discussed in the context of neurocognitive disorders. Preclinical and clinical evidence suggest that targeting specific HDAC isoforms (HDAC3, HDAC6, etc) can restore synaptic plasticity and improve cognitive function. This chapter further discusses recent advances, challenges in drug specificity, BBB permeability, and off-target epigenetic effects, which remain barriers to clinical translation. Case studies highlighting successful epigenetic interventions in AD models were presented to demonstrate therapeutic feasibility. Overall, epigenetic modulators present a promising therapeutic approach for neurodegeneration, and continued research integrating various assays like DNA methylation analysis, histone modification analysis, non-coding RNA analysis, neuroinflammation analysis, and functional and behavioral assays in AD models is significant in harnessing the full potential for AD treatment.

1. Introduction

Epigenetic modulators have emerged as a promising therapeutic for neurodegenerative disorders, especially Alzheimer's disease (AD), in which cognitive decline is closely linked to dysregulated gene expression. Epigenetics refers to heritable, reversible modifications in gene expression that occur without changes in the DNA sequence and consist of histone modifications (acetylation, methylation), DNA methylation, and chromatin remodeling. In the central nervous system (CNS), epigenetic regulation is crucial for synaptic plasticity, learning, and memory formation (Ostaiza-Cardenas et al., 2025). In Alzheimer's disease, aberrant epigenetic patterns- such as altered histone acetylation and DNA methylation lead to transcriptional repression of neuroprotective genes and activation of neurotoxic pathways. In AD and related dementias, there is convergent evidence for abnormal DNA methylation, disrupted non-coding RNA networks in neurons and glia, and global and locus-specific histone hypo or hyperacetylation. These epigenetic changes alter the expression of genes involved in amyloid processing, tau phosphorylation, synaptic plasticity, neuroinflammation, and oxidative stress. The enzymes that write, read, or erase these epigenetic marks are reversible, making them a crucial therapeutic target for disease treatment, rather than symptomatic relief (Sharma et al., 2020). The global burden of neurodegenerative diseases is rising due to aging populations. Alzheimer's disease accounts for approximately 60–70 % of dementia cases worldwide, which affects over 5.5 crore people, with an expected rise of 13 crore by 2050. Current therapies offer only symptomatic relief, which are unable to address the therapeutic need for a permanent solution (Halder et al., 2023; https://www.who.int/, 2026). Therefore, targeting the core pathogenic mechanisms of AD, epigenetic dysregulation is an important cause. Epigenetic modulators provide a significant avenue because they can reversibly restore gene expression patterns associated with cognitive impairments and neuronal survival. Recent preclinical and clinical studies provided evidence that targeting epigenetic enzymes such as HDAC, DNMT, and BET by promising candidates can improve neuronal plasticity, memory, reduce amyloid-β (Aβ) aggregation, and attenuate neuroinflammation and oxidative stress in AD models, which highlights their therapeutic potential (Xu et al., 2026).

Epigenetic modulation operates through enzymatic 'writers', 'erasers', and 'readers' at a molecular level. DNA methyltransferases (DNMTs) are known as writers, adding or catalyzing methyl groups to cytosine in CpG dinucleotide islands, leading to gene silencing, whereas the ten-eleven translocation (TET) enzymes mediate active demethylation. It was reported that locus-specific promoter hypermethylation of neuroprotective genes, as well as hypomethylation of pro-amyloid or pro-inflammatory genes, occurs in AD brains (Honer et al., 2024). Modulating these methylation patterns may restore them to a normal condition. Histone acetyltransferases (HATs) add acetyl groups to histones, promoting chromatin relaxation and transcriptional activation. However, erasers, such as histone deacetylases (HDACs), remove acetyl groups, which leads to chromatin condensation and transcriptional repression. Abnormal HDAC activity, including HDAC2, HDAC3, and HDAC6, was directly associated with cognitive impairment and tau hyperphosphorylation in AD models (Pereira et al., 2024). Bromodomain and extra-terminal (BET) proteins, which consist of BRD2, BRD3, BRD4, and BRDT, act as epigenetic readers that identify acetyl-lysine residues on histones and recruit transcriptional domains. BET-dependent transcriptional programs regulate pro-inflammatory cytokines in microglia and astrocytes, thereby influencing neurodegeneration and synaptic injury (Liu et al., 2021). Non-coding RNAs integrate with DNA and histone modifications for modulation of neuronal and glial gene expression. Epigenetic modulators or drugs often modulate these RNA networks indirectly. Impairment of these processes in AD contributes to dysfunctional gene expression in synaptic transmission and neurodegeneration (Olufunmilayo & Holsinger, 2023).

The epigenetic modulators function as inhibitors or alterations of the activity of writers (DNMTs, HATs), erasers (HDACs), or readers (BETs), which reshape the chromatin structure and transcriptional outputs. Inhibition of HDACs by HDAC inhibitors increases histone acetylation, promoting genes related to synaptic plasticity, upregulating brain-derived neurotrophic factor (BDNF), and improving memory by enhancing gene transcription in AD models (Ateya et al., 2025). DNMT inhibitors can demethylate aberrant DNA methylation, thereby reactivating silenced neuroprotective genes. Although global demethylation causes oncogenic and off-target effects, this has driven interest in more selective targeted therapeutics. BET inhibitors block the interaction of reader proteins with acetylated histones, further modulating transcription of inflammatory and disease-related genes (Qin et al., 2023). BET inhibitors attenuate the expression of inflammatory mediators and break chronic neuroinflammatory cycles in AD (Liu et al., 2021). The therapeutic relevance of these epigenetic modulators lies not only in their symptomatic relief of cognitive impairment but also in their ability to simultaneously influence multiple pathways, offering a system-level therapeutic effect.

The major classes of epigenetic modulators can be broadly grouped into three categories: HDAC inhibitors, DNMT inhibitors, and BET inhibitors. Other classes include HAT activators and other histone-mark modulators (EZH2 inhibitors, LSD1 inhibitors), and finally Non-coding RNA-targeted approaches in neurodegenerative disorders. In the exploration of AD treatment, HDAC, DNMT, and BET modulators are currently the most intensively studied pharmacological classes, which not only accumulate mechanistic and preclinical data but also explore the clinical perspective (Olawade et al., 2025). Among the HDAC isoforms, HDAC3 and HDAC6 have received particular attention for their crucial roles in memory suppression and neurodegeneration. HDAC3 inhibitors such as RGFP966, RGFP963, PT3, and the HDAC6 inhibitor Tubastatin A were explored in various AD and other neurodegenerative and neuroinflammatory models (Pulya et al., 2021; Rumbaugh et al., 2015; Zhang et al., 2014). Also, overexpression of HDAC2 in hippocampus and cortical neurons correlates with decreasing dendritic spine density, causes long-term impairment, and memory deficits. Genetic or pharmacological downregulation of HDAC2 can reverse these phenotypes in AD models (Guan et al., 2009). DNMT inhibitors, such as 5-azacytidine and decitabine, act by reversing hypermethylation-induced gene silencing, though their application in the treatment of neurocognitive disorders is still under investigation due to potential toxicity (Lund et al., 2014). BET inhibitors, such as JQ1, represent a novel class that targets epigenetic readers, further modulating transcriptional programs linked to neuroprotection and attenuating neuroinflammation (Singh & Sartor, 2020). Recent reviews emphasize that epigenetic modulators represent a transformative approach in the treatment of AD by targeting reversible mechanisms underlying genetic dysregulation. The ability to restore cognitive functions and modify disease progression systematically makes epigenetic modulators a highly promising class of therapeutics, although clinical validations and specific optimization are required.

2. Epigenetic regulation of the nervous system

The epigenetic regulation is fundamental to the development, function, and plasticity of the nervous system, in which precise temporal and spatial control of gene expression is required without altering the underlying DNA sequence. Neurons are specialized, post-mitotic cells that rely on epigenetic mechanisms to maintain identity, regulate synaptic activity, and support long-term memory formation. The major epigenetic mechanisms involved are DNA methylation, histone modifications, chromatin remodeling, and non-coding RNA-mediated regulation, all of which interact to modulate and adjust neuronal gene expression (Peña, 2026).

2.1 DNA methylation in neural regulation

DNA methylation is among the most extensively studied epigenetic processes in the CNS and is mediated by DNA methyltransferases (DNMT1, DNMT3A, and DNMT3B). These enzymes catalyze the transfer of a methyl group to the 5′ position of cytosine residues in CpG dinucleotides, generating 5-methylcytosine (5-mC). DNA methylation in neurons functions as a dynamic regulatory mechanism rather than a static silencing process. Promoter methylation suppresses gene transcription by:

- Prevention of transcription factor binding
- Recruitment of methyl-CpG binding proteins such as MeCP2
- Facilitati9ng HDAC-containing co-repressor complex formation

Although, active demethylation can occur through ten-eleven translocation (TET) enzymes that convert 5-mC to 5-hydroxymethylcytosine (5-hmC), thereby promoting activity-dependent gene expression associated with neural plasticity and cognitive function (Kim & Costello, 2017; Lund et al., 2014).

2.2 Histone modifications and transcriptional regulation

Histone proteins (H2A, H2B, H3, and H4) from nucleosome structures that package genomic DNA into chromatin. Histone tails undergo diverse post-translational modifications, including:

- Acetylation
- Methylation
- Phosphorylation
- Ubiquitination

Histone acetyltransferases (HATs), including CBP/p300, catalyze acetylation reactions that neutralize histone positive charge and reduce DNA-histone affinity, resulting in transcriptionally active chromatin. In contrast, histone deacetylases (HDACs), particularly HDAC2 and HDAC3 in neurons, remove acetyl groups, leading to chromatin condensation and transcriptional repression (Park et al., 2022).

Activity-dependent increases in histone acetylation at promoters of immediate early genes, such as:

- c-Fos
- Brain-derived neurotrophic factor (BDNF)

It facilitates rapid transcriptional responses necessary for synaptic strengthening and memory consolidation. Histone methylation produces context-dependent effects on transcription:

- Transcriptionally activating marks – H3K4me3
- Transcriptionally repressing marks – H3K27me3

The balance among these modifications regulates neuronal differentiation, neurogenesis, and synaptic plasticity (Geng et al., 2021; Khan et al., 2025).

2.3 Chromatin remodeling in neuronal function

Chromatin remodeling complexes alter nucleosome positioning through ATP-dependent mechanisms to regulate DNA accessibility.
The SWI/SNF (BAF) complex contributes to:

- Dendritic development
- Neuronal differentiation
- Activity-dependent gene expression
- Synaptic plasticity

Dysregulation of chromatin remodeling pathways has been implicated in multiple neurodegenerative disorders, including AD (Sokpor et al., 2017).

2.4 Non-coding RNA-mediated regulation

Non-coding RNAs (ncRNAs) constitute an additional regulatory layer of epigenetic control. Major ncRNA classes include:

- MicroRNAs (miRNAs)
  Functions:
  - Post-transcriptional regulation
  - mRNA degradation
  - Translational repression

  Examples:
  - miR−132
  - miR−124
- Long non-coding RNAs (lncRNAs)
  Functions:
  - Recruitment of chromatin-modifying complexes
  - Locus-specific transcriptional regulation
  - Regulation of neuronal gene expression networks (Tang et al., 2010)

2.5 Integration of epigenetic mechanisms in synaptic plasticity

Epigenetic pathways function as highly interconnected regulatory systems during both neuronal development and adult brain function.

During embryonic development:

- Neural stem cells differentiate into neuronal and glial lineages
- DNA methylation patterns become progressively specialized
- Histone modifications stabilize lineage commitment

During adult brain function:

- Neuronal stimulation activates intracellular signaling pathways
- MAPK/ERK and CaMKII/IV pathways activate CREB
- CREB recruits CBP/p300
- Histone acetylation increases expression of plasticity-related genes, including BDNF

These coordinated processes support:

- Long-term potentiation
- Learning
- Memory formation
- Synaptic remodeling

The epigenetic mechanisms involved in neurodegeneration are reported in Fig. 1.

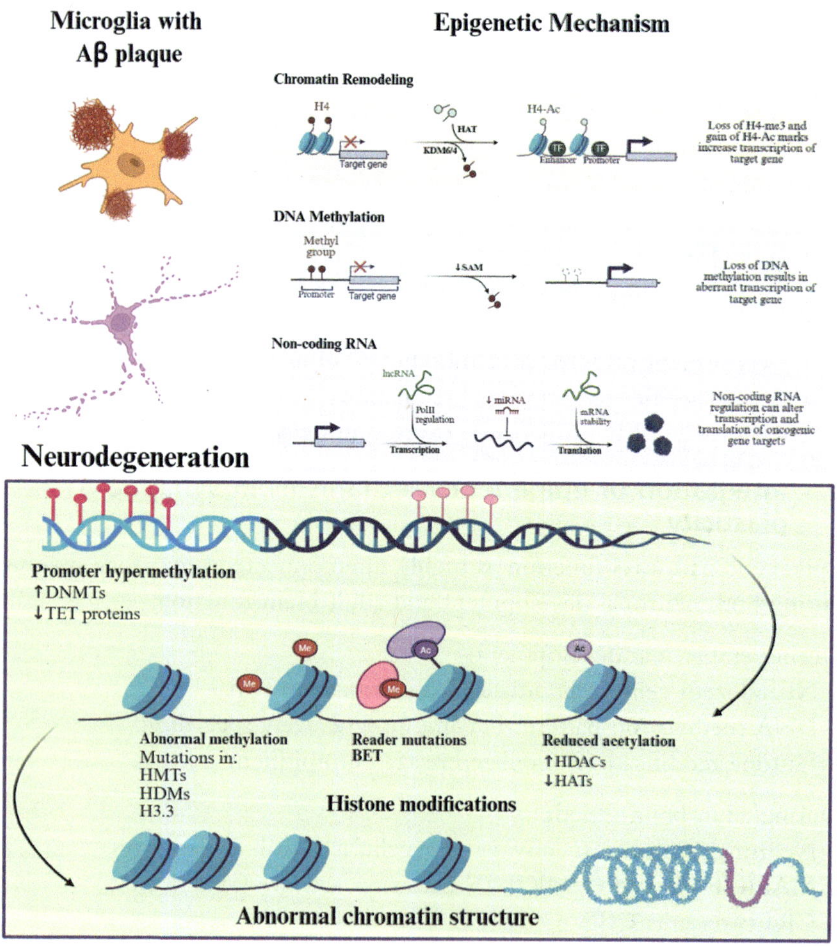

Fig. 1 Epigenetic mechanisms of neurodegeneration.

2.6 Epigenetic dysregulation in neurodegenerative disorders

Aging and environmental stressors progressively alter epigenetic profiles and increase susceptibility to neurodegenerative diseases. Early-life environmental exposures may induce persistent epigenetic modifications through "epigenetic priming," which can remain latent until later life events trigger pathological outcomes.
In AD, dysregulation of epigenetic pathways produces:

- Histone hypoacetylation
- Altered DNA methylation patterns
- Abnormal chromatin structure
- Disrupted transcription of synaptic genes

These changes silence genes required for synaptic transmission, long-term potentiation, and neuronal survival. Ultimately, these molecular alterations contribute to progressive cognitive decline and memory impairment (Gräff et al., 2011). Epigenetic regulation within the CNS involves a dynamic interaction among DNA methylation, histone modifications, chromatin remodeling, and non-coding RNA networks. Together, these mechanisms regulate neuronal development, synaptic plasticity, and cognitive function. Dysregulation of these pathways contributes significantly to the pathogenesis of neurocognitive disorders, which highlights epigenetic modulation as an important therapeutic strategy.

3. Epigenetic dysfunction in AD

AD is a progressive neurodegenerative disorder characterized by cognitive decline, memory impairment, and behavioral dysfunction. The principal neuropathological hallmarks of AD include:

- Extracellular deposition of amyloid-β (Aβ) plaques
- Intracellular aggregation of hyperphosphorylated tau protein in the form of neurofibrillary tangles (NFT) (Woźniak et al., 2024).

Although these pathological features have been extensively investigated, the precise molecular mechanisms underlying AD remain incompletely understood because of the multifactorial nature of the disease and the complex interplay between genetic and environmental factors. Increasing evidence indicates that epigenetic dysregulation is a central contributor to

AD pathogenesis. Aberrant alterations in DNA methylation, histone modifications, and non-coding RNA (ncRNA)-mediated gene upregulation disrupt transcriptional homeostasis in neuronal and glial cells, thereby influencing pathways associated with amyloid precursor protein (APP) processing, tau phosphorylation, synaptic plasticity, neuroinflammation, and neuronal survival (Behl et al., 2024). Consequently, epigenetic mechanisms have emerged as promising therapeutic targets for neurodegenerative disorders.

3.1 DNA methylation in AD

DNA methylation is one of the most extensively studied epigenetic mechanisms implicated in AD. It typically occurs at CpG dinucleotides and is mediated by DNA methyltransferases (DNMTs), resulting in transcriptional repression when present within promoter regions. In contrast, DNA hypomethylation is commonly associated with increased transcriptional activity. Studies of AD brains have demonstrated global DNA hypomethylation together with locus-specific hypermethylation, leading to widespread transcriptional dysregulation (Gao et al., 2022).

Several AD-associated genes exhibit aberrant methylation profiles. Hypermethylation within the promoter and intronic regions of the APP gene alters APP processing and promotes increased production of amyloidogenic beta peptides (Behl et al., 2024). Similarly, hypomethylation of the presenilin 1 (PSEN1) and presenilin 2 (PSEN2) genes encodes catalytic components of the γ-secretase complex, enhances gene expression, and contributes to excessive Aβ generation and plaque deposition. Epigenetic alterations are also evident in the β-site APP-cleaving enzyme 1 (BACE1) gene. DNA hypermethylation at CpG sites located within intron 1 disrupts U1 small nuclear ribonucleoprotein (U1 snRNP) binding and interferers with normal splicing, increasing BACE1 expression and exacerbating amyloid pathology (Tiwari et al., 2025).

The apolipoprotein E (APO E) gene represents another major target of epigenetic dysregulation in AD. The APOε4 allele is the strongest genetic risk factor for sporadic AD, and hypomethylation of the APO E promoter is associated with elevated APO E expression, impaired Aβ clearance, and enhanced neuroinflammatory responses (Behl et al., 2024; Gao et al., 2022). Also, age-related changes and environmental factors further contribute to these methylation abnormalities, emphasizing the dynamic nature of epigenetic regulation in AD.

3.2 Histone modifications in AD

Histone modifications constitute another critical layer of epigenetic regulation disrupted in AD. Histones are octameric proteins around which DNA is wrapped to form nucleosomes, the fundamental structural units of chromatin. Post-translational modifications of histone N-terminal tails, including acetylation, methylation, phosphorylation, and ubiquitination, regulate chromatin accessibility and gene transcription (Kitazawa et al., 2023). These modifications determine whether chromatin adopts a transcriptionally permissive euchromatin state or a condensed heterochromatin configuration (Payne, 2014).

Histone acetylation, mediated by histone acetyltransferases (HATs) and histone deacetylases (HDACs), is essential for synaptic plasticity, learning, and memory formation. In AD, dysregulation of histone acetylation contributes significantly to cognitive impairment. Elevated expression of HDAC2 has been identified in both AD patients and AD mouse models (Chouliaras et al., 2013; Gräff et al., 2012). HDAC2 overexpression promotes deacetylation of histones H3 and H4 at lysine residues such as H3K9 and H4K12, resulting in transcriptional repression of genes associated with synaptic function and memory (Asadi et al., 2025; Guan et al., 2009). Genes affected include brain-derived neurotrophic factor (BDNF), early growth response protein 1 (EGR1), glutamate ionotropic receptor AMPA type subunit 1 (GLUR1), and activity-regulated cytoskeleton-associated protein (ARC). Reduced expression of these genes contributes directly to synaptic dysfunction and progressive cognitive decline in AD (Gräff et al., 2012).

Histone methylation also plays an important role in AD pathogenesis. Depending on the modified lysine residue and the degree of methylation, histone methylation can either activate or repress transcription (Behl et al., 2024). Trimethylation of histone H3 lysine 4 (H3K4me3), an activating transcriptional mark, is elevated in the prefrontal cortex of AD patients. Increased H3K4me3 mediated by methyltransferases such as SETD1A/B and mixed-lineage leukemia proteins (MLL1–4) enhances serum- and glucocorticoid-regulated kinase 1 (SGK1) signalling, thereby promoting tau hyperphosphorylation and neurofibrillary tangle formation (Saleem et al., 2026). In contrast, increased dimethylation of histone H3 lysine 9 (H3K9me2), a transcriptionally repressive modification catalyzed by the G9a/GLP methyltransferase complex, suppresses glutamate receptor genes, including GRIA2 and GRIN2B, impairing glutamatergic neurotransmission and memory processes (Coneys & Wood, 2020; Park et al., 2022).

Recent studies have identified additional histone modifications implicated in AD, including crotonylation and lactylation. Reduced histone H3 lysine 27 crotonylation (H3K27cr) disrupts the expression of genes involved in autophagy and endocytosis, thereby impairing Aβ clearance by glial cells (Park et al., 2022; Wang et al., 2021). Conversely, elevated histone H4 lysine 12 lactylation (H4K12la) in microglia activates glycolytic pathways and nuclear factor kappa B (NF-κB) signalling, contributing to sustained neuroinflammatory responses and neuronal damage (Li et al., 2025).

3.3 Non-coding RNAs in AD

Non-coding RNAs, including microRNAs (miRNAs) and long non-coding RNAs (lncRNAs), are important regulators of post-transcriptional gene expression and epigenetic signaling pathways in AD (Tiwari et al., 2025). MiRNAs are short RNA molecules that bind complementary sequences within the 3′ untranslated regions of target messenger RNAs, resulting in translational repression or mRNA degradation. Dysregulated miRNA expression profiles have been strongly associated with both amyloid and tau pathology in AD (Santa-Maria et al., 2015).

Downregulation of miR-29a, miR-29b, and miR-107 removes inhibitory regulation of BACE1 expression, thereby increasing Aβ production (Zhang et al., 2021). Similarly, reduced expression of miR-132 and miR-212 contributes to sustained activation of glycogen synthase kinase 3β (GSK3β) and cyclin-dependent kinase 5 (CDK5), two major kinases involved in tau hyperphosphorylation. Dysregulated miRNAs also influence neuroinflammatory pathways. Upregulation of miR-146a and miR-125b suppresses complement factor H expression, thereby promoting persistent inflammatory responses and neuronal injury.

LncRNAs function as molecular scaffolds for chromatin-modifying complexes and play essential roles in transcriptional regulation (Zhang et al., 2021). The lncRNA BACE1 antisense transcript (BACE1-AS) is significantly upregulated in AD and stabilizes BACE1 mRNA, thereby amplifying amyloidogenic processing. In contrast, reduced expression of nuclear enriched abundant transcript 1 (NEAT1) decreases P300-mediated histone acetylation at the frizzled class receptor 3 (FZD3) promoter, suppressing Wnt signaling and indirectly facilitating tau hyperphosphorylation (Li et al., 2019).

3.4 Epigenetic regulation of peripheral immune responses in AD

Emerging evidence indicates that the peripheral immune system plays an important role in AD pathogenesis and is subject to extensive epigenetic regulation. Advanced single-cell sequencing studies have revealed widespread epigenetic reprogramming of peripheral monocytes and CD8+ T cells in AD patients (Xu & Jia, 2021). CD8+ memory T cells exhibit altered chromatin accessibility at regulatory regions associated with the CXC chemokine receptor 3 (CXCR3) gene. Increased CXCR3 expression promotes migration of cytotoxic T cells across the blood–brain barrier into the AD brain, where they interact with Aβ plaque-associated microglia and modulate neuroinflammatory responses (Constantin et al., 2024).

Peripheral monocytes also display disease-specific changes in chromatin accessibility. In particular, an AD-associated RELA transcription factor-binding site adjacent to an open chromatin region within the NF-κB pathway enhances pro-inflammatory signaling cascades. Importantly, these epigenetic abnormalities demonstrate gene-dose dependence on the APOEε4 allele. AD patients homozygous for APOEε4 exhibit increased chromatin accessibility at inflammatory cytokine genes, including CCL4L2 and CXCL2, as well as at the ATP-binding cassette transporter A1 (ABCA1) gene, which is involved in apolipoprotein lipidation (Ramakrishnan et al., 2024). These findings highlight the synergistic interaction between genetic susceptibility factors and epigenetic dysregulation in promoting chronic neuroinflammation and neurotoxicity.

In conclusion, the pathogenesis of AD is unequivocally intertwined with pervasive epigenetic dysfunction. The intricate disruptions across DNA methylation landscapes, histone modification profiles, and non-coding RNA regulatory networks do not operate in isolation; rather, they form a highly interconnected web that actively drives the molecular hallmarks of the disease (Tiwari et al., 2025). Methylation and histone abnormalities directly upregulate the machinery responsible for Aβ and tau accumulation, while simultaneously repressing the genes required for synaptic plasticity, memory consolidation, and neuronal survival (Gao et al., 2022). Concurrently, dysregulated non-coding RNAs post-transcriptionally amplify these toxic pathways and suppress neuroprotective mechanisms. The expanding understanding of how these central and peripheral epigenetic changes synergize with established genetic risk factors, such as the APOEε4 allele, provides a transformative perspective on disease etiology. Pharmacological agents targeting epigenetic enzymes hold immense potential to halt neurodegeneration. The schematic representation of the role of Epigenetic regulation in the AD brain was reported in Fig. 2.

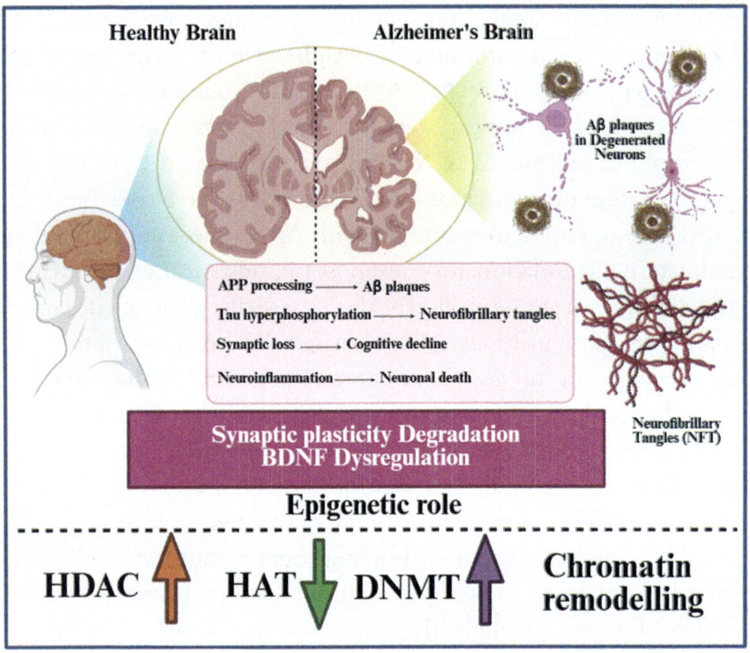

Fig. 2 Schematic representation of the role of Epigenetic regulation in the AD brain.

3.5 Experimental assays and analytical methods used in epigenetic studies of AD

The following assays and analytical methods are commonly employed to investigate epigenetic dysfunction and evaluate therapeutic interventions in AD models. Epigenetic abnormalities have been recognized as important in the pathogenesis and progression of Alzheimer's disease (AD). To explore DNA methylation, histone modifications, and non-coding RNA-mediated regulation in neurodegeneration, various experimental assays and analytical approaches are utilized. These methods can give insight into changes in gene expression, chromatin remodeling, and epigenetic markers related to the pathobiology of AD.

3.5.1 DNA methylation analysis

3.5.1.1 Bisulfite sequencing

Principle:

The principle behind bisulfite sequencing is that unmethylated cytosines are converted to uracil by sodium bisulfite, whereas methylated cytosines are unaffected. Uracils are converted to thymine during PCR amplification, allowing differentiation between methylated and unmethylated cytosines by DNA sequencing. This method allows for single-base resolution analysis of DNA methylation of genes involved in Alzheimer's disease [amyloid precursor protein (APP), presenilin 1 (PSEN1), presenilin 2 (PSEN2), beta-site amyloid precursor protein cleaving enzyme 1 (BACE1) and apolipoprotein E (APOE)].

Reagents Required:

- Isolation kit for genomic DNA
- Sodium bisulfite conversion kit
- Sodium bisulfite reagent
- Hydroquinone
- Sodium hydroxide (NaOH)
- Ethanol
- PCR master mix
- Bisulfite-specific primers
- Agarose gel electrophoresis chemicals.
- DNA sequencing reagents

Process.

1. **DNA Isolation and Fragmentation**
   - Extraction: Pure genomic DNA is isolated from the target biological sample (e.g., tissue, cells, or blood).
   - Shearing: DNA is physically or enzymatically fragmented into lengths optimal for sequencing.
   - Adapter Ligation: For Next-Generation Sequencing (NGS) workflows, specific sequencing adapters are ligated to the ends of the fragments.
2. **Bisulfite Conversion (The Core Chemical Step)**

   The fragmented DNA undergoes a chemical conversion protocol consisting of three main phases:
   - Denaturation: The double-stranded DNA is melted using heat or a chemical agent (like NaOH) into single strands, as bisulfite can only react with single-stranded DNA.
   - Sulfonation & Deamination: The single-stranded DNA is incubated with sodium bisulfite under hot, acidic conditions. This converts unmethylated cytosines into uracil-sulfonate intermediates. Methylated cytosines (5-mC or 5-hmC) are sterically protected and remain completely intact.

- Desulfonation: An alkaline solution is introduced to remove the sulfonate groups, finalizing the conversion of the unmethylated bases into standard uracil.

3. **PCR Amplification**
   - Uracil to Thymine Transformation: The converted single-stranded DNA is amplified using Polymerase Chain Reaction (PCR). During this amplification, DNA polymerase treats the chemical uracils as thymines (T), synthesizing a complementary adenine (A) strand.
   - Final Sequence Result: By the end of PCR, original unmethylated cytosines (C) have completely mutated into thymines (T), while original methylated cytosines remain as cytosines (C). Specialized polymerases are typically used to successfully bypass uracil and manage the high AT-content of the converted DNA.
4. **Sequencing**
   - Platform Loading: The amplified library is loaded onto a high-throughput NGS platform or processed via Sanger sequencing for targeted single-gene regions.
   - Data Output: The sequencer reads out millions of short-read fragments consisting of (A), (G), (T), and only the modified (C) bases.
5. **Bioinformatic Alignment and "Methylation Calling"**
   - Reduced Complexity: Because most cytosines convert to thymines, the physical genomic sequence loses complexity, converting from a 4-base language (A, T, C, G) into a mostly 3-base language (A, T, G).
   - Algorithmic Alignment: Specialized alignment programs, such as Bismark, are used to map these altered sequences back to an original, unconverted reference genome.
   - Methylation Score Calculation: The software evaluates every single cytosine position. The percentage of methylation at any given locus is determined using a simple formula:

Methylation (%) = (Number of unconverted C-reads/Total C+T reads at that position)*100 (Darst et al., 2010).

3.5.1.2 Methylation-specific PCR (MSP)

Principle:

The principle behind methylation-Specific PCR is to modify the DNA by bisulfite treatment and then selectively amplify DNA containing methylated or unmethylated sequences with specific primer sets. Differential amplification patterns show the state of methylation of target genes associated with AD progression.

Reagents Required:
- Bisulfite-treated DNA
- Methylated-specific primers
- Unmethylated-specific primers
- PCR master mix
- Taq DNA polymerase
- dNTPs
- Agarose gel reagents
- DNA ladder

Process:
1. The genomic DNA is isolated and bisulfite converted.
2. There are different primer sets for methylated and unmethylated DNA.
3. Both sets of primers are used to amplify the material by PCR.
4. Separation of amplified products is done by agarose gel electrophoresis.
5. Bands are detected by the presence or absence of bands in the gel (Herman et al., 1996).

3.5.1.3 Pyrosequencing

Principle:

Pyrosequencing is a 'sequencing by synthesis' technique for quantitative analysis of CpG methylation. During the synthesis of DNA, pyrophosphate is released, which initiates an enzyme cascade resulting in light signals that are detectable and correspond to the amount of nucleotide that was incorporated.

Reagents Required:
1. Bisulfite-converted DNA
2. PCR reagents
3. Biotinylated primers
4. Streptavidin-coated beads
5. Sequencing primers
6. Pyrosequencing enzyme mixture
7. Luciferase substrate reagents

Process:
1. Biotinylated primers are used to amplify bisulfite-converted DNA.
2. The PCR products are bound to the streptavidin-coated beads.

3. To obtain single-stranded templates, double-stranded DNA is denatured.
4. Primers are attached by annealing.
5. Pyrosequencing is the sequential addition of nucleotides.
6. The incorporation of a nucleotide within the molecules produces light signals, which are measured.
7. The percentage of CpG methylation is determined quantitatively (Ramon et al., 2003).

3.5.1.4 Whole-genome bisulfite sequencing (WGBS)

Principle:

Whole-Genome Bisulfite Sequencing is a combination of sodium bisulfite conversion and next-generation sequencing that enables genome-wide DNA methylation profiling at single-base resolution. This approach allows the identification of differentially methylated regions and new epigenetic markers for AD.

Reagents Required:

- High-quality genomic DNA
- DNA fragmentation reagents
- Bisulfite conversion kit
- Library preparation kit
- PCR amplification reagents
- Next-generation sequencing reagents

Process:

1. Genomic DNA is isolated and cut up.
2. DNA fragments are treated with bisulfite.
3. Sequencing libraries are prepared.
4. Genes are amplified in a Polymerase Chain Reaction.
5. High-throughput sequencing is carried out.
6. Genome-wide methylation pattern analysis is performed by bioinformatics analysis (Vargas-Landin et al., 2024).

3.5.1.5 Methylation assays using chromatin immunoprecipitation (ChIP)

Principle:

ChIP-based methylation assays involve taking advantage of antibodies against methylated histones, or chromatin-binding proteins, to selectively immunoprecipitate chromatin-protein complexes. This is followed by qPCR or sequencing for the identification of regions of chromatin that are methylated and the epigenetic regulation of AD.

Reagents Required:
- Formaldehyde
- Glycine
- Cell lysis buffer
- Sonication buffer
- Antibodies to methylated histones
- MP-Beads Protein A/G
- DNA purification kit
- The reagents used for qPCR or sequencing.

Process:
1. Formaldehyde is used to cross-link cells or tissues.
2. Isolation and fragmentation of chromatin by sonication.
3. Specific antibodies are used to incubate chromatin fragments.
4. Protein A/G beads can be used for the capture of immune complexes.
5. Reversing cross-links and purification of DNA.
6. The enriched DNA fragments are then sequenced or analyzed by qPCR (Li & Tollefsbol, 2011).

3.5.2 Histone modification analysis

3.5.2.1 Chromatin immunoprecipitation (ChIP) assays

Principle:

Chromatin immunoprecipitation assays enrich for specific DNA regions using antibodies against acetylated or methylated histone residues, such as H3K9ac, H4K12ac, H3K4me3, and H3K9me2. This technique can be used to assess chromatin remodeling and transcriptional regulation in AD.

Reagents Required:
- Formaldehyde
- ChIP lysis buffer
- Sonication reagents
- Histone modification-specific antibodies
- Protein A/G beads
- DNA purification kit
- qPCR reagents

Process:
1. Chromatin is broken and crosslinked.
2. Immunoprecipitation is performed with antibodies specific for.
3. Isolation of complexes of DNA with histones.

4. Purification of DNA fragments.
5. The ChIP-qPCR or ChIP-seq is carried out (Milne et al., 2009).

3.5.2.2 Western blotting

Principle:

Western blotting involves the separation of proteins by SDS-PAGE and the immunodetection of the separated proteins by specific antibodies. It is commonly employed to identify the acetylation, methylation, and HDAC expression in AD models.

Reagents Required:
1. Protein extraction buffer
2. SDS-PAGE reagents
3. PVDF membrane
4. Blocking buffer
5. Primary antibodies
6. Secondary antibodies
7. ECL detection reagent

Process:
1. Samples are used to isolate proteins.
2. SDS-PAGE is used to separate protein samples.
3. Proteins are transferred onto PVDF membranes.
4. Primary antibodies are blocked and incubated with the membrane(s).
5. Secondary antibodies are added.
6. Chemiluminescence is used to see the protein bands (Pulya et al., 2021).

3.5.2.3 Immunohistochemistry and immunofluorescence

Principle:

These antibody-based methods identify and map the histone modifications or the epigenetic proteins in tissue sections with chromogenic or fluorescent tags.

Reagents Required:
- Paraformaldehyde
- Tissue embedding reagents
- Blocking serum
- Primary antibodies
- Secondary antibodies (either fluorescent or HRP-conjugated)
- DAPI
- Mounting medium

Process:

1. Fixed and cut brain tissue/neuronal cells – SHSY5Y, primary neuronal cells, Neuro2A.
2. Tissue sections are permeabilized and blocked.
3. Primary antibodies are added and incubated.
4. Secondary antibodies are added.
5. Microscopy is used to visualize signals (Zhang et al., 2022).

3.5.2.4 HDAC activity assays

Principle:

HDAC activity assays are based on the enzymatic deacetylation of acetylated substrates by HDAC, which results in a fluorescent or colorimetric signal that is directly proportional to HDAC activity.

Reagents Required:

1. HDAC assay kit
2. A fluorogenic or colorimetric substrate.
3. Assay buffer
4. HDAC inhibitors
5. Microplate reader

Process:

1. Enzymes of HDACs are isolated as protein lysates.
2. Add substrates that have been acetylated.Add acetylated substrates.
3. Reaction mixtures are allowed to incubate.
4. The fluorescence or absorbance is measured.
5. The activity of HDAC is measured and compared (Pulya et al., 2021).

3.5.2.5 Mass spectrometry-based proteomics

Principle:

Using mass spectrometry-based proteomics, the mass-to-charge ratios of peptides derived from enzyme digestion are used to identify post-translational modifications of histones (Zhao et al., 2025).

Reagents Required:

1. Histone extraction buffer
2. Trypsin enzyme
3. LC-MS grade solvents
4. Chromatography columns
5. Proteomics analysis software

Process:

1. The proteins (histone proteins) are extracted and purified.
2. Trypsin is an enzyme that breaks down proteins to peptides.
3. Liquid chromatography is used to separate peptides.
4. Analyses by mass spectrometry are carried out.
5. Histone modifications are detected and measured (Huang et al., 2015).

3.5.3 Non-coding RNA analysis

3.5.3.1 Real-time PCR (qRT-PCR)

Principle:

Quantitative Real-Time PCR (qRT-PCR) is a technique that includes reverse transcription of RNA to cDNA and amplification with fluorescent probes or fluorescent dyes in real time. Fluor intensity is proportional to levels of target ncRNAs.

Reagents Required:

- RNA isolation kit
- Reverse transcription kit
- SYBR Green or TaqMan master mix
- miRNA/lncRNA-specific primers
- RNase-free water

Process:

1. Tissues or cells are used for isolating total RNA.
2. The cDNA is created from RNA.
3. Specific primers are used for qPCR amplification.
4. Real-time measurement of fluorescence signals.
5. The Ct values are used for relative gene expression (Zhou et al., 2021).

3.5.3.2 RNA sequencing (RNA-seq)

Principle:

A high throughput sequencing approach that is used to study the whole transcriptome expression of coding and non-coding RNAs (Kaur et al., 2025).

Reagents Required:

1. RNA extraction kit
2. Reagents for assessing the quality of RNA.
3. Library preparation kit
4. Sequencing reagents
5. Bioinformatics software

Process:

1. Good quality RNA is recovered.
2. Integrity of the RNA is determined.
3. cDNA libraries are made.
4. NGS platforms are used for sequencing.
5. Dysregulated ncRNAs are identified using bioinformatics analysis (Ng et al., 2024).

3.5.4 Neuroinflammation and immune profiling assays

3.5.4.1 Flow cytometry

During neuro-inflammation, immune cell subsets, inflammatory mediators, and cellular activation profiles are characterized using highly sensitive and quantitative flow cytometry. The assay allows multiparametric analysis of surface and intracellular markers in heterogeneous cell populations from brain tissues, peripheral blood, spleen, or cultured cells.

Principle.

The principle of flow cytometry is hydrodynamic focusing, which allows single-cell suspensions to flow one cell at a time through the laser beam. The antibodies are labeled with fluorophores, and the antibodies bind to specific antigens present on the cells, which will generate a fluorescent signal that is detected and quantified. The technique enables the simultaneous detection of immune cell phenotypes, cytokine expression, apoptotic markers, oxidative stress indicators, and microglial and astrocytic activation sites.

Reagents and Materials.

A. Biological samples

1. Brain tissues: hippocampus, cortex, or whole brain homogenates.
2. Mesenchymal stem cells (MSCs)
3. Splenocytes
4. Neuronal or glial cells in culture
5. Cerebrospinal fluid (optional)

B. Buffers and Solutions

1. Phosphate Buffered Saline (PBS)
   - pH: 7.2–7.4
   - Sterile and calcium/magnesium-free
2. FACS Buffer

   Composition:
   - PBS
   - 1–2 % fetal bovine serum (FBS)

- 0.1 % sodium azide

  Purpose:

- Free from nonspecific antibody binding and maintains cell integrity

3. RBC Lysis Buffer

   Commercial or prepared using:

   - Ammonium chloride
   - Potassium bicarbonate
   - EDTA
   - Purpose:
   - Blood sample erythrocytes removed

4. Cell Fixation Buffer

   - 2–4 % paraformaldehyde in PBS
   - Purpose:
   - Maintenance of cell shape and integrity

5. Permeabilization Buffer

   Common reagents:

   - Saponin
   - Triton X-100
   - Tween-20
   - Purpose:
   - Intracellular staining for cytokines/transcription factors

C. Antibodies

Surface Markers of Neuroinflammation

Marker Cell Type/Function

CD11b Microglia/macrophages

CD45 Leucocyte marker

CD68 Activated microglia

CD206 M1 inflammatory phenotype

Marker Cell Type/Function

GFAP Astrocyte activation

CD3 T lymphocytes

There are cells called CD4 + helper T cells.

CD8 – Cytotoxic T cells

CD19 B lymphocytes

Intracellular Cytokine Markers

Marker Function

TNF-α Pro-inflammatory cytokine

IL-1β Neuroinflammatory mediator

IL-6 Inflammatory cytokine

IFN-γ Immune activation

IL-10 Anti-inflammatory cytokine

Fluorophores Commonly Used:

- FITC
- PE
- APC
- PerCP
- Alexa Fluor dyes

D. Additional Reagents

1. Fc receptor blocking agent
2. Viability dyes (7-AAD, propidium iodide, Zombie dyes)
3. Brefeldin A or Monensin (intracellular cytokine accumulation) equipped
4. Trypan blue
5. Collagenase and DNase I (for brain tissue dissociation)

E. Equipment and Consumables.

1. Flow cytometer with multiple lasers
2. Centrifuge
3. Cell strainer (40–70 μm)
4. Refrigerated microcentrifuge
5. Haemocytometer or other automated cell counter
6. Polystyrene FACS tubes or 96 well plates.
7. Biosafety cabinet
8. Vortex mixer
9. Ice bucket

Experimental Procedure.

A. Preparation of Single-Cell Suspension

1. Brain Tissue Isolation
   a) Animals are euthanized according to institutional ethical guidelines
   b) Brain tissues are preferably dissected under sterile conditions.
   c) Hippocampus and cortex are separated and cut apart into smaller bits.
2. Tissue Dissociation
   a) In the digestion buffer: Tissue fragments are incubated in a digestion buffer that contains:
      - Collagenase (1–2 mg/mL)
      - DNase I (50–100 μg/mL)

b) The samples are then incubated for 20–45 min with gentle shaking at 37 °C.
c) Digested tissue mechanically tritrated by pipettes

3. Filtration
   a) Suspension of cells is filtered using a 70 μm cell strainer.
   b) Filtrate is collected in the chilled centrifuge tubes.
4. Centrifugation
   a) Cells are centrifuged at 300–400 X *g* for 5–10 min at 4 °C.
   b) Supernatant is removed
   c) The cell pellet is then re-suspended in FACS buffer.

   After collection, the number of cells and their viability are determined. Trypan blue is added to a mixture of cells. The number of viable cells is determined by hemocytometer

   1. The concentration of cells is adjusted to about 1 × 106 cells/mL.

B. Fc Blocking

Blocking reagent for FC receptor is added to the cells and incubated for 10–15 min at 4 °C. This reduces non-specific antibody binding as a step.

C. Surface Marker Staining

A cell suspension is mixed with:

1. Fluorophore-conjugated antibody
2. Antibody concentrations are appropriately applied according to manufacturer's instructions.
3. Samples are then kept in the dark for 20–40 min at 4 °C.

Cells have to be washed twice with FACS buffer. Centrifugation is done at 300 X *g* for 5 min.

D. Intracellular Cytokine Staining

1. Cytokine Induction
   a) Cells are stimulated using:
      - PGE2 (prostaglandin E2)
      - Ionomycin
   b) Cytokine secretion is blocked by adding brefeldin A or monoensin.
   c) The incubation lasts 4–6 h at 37 °C.
2. Fixation and Permeabilization
   a) Cells are paraformaldehyde fixed for 15–20 min.
   b) Washing and ermeabilization of cells.

3. Intracellular Staining
   a) Cytokine-specific antibodies are added.
   b) Incubation takes place for 30 min at 4 °C and in the dark.
   c) Cells are washed and resuspended in PBS.

E. Acquisition using Flow Cytometry

1. Single-stained controls are used for Instrument compensation.
2. Unstained and fluorescence-minus-one (FMO) controls are included.
3. At least 10,000–100,000 events/sample are acquired.
4. Forward scatter (FSC) and side scatter (SSC) parameters are used to exclude debris and aggregates.

Data Analysis

Gating Strategy:

1. Using FSC/SSC to exclude debris.
2. Doublet discrimination.
3. Live-cell gating.
4. Characterization of immune cells:
   - CD11b + microglia
   - Leukocytes are infiltrating (CD45 high)
   - CD3 + T cells
5. Quantification of cytokine positive cells

Interpretation of Results

Increased Neuroinflammation can be observed as:

- Higher CD68, CD86, TNF-α, IL-1β, and IL6 levels of expression.
- Peripheral immune cells infiltrated more.
- Enhanced microglial activation.
- Anti-inflammatory and/or Neuroprotective Effects can be observed as:
- Up-regulated expression of CD206 and IL-10.
- Reduced pro-inflammatory cytokines.
- Reduced numbers of activated microglia.

Precautions.

1. Keep samples cool as much as possible.
2. Keep fluorophore-conjugated antibodies out of the light.
3. Do not mechanically dissociate too much, as it will cause cell death.
4. Properly use compensation controls to reduce spectral overlap.
5. Prepare and analyse fresh samples for optimum results. (Siddiqui & Liválk, 2023)

3.5.4.2 single-cell RNA sequencing (scRNA-seq)

Principle:

Single-cell RNA sequencing (scRNA-seq) is a high-throughput transcriptomic method to study cellular gene expression. It includes single viable cell isolation, cellular mRNA isolation, and reverse transcription to complementary DNA (cDNA), amplification, and sequencing with cell-specific barcodes, allowing the identification of cellular heterogeneity, rare cell populations, and transcriptional changes linked to Alzheimer's disease.

Reagents Required:

1. Fresh brain tissue, cultured neural cells, or isolated nuclei
2. Sandwich assembly of the plates and membranes to create a microgravity environment
3. Cell culture media and solutions for cell suspension and culture.
4. Sodium citrate buffer
5. Barcoded beads for a single-cell capture platform
6. Reverse transcription and cDNA amplification reagents are used to convert RNA into DNA.
7. Library preparation kit, sequencing reagents and software for bioinformatics

Process:

Single-cell suspensions of viable brain tissue or cells are obtained by dissociation. Single cells are isolated and "barcoded". The cellular mRNA is reverse transcribed to form cDNA and amplified. Sequencing libraries are created, and next-generation sequencing is performed. Bioinformatics analysis reveals cellular sub-populations associated with disease and gene expression changes specific to each cell type in AD (Jovic et al., 2022).

3.5.4.3 Assay for transposase-accessible chromatin using sequencing (ATAC-seq)

Principle:

ATAC-seq is a fast epigenomic method for genome-wide mapping of open chromatin regions that are linked to active promoters, enhancers, and regulatory elements. Uses hyperactive Tn5 transposase to both fragment accessible chromatin and to insert sequencing adapters into exposed DNA regions to analyze chromatin accessibility and gene regulatory changes important to Alzheimer's disease.

Reagents Required:

1. Brain tissue, cultured neural cells or isolated nuclei that are not fresh.
2. DNA isolation buffer and kit.

3. The hyperactive Tn5 transposase and the Tn5 transposomes that are loaded with adapters.
4. PCR master mix and indexing primers
5. Magnetic beads or DNA purification kit
6. Reagents and bioinformatics software for next generation sequencing

Process:

Brain tissue or cells are taken, and with care taken, nuclei are isolated and chromatin maintained. Tagmentation of accessible chromatin with Tn5 transposase in isolated nuclei. Adapter tags are used to purify and PCR-amplify DNA fragments. All libraries are quality-checked and subjected to high-throughput sequencing. Sequencing data analyzed for open chromatin regions and epigenetic changes related to Alzheimer's disease (Miskimen et al., 2017).

3.5.4.4 Enzyme-linked immunosorbent assay (ELISA)

Principle:

ELISA is an antibody-based immunoassay that involves the binding of antigen and antibody on the solid surface of a microplate for the detection and quantification of a specific protein or biomarker. A detection antibody with an enzyme linked to the antibody reacts with a chromogenic substrate to generate a measurable color signal that is proportional to the amount of target analyte. ELISA is widely applied to estimate biomarkers in Alzheimer's disease studies, like amyloid-beta, tau (total and phosphorylated), inflammatory mediators related to progression of the disease, and epigenetic regulation.

Reagents Required:

1. Serum, plasma, cerebrospinal fluid or tissue lysate samples.
2. ELISA microtiter plate.
3. Capture and detection antibodies.
4. Blocking buffer.
5. Washing buffer (PBS-T).
6. Enzyme substrate, e.g. TMB.
7. Stop solution and microplate, reader.

Process:

Samples and standards were placed in antibody-coated wells. Unbound material is washed off after incubation. A detection antibody is added, which is linked to an enzyme. Substrate is added for color development. The biomarker of interest is quantified by measurement of absorbance (Metzendorf et al., 2025).

3.5.5 Functional and behavioral assays in AD models

Behavioral and functional assays in rodent AD models are crucial to assess the therapeutic potential of epigenetic modulators, nanomedicines, and immunotherapeutic interventions, linking them to the mechanistic findings and cognitive and synaptic outcomes.

3.5.5.1 Morris water maze test (MWM)

Principle:

The Morris Water Maze is a behavioral test to assess spatial learning and memory in rats. It is based on how readily the animal learns and remembers the position of a not visible escape platform in cloudy water, and is widely used in Alzheimer's disease research as a model of cognitive dysfunction and epigenetic memory changes.

Reagents Required:

1. Circular water tank.
2. Hidden escape platform.
3. Water is kept at a controlled temperature (22–25 °C).
4. Water opaque with non-toxic white tempera paint/milk powder.
5. Visual cues around the testing room.
6. Stopwatch or automatic video tracking device to time the duration of the event.

Process:

The tank is filled with opaque water, and the escape platform is in the water below the surface. Rodents are put into the pool from various start locations and are given time to find the hidden platform. Escape latency and swim path are measured from repeated training sessions. During the probe trial, the platform is not placed, and the time spent in the target quadrant is measured. Preserved or improved spatial memory as shown by reduced escape latency and increased time spent in the target quadrant.

3.5.5.2 Novel object recognition test (NOR)

Principle:

The Novel Object Recognition test is a non-invasive behavioral assay that can be used to assess recognition memory and cognitive function in rodents. It relies on rodents' natural preference for a new object over a familiar object and is commonly used in Alzheimer's disease research to identify memory deficits that are epigenetically mediated without the use of aversive stimuli or food deprivation.

Reagents Required:

1. Open field arena or test box with opaque walls.
2. Two similar objects that the subject rodent is already familiar with for the habituation phase.
3. One new object for the test phase.
4. 70% ethanol between trials as a cleaning solution.
5. Stopwatch or automatic video tracking program.

Process:

Rodents are acclimated to the testing arena prior to the experiment for 10 min. In the sample phase, two similar objects are introduced in the arena, and the animal is given 10 min of free exploration time. One familiar object is swapped for a novel object after a set defined retention period. During the test phase, the time spent exploring each object is recorded. To assess recognition memory, a discrimination index is calculated, and a lower preference for the novel object means that there is some cognitive impairment.

3.5.5.3 Y-Maze and T-Maze Tests

Principle:

Behavioral assessment of spatial working memory and spontaneous alternation behavior in rodents is carried out by the Y-Maze and T-Maze tests, respectively. They rely on rodents' natural preference for not going back down arms of the maze they have already visited or for exploring less familiar/novel arms of the maze. The tests are commonly used in Alzheimer's disease research for the evaluation of impairments in working memory dependent on the function of the hippocampus and in short-term memory of spatial location caused by epigenetic aberrations.

Reagents Required:

1. Opaque arms Y- or T-shaped maze.
2. Visual spatial cues around the testing room.
3. For cleaning between trials: 3. 70 % ethanol or acetic acid.
4. Stopwatch or automatic video tracking system.

Process:

Rodents are made familiar with the testing room and maze prior to experimentation. In the sample phase, one arm is closed, and the animal is given access to the two open arms. Following the intertrial interval, the blocked arm is released, and all arms are available. Time spent in each arm and the number of entries into each arm are recorded. The novel arm is less explored due to the spatial working memory deficits characteristic of Alzheimer's disease.

3.5.5.4 Electrophysiological recordings

Principle:

The electrical activity and the synaptic transmissions of a neuron are measured by electrophysiological recording to assess functional changes related to Alzheimer's disease. These techniques assess parameters like long-term potentiation (LTP), field excitatory postsynaptic potentials (fEPSPs), and action potentials and deliver direct evidence of how epigenetic dysregulation can result in a lack of synaptic plasticity and memory-related neuronal dysfunction.

Reagents Required:

1. Acute brain slice or cultured neurons from the hippocampus.
2. Artificial cerebrospinal fluid (aCSF) gassed with 95 % O_2/5 % CO_2.
3. Glass recording and stimulating microelectrodes.
4. Patch Clamp amplifier and signal digitizer.
5. Setup for protocol of high-frequency stimulation.
6. Software for data acquisition and analysis.

Process:

Brain tissue is cut, and acute hippocampal slices are generated and kept in oxygenated aCSF. The electrodes are inserted in the CA1 or CA3 area of the hippocampus for recording and stimulating. Synaptic responses are measured after an electric stimulus has been delivered in order to get baseline measurements. LTP is induced by high-frequency stimulation. The amplitude and slope of the post-stimulation fEPSP are measured to assess synaptic plasticity and detect epigenetically-mediated deficits in Alzheimer's disease models.

3.5.5.5 Amyloid and tau quantification assays

Principle:

Assays for the detection and quantification of amyloid-beta (Aβ40, Aβ42) peptides and total and phosphorylated tau proteins in the brain tissue, cerebrospinal fluid, or plasma samples. These assays are antibody-based and detect using sandwich ELISA or western blotting and are essential to the evaluation of the pathological hallmarks of Alzheimer's disease that are epigenetically regulated.

Reagents Required:

1. Homogenates of brain tissue, CSF, or plasma samples.
2. Sandwich ELISA kits for Aβ40, Aβ42, total tau and phosphorylated tau (p-tau).
3. Anti-Amyloid and anti-Tau primary and secondary antibodies.
4. Protein extraction and lysis buffer.
5. SDS-PAGE and PVDF membrane for Western blot.
6. Microplate reader and ECL detection reagents.

Process:

Brain tissue or biological fluids are collected and processed under controlled conditions. Protein extracts are made and determined by standard procedure. Sandwich ELISA or western blot of the samples. Specific antibodies against Aβ and tau isoforms are used for detection. Signal intensity is measured and used to assess the contribution of epigenetics to the pathology of AD with respect to levels of Aβ and tau (Puzzo et al., 2014).

The experimental assay and its functions were reported in Fig. 3 along with a diagrammatic representation.

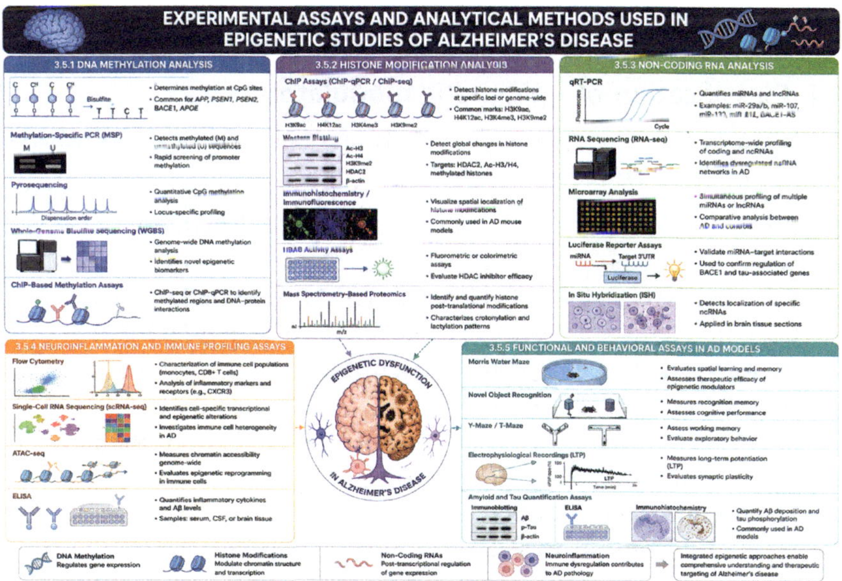

Fig. 3 Experimental process of assays and their functions.

4. Epigenetic modulators for neurodegenerative disorders

Epigenetic modulators constitute an adaptive approach for disease modification in neurodegenerative disorders, by reprogramming transcriptional pathways that control proteostasis, mitochondrial function, synaptic plasticity, and neuroinflammation. Alzheimer's disease (AD), Parkinson's disease (PD), Huntington's disease (HD), and ALS are such neurodegenerative disorders that share epigenetic abnormalities linked to DNA methylation, histone acetylation or methylation, and chromatin structure in neurons and glia. Dysregulation of HDACs, DNMTs, and the activity of bromodomain readers contributes to reduced plasticity-related gene expression, abnormalities in inflammatory gene expression, impairment of autophagy, and altered mitochondrial homeostasis. Small molecule epigenetic modulators- HDAC inhibitors, DNMT inhibitors, BET inhibitors, and HAT modulators can restore neuroprotective expression profiles, which often produce robust cognitive benefits and neuroprotection in preclinical models. The primary challenge in translation is to achieve isoform-, cell-, and brain-selective modulation with acceptable long-term safety, which motivates the development of multi-target ligands, brain-biased chemotypes, and novel approaches such as PROTACs/degraders (Prasanth et al., 2024).

4.1 Classification of epigenetic modulators

The primary classes of epigenetic modulators explored in neurodegeneration are classified as follows:

4.1.1 HDAC inhibitors (Class I, II, and IV) – Zinc-dependent

- Pharmacological targets include:
  - Class I HDACs: HDAC1, HDAC2, HDAC3, and HDAC8
  - Class IIa HDACs: HDAC4, HDAC5, HDAC7, and HDAC9
  - Class IIb HDACs: HDAC6 and HDAC10
  - Class IV HDACs: HDAC11
- HDAC inhibitors are chemically classified on the basis of the pharmacophore of the zinc-binding group (ZBG) into:
  - Hydroxamic acids: Vorinostat, Tubastatin A, Givinostat
  - Short-chain fatty acids: Valproate, Sodium butyrate
  - Benzamides: Entinostat, CI-994, RGFP966
  - Hydrazides: UF-010, SR-4370
  - Miscellaneous compounds: TYA-018, CG-34B, Santacruzamate A

- Mechanisms of action in neurodegenerative disorders:
  - Increase histone acetylation
  - Enhance expression of memory-related genes such as BDNF
  - Improve synaptic dysregulation
  - Inhibit tau hyperphosphorylation
  - Reduce Aβ plaque aggregation through autophagy
  - Downregulate NF-κB-mediated neuroinflammation
  - Promote neurogenesis
- Therapeutic relevance:
  - Demonstrated neuroprotective effects in Alzheimer's disease (AD) models
  - Improved cognitive and synaptic functions in preclinical studies (Hawash, 2025; Ji et al., 2025)

4.1.2 Sirtuin modulators (class III HDAC) – NAD + -dependent

- Sirtuins comprise:
  - SIRT1–SIRT7 enzymes
- Biological functions:
  - Regulation of mitochondrial function
  - Modulation of metabolism and oxidative stress
  - Maintenance of synaptic connectivity
- Sirtuin inhibitors are classified into:
  - Synthetic inhibitors
  - Natural inhibitors
  - NAD + -competitive or mechanism-based inhibitors

- **Synthetic Sirtuin Inhibitors**
  - Designed for selective inhibition of sirtuin isoforms
  - Examples: AK-7 and AGK-2: Selective SIRT2 inhibitors, EX527: Selective SIRT1 inhibitor.
    - Pharmacological effects:
      - Reduction of Aβ accumulation
      - Improvement of cognitive decline
      - Modulation of transcription and autophagy
      - Reduction of pathological protein accumulation

- **NAD + -Competitive and Mechanism-Based Inhibitors**
- Function as pan-SIRT inhibitors
  - Example: Nicotinamide
- Therapeutic outcomes: Amelioration of tau phosphorylation in triple-transgenic AD mouse models

- **Natural Sirtuin Inhibitors**
  - Example: Splitomicin (SIRT1 inhibitor)
    - Advantages: Lower toxicity, multi-target neuroprotective effects

- **Sirtuin Activators**
  - Examples: Resveratrol, SRT2104.
- Mechanisms:
  - Activation of SIRT1
  - Promotion of neuroprotection and cognitive benefits

- **Current research focus:**
  - Development of isoform-specific modulators
  - Exploration of SIRT1 and SIRT2 selective inhibitors in AD and PD models. (Bonomi et al., 2024; Razick et al., 2023; Thapa et al., 2024).

4.1.3 DNMT inhibitors

DNA methyltransferase (DNMT) inhibitors target abnormal DNA methylation associated with gene silencing in neurodegenerative disorders.

- DNMT classes include:
  - **DNMT1:** Maintenance methylation
  - **DNMT3A and DNMT3B:** *De novo* methylation
- Mechanism:
  - Catalyze transfer of methyl groups to cytosine residues in CpG islands
  - Result in transcriptional repression
- Therapeutic effects of DNMT inhibition:
  - Reverse abnormal hypermethylation
  - Restore genes associated with:
    - Synaptic plasticity
    - Neuronal survival
    - Memory formation
- DNMT inhibitors are classified into:

- **Nucleoside Analog Inhibitors**
  - Contain cytidine moieties incorporated into DNA during replication
  - Form covalent complexes with DNMT enzymes
  - Causes passive demethylation and DNMT degradation
  - Examples: 5-Azacytidine, Decitabine, Zebularine.
  - Therapeutic significance: Restoration of neuronal survival-associated genes

- **Non-Nucleoside DNMT Inhibitors**
  - Do not incorporate into DNA
  - More suitable for non-dividing neuronal cells
  - Safer for long-term administration
  - Examples: RG108, Hydralazine, Procainamide.
- **Natural DNMT Inhibitors**
  - Examples: Epigallocatechin-3-gallate (EGCG), Curcumin, Genistein
  - Neuroprotective effects:
    - Restoration of BDNF expression
    - Rebalancing of APP processing
    - Reduction of neuroinflammation (DNA methylation inhibitors: Retrospective & perspective view, 2021)

4.1.4 BET inhibitors

Bromodomain and Extra-Terminal domain (BET) inhibitors target BET family proteins:

- BRD2
- BRD3
- BRD4
- BRDT

- BET inhibitors are classified into:
  - Pan-BET inhibitors
  - Selective bromodomain inhibitors (BD1 or BD2 selective)
  - Bivalent inhibitors
- **Pan-BET Inhibitors**
  - Target both BD1 and BD2 bromodomains across all BET proteins
  - Examples: JQ1, I-BET762 (Molibresib), OTX015 (Birabresib)
    - Pharmacological effects:
      - Suppression of pro-inflammatory genes
      - Reduction of neuroinflammation
      - Improvement of cognitive function
- **BD1-Selective Inhibitors**
  - Improve specificity toward BET protein domains
  - Examples: MS436, Olinone, GSK789, GSK778
- Features:
  - Selective targeting of BRD4 BD1 domain
  - Olinone exhibits approximately 100-fold selectivity for BD1 over BD2

- **BD2-Selective Inhibitors**
  - Examples: RVX-208, GSK046, ABBV-744
  - Investigated in:
    - Inflammatory diseases
    - Cardiovascular disorders
    - Preclinical inflammation models
- The role of BD2 selectivity in neurodegenerative disorders remains insufficiently explored.

- **Bivalent BET Inhibitors**
  - Simultaneously engage BD1 and BD2 domains within the same BET protein
  - Example: AZD5153
  - Effects:
    - Stronger transcriptional repression,
    - Reduction of transcriptional dysregulation
  - Overall neuroprotective effects:
    - Reduction of microglial activation
    - Suppression of neuroinflammation
    - Modulation of aberrant transcription (Singh & Sartor, 2020)

4.1.5 HAT activators or modulators

Histone acetyltransferase (HAT) activators/modulators remain relatively underdeveloped but represent a promising class of epigenetic therapeutics.

- Reduced histone acetylation is associated with:
  - Impaired memory
  - Synaptic dysfunction
  - Neuronal loss

HAT activators promote histone acetylation and transcription of neuroprotective genes.

HAT modulators are classified into:

- Direct HAT activators
- Indirect activators
- Natural compounds
- Dual epigenetic modulators
- **Direct HAT Activators**
  - Example: CTPB [N-(4-chloro-3-trifluoromethylphenyl)-2-ethoxybenzamide]
    - Mechanism: Selective activation of p300/CBP, Enhancement of histone acetylation and transcriptional activity

Table 1 Epigenetic modulators, along with their structure, target selectivity, and disease/disorder, are explored in neurodegeneration.

HDAC inhibitors

Class	Compound	Structure	Target selectivity	Disease/Target (Explored)
I	CI994	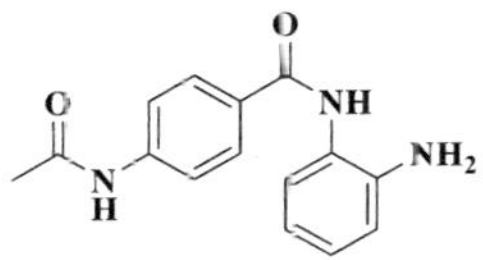	HDAC1/2/3	AD, Wnt/β-catenin signalling (Lu et al., 2026)
	BRD6929	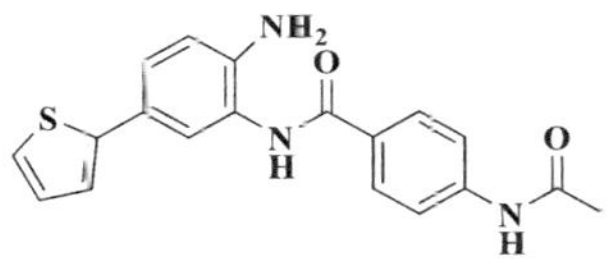	HDAC1/2	AD, Increase histone acetylation (Tammaro et al., 2024)
	BRD4884	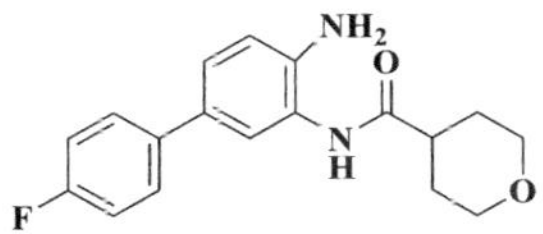	HDAC1/2	Cognitive enhancer (Wagner et al., 2015)
	BRD3321	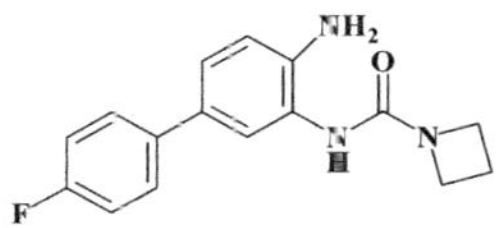	HDAC1/2	Cognitive enhancer (Wagner et al., 2015)
	BRD3349	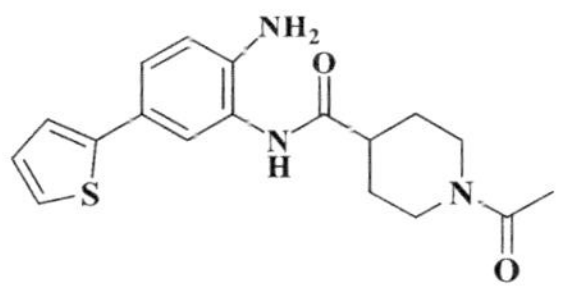	HDAC1/2	Cognitive enhancer (Wagner et al., 2015)

(continued)

Table 1 Epigenetic modulators, along with their structure, target selectivity, and disease/disorder, are explored in neurodegeneration. (*cont'd*)

HDAC inhibitors				
	BG45	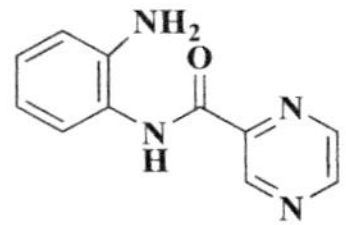	HDAC1/2/3	AD, Reducing microglial inflammation in APP/PS1 mice (Wang et al., 2023)
	MS-275	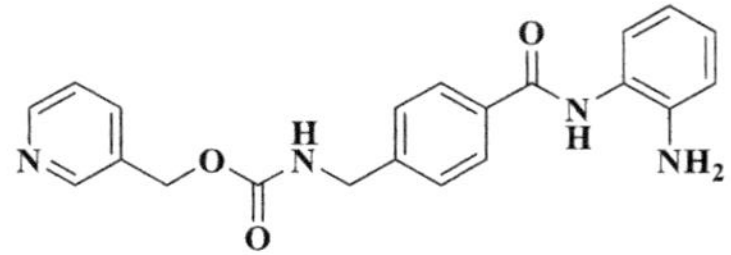	HDAC1/2/3	Cognitive enhancer, AD by reducing Aβ aggregation (Mitsuishi et al., 2024; Zhang & Schluesener, 2013)
	RGFP963	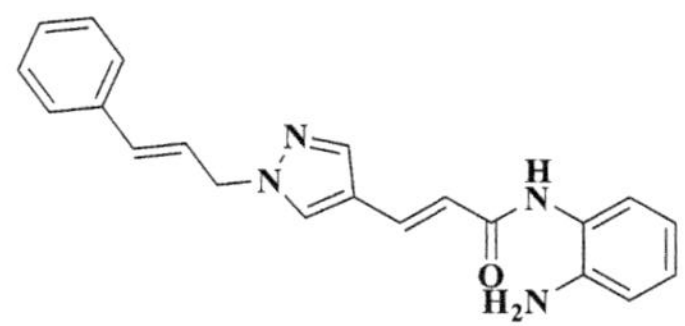	HDAC1/2/3	Memory enhancer, Synaptic plasticity (Rumbaugh et al., 2015)
	RG2833	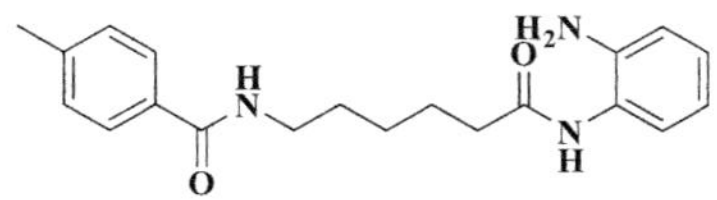	HDAC1/3	AD, Memory enhancer (Ndukwe et al., 2025)
	RGFP966	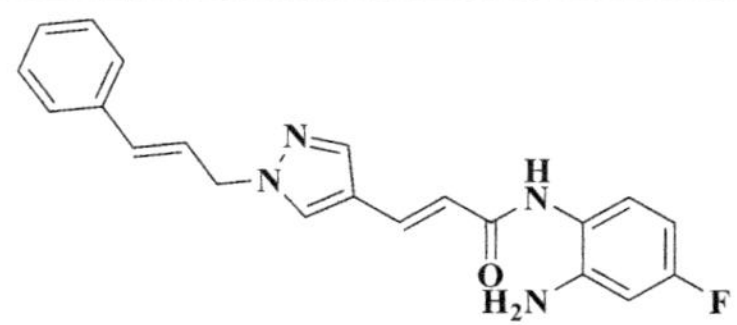	HDAC3	AD, Neuroprotection, Cognitive enhancer (Davis et al., 2024; Gu et al., 2023; Sun et al., 2022)

	PT3	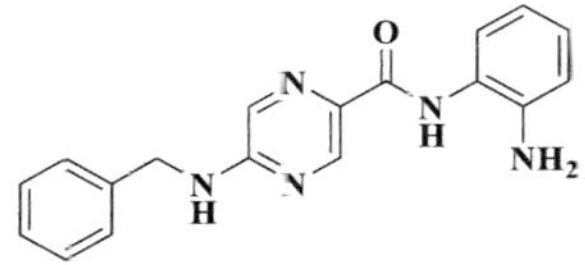	HDAC1/2/3	Memory enhancer (Pulya et al., 2021)
	Santacruzamate A	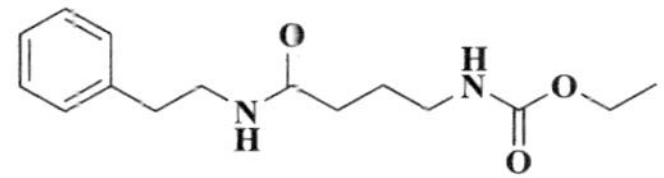	HDAC2	AD, Neuroinflammation (Qin et al., 2024)
	WK2-16	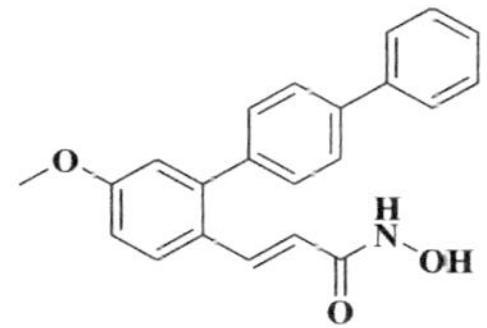	HDAC8	Cognitive enhancer (Lin et al., 2019)
	PCI-34051	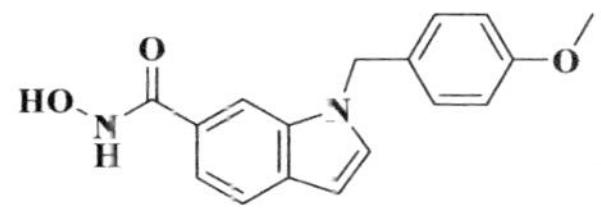	HDAC8	Neuroprotection (He et al., 2025)
Class IIa	TMP269	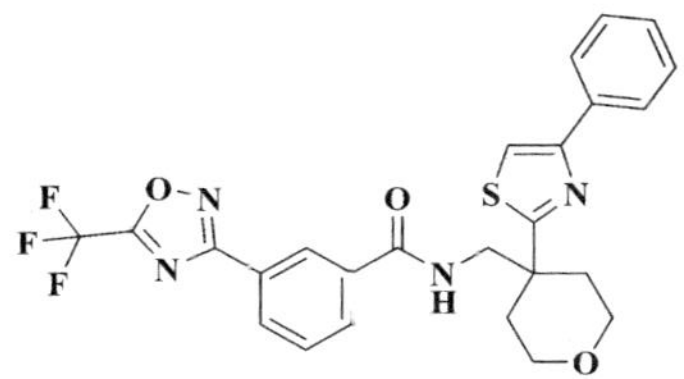	HDAC4/5/7/9	Cerebral ischemia, neuroprotection (Su et al., 2020)

(continued)

Table 1 Epigenetic modulators, along with their structure, target selectivity, and disease/disorder, are explored in neurodegeneration. (*cont'd*)

HDAC inhibitors

	TMP195	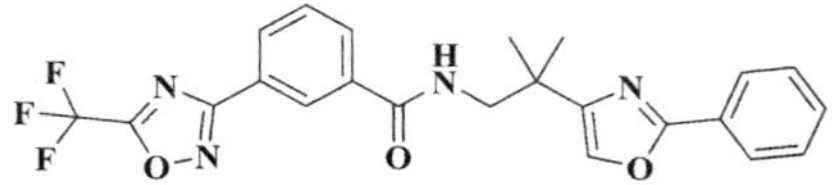	HDAC4/5/7/9	AD, Tau phosphorylation (Ye et al., 2025)
	MC1568	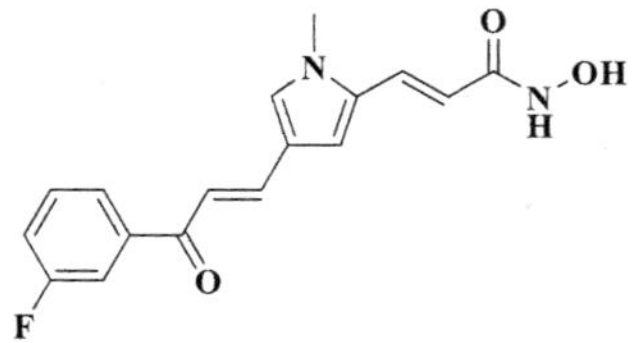	HDAC4/5/7/9	PD, Neuroprotection (Mazzocchi et al., 2022)
	LMK235	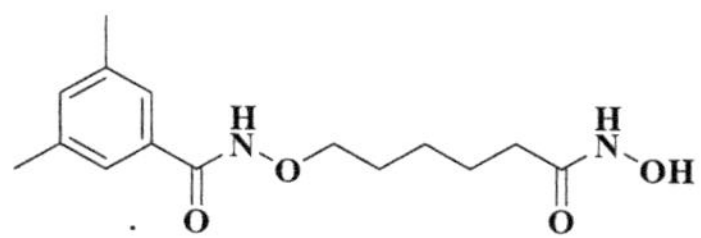	HDAC4/5	Neurite outgrowth, Neuroprotection (Mazzocchi et al., 2021)
Class IIb	Tubastatin A	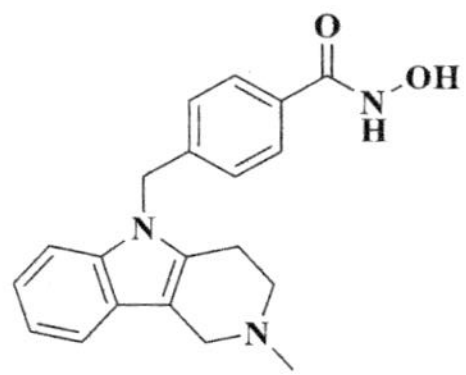	HDAC6	AD, reduces Aβ aggregation, Tau phosphorylation (Selenica et al., 2014; Zhang et al., 2014)
	Ricolinostat	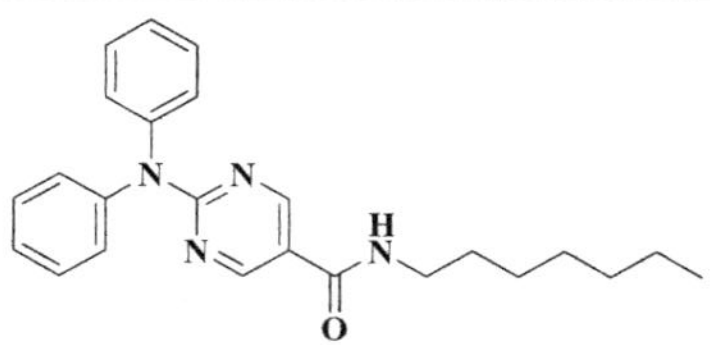	HDAC6	AD, reduces Aβ aggregation, Tau phosphorylation (Zhang et al., 2014)

	ACY-738	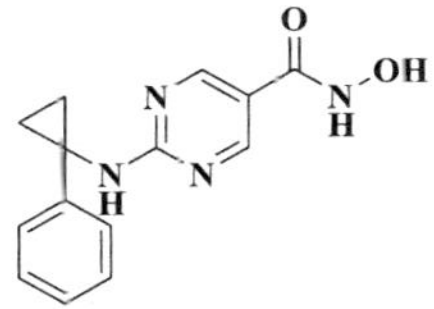	HDAC6	AD, APP processing (Majid et al., 2015)
	ACY-1083	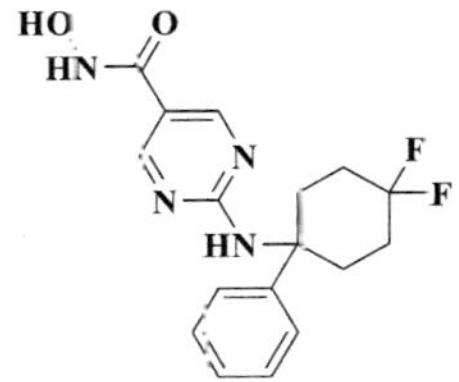	HDAC6	Cognitive enhancer, synaptic plasticity (McAlpin et al., 2022)
	MPT0G413 (Compound 6)	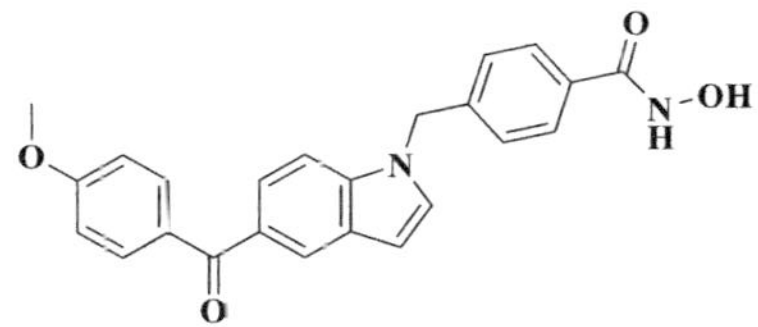	HDAC6	AD, reduces Aβ aggregation (Lee et al., 2018)
	MPT0G211	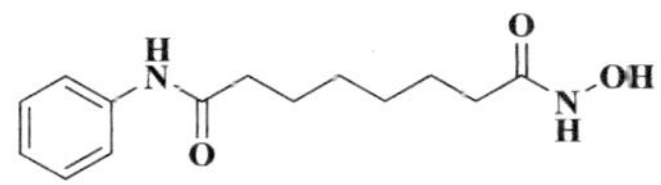	HDAC6	AD, reduces Aβ aggregation, tau phosphorylation (Fan et al., 2018)
Class IV	FT895	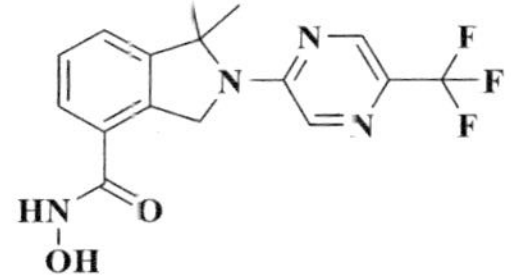	HDAC11	Neuroprotection (Huang et al., 2023; Xie et al., 2023)

(continued)

Table 1 Epigenetic modulators, along with their structure, target selectivity, and disease/disorder, are explored in neurodegeneration. (*cont'd*)

HDAC inhibitors

	SIS17	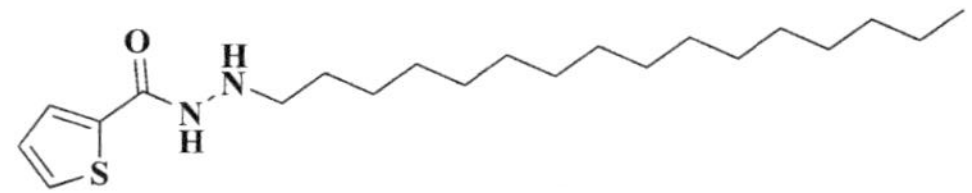	HDAC11	Neuroprotection (Son et al., 2019)
Pan-HDAC	Vorinostat	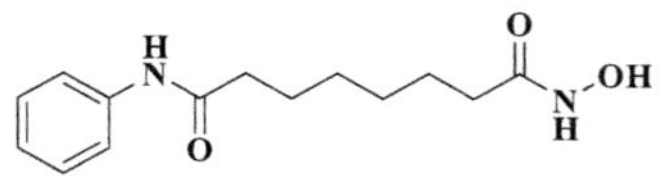	HDAC1/2/3/4/5/6/8/10	AD, histone acetylation (Bose et al., 2025)
	Tucidinostat	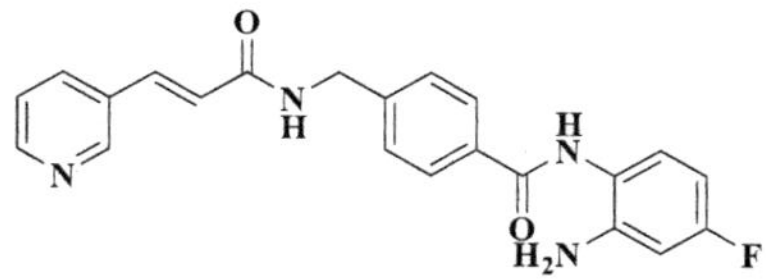		AD, APP/PS1 mice (Zhang et al., 2024)
	Domatinostat	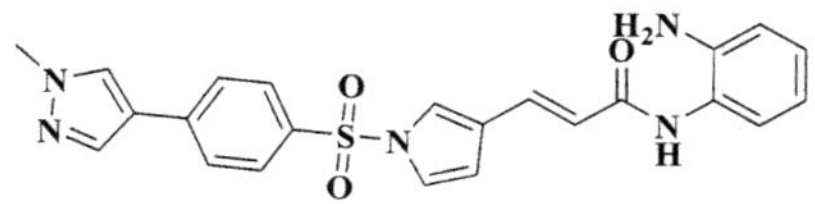		Neuroprotection (Nakagawa-Saito et al., 2022)
	Romidepsin	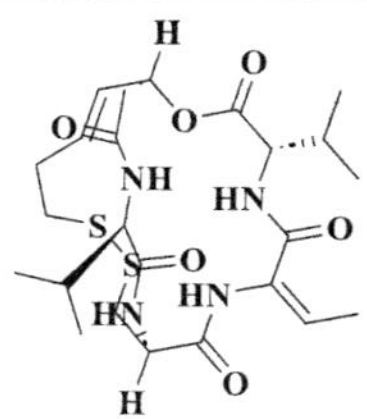		AD, Dementia (Das et al., 2024)

	Crebinostat	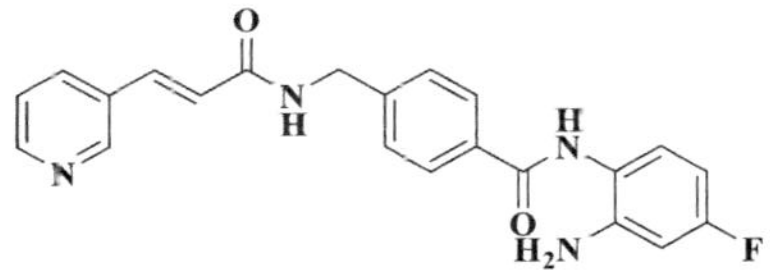		AD, Neuroplasticity (Fass et al., 2013)
Sirtuin modulator				
Inhibitor	EX-527 (Selisistat)	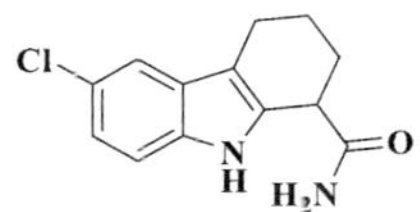	SIRT1	Neuronal differentiation, Neuroinflammation, Cellular model of Huntington`s disease (Nikseresht et al., 2019) (Kim et al., 2016) (Smith et al., 2014)
	AK7	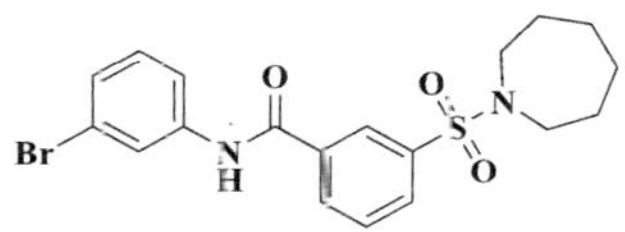	SIRT2	ALS, Cerebral ischemia (Chen et al., 2015)
	AGK2	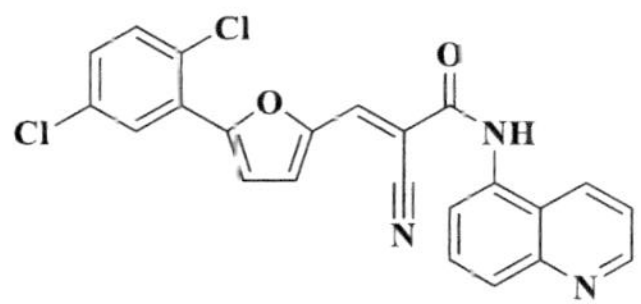	SIRT2	AD, Aβ plaque (Biella et al., 2016)

(continued)

Table 1 Epigenetic modulators, along with their structure, target selectivity, and disease/disorder, are explored in neurodegeneration. (*cont'd*)
HDAC inhibitors

	AK-1	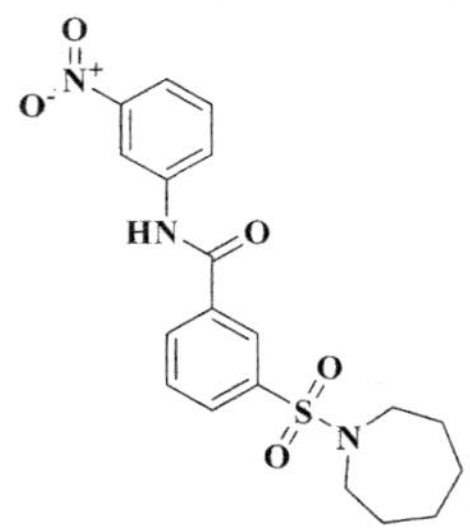	SIRT2	Dementia, Neuroprotection (Spires-Jones et al., 2012)
	Nicotinamide	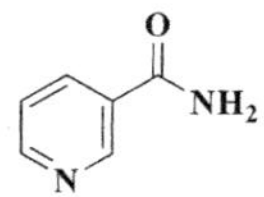	Pan SIRT	AD, Dementia, Memory (Green et al., 2008)
—	Resveratrol	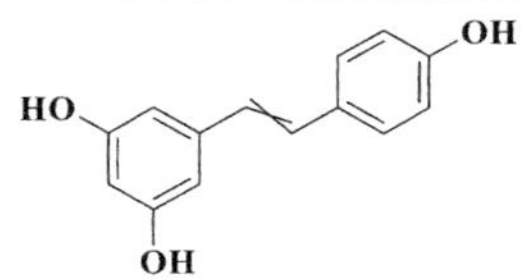	Activator	AD, Neuroprotection (Gomes et al., 2018)
DNMT inhibitors				
—	Decitabine	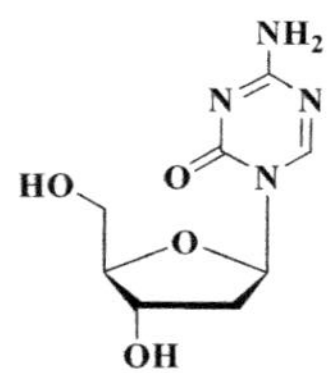	Nucleoside, DNMT1	AD, Neuroprotection (Olawade et al., 2025)

—	RG108	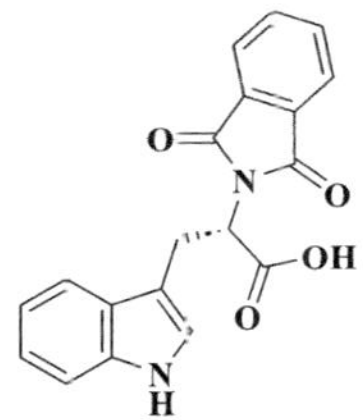	Non-nucleoside, DNMT1	Neuroprotection (Zheng et al., 2021)
BET inhibitors				
—	JQ1	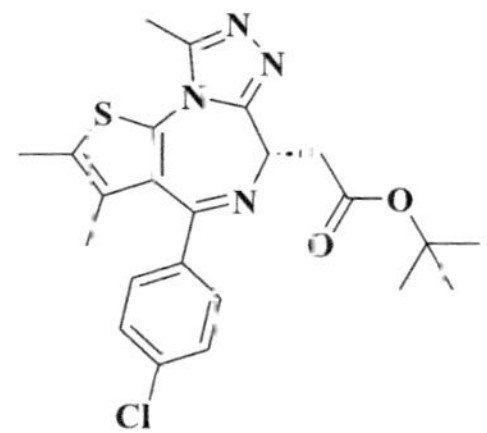	—	AD, Neuroinflammation (Magistri et al., 2016)

Table 2 Approved drugs/ compounds in clinical trials (https://clinicaltrials.gov/, 2026).

Compounds	Target classes	Conditions	Combination with	Clinical trial status
Vorinostat	Pan-HDACi	Mild AD	Alone	Phase I (NCT03056495 /April 10, 2024/Terminated)
		Niemann-pick disease, Type C1	Alone	Phase I/II (NCT02124083/Feb. 22, 2018/ Completed)
Phenylbutyrate	Pan-HDACi	AD	Taurursodiol (AMX0035)	Phase II (NCT03533257/March 10, 2025/Completed)
	Pan-HDACi	ALS	Taurursodiol (AMX0035)	Phase II/III (NCT04516096/ April 07, 2023/Completed)
	Pan-HDACi	HD	Alone	Phase II (NCT00212316/Aug 15, 2012/Completed)
CKD-504	HDAC6	HD	Alone	Phase II (NCT03713892/Feb 24, 2020/Unknown status)
Valproic Acid	Pan-HDACi	Dementia	Alone	Phase III (NCT00071721/Sep 25, 2014/Completed)
		AD	Alone	Early Phase I (NCT01729598/Oct 09, 2019/ Completed)

Ricolinostat	HDAC6	Painful diabetic peripheral neuropathy	Alone	Phase II (NCT03176472/Sep 21, 2023/Completed)
MBF-015	HDAC1/2	HD	Alone	Phase II (NCT06469853/March 03, 2025/Completed)
Nicotinamide, Riboside	Sirtuin	AD	Alone	Phase II (NCT05617508/Dec. 01, 2025/Completed)
		Improving memory and mild cognitive impairment	Alone	Phase I/II (NCT03482167/April 04, 2025/Completed)
SEN0014196 (EX-527, Selisistat)	Sirtuin	HD	Alone	Phase II (NCT01521585/Nov. 25, 2015/Completed)

- Example: TTK21
  - Features: Small-molecule activator conjugated to carbon nanospheres, Enhances CBP/p300 activity, Promotes neurogenesis
- Indirect HAT Activators
  - Example: Forskolin (Activates adenylate cyclase, Increases intracellular cAMP, stimulates the CREB signaling pathway)
  - Rolipram (Inhibits phosphodiesterase-4 (PDE-4), Elevates cAMP levels, Indirectly enhances CREB-mediated transcription)

- **Natural HAT Modulators**
  - Examples: Curcumin, Resveratrol, Garcinol
  - Functions: Modulate p300/CBP activity, restore histone acetylation balance
    - Therapeutic implications in neurodegeneration:
      - Restoration of synaptic plasticity-associated genes
      - Reduction of neuroinflammation
      - Enhancement of neuronal survival (Piaz et al., 2011)

Various challenges include poor pharmacokinetics, limited brain penetration, and a lack of specificity for the small-molecule epigenetic modulators. Hence, novel technologies like targeted degradation by PROTAC technology, nanomedicine, personalized medicine, etc., are emerging in the epigenetic therapeutics of neurodegenerative disorders like AD. The list of epigenetic modulators, along with their structures, target selectivity, and the diseases/disorders explored in preclinical studies, is reported in Table 1. The approved drugs/Compounds reported in clinical trials as epigenetic modulators in neurodegeneration are reported in Table 2.

5. Prospects and emerging therapeutics

Epigenetic modulators hold promise as a therapeutic approach to neurodegenerative diseases, such as AD, owing to recent developments in isoform-selective inhibitors and medicinal chemistry advancements (Alkhammash & Alotaibi, 2025). For instance, selective histone deacetylase (HDAC) inhibitors, and more specifically HDAC6 inhibitors, have shown significant preclinical promise in improving protein clearance mechanisms, reducing tau protein aggregation and amyloid-beta formation, and minimizing the side effects observed with pan-HDAC inhibitors.

Recent developments in proteolysis-targeting chimeras (PROTACs) are revolutionizing the epigenetic therapy space, as they enable the selective and sustained degradation of disease-relevant targets, including HDACs and sirtuins (Himaja et al., 2025). Recent advancements in antibody-proteolysis-targeting chimeric conjugates (Ab-PROTACs) provide further precision in epigenetic therapy, as they integrate the binding and degrading capabilities of both antibodies and PROTACs. These approaches have the potential to overcome the limitations associated with small-molecule inhibitors, including tissue distribution and toxicity. At the same time, advancements in DNA methyltransferases (DNMTs) and bromodomain and extra-terminal (BET) protein modulation are providing more options for epigenetic therapy, including combination therapy (Dai et al., 2024).

Recent developments in drug delivery and gene modulation are significantly enhancing the prospects for the next generation of therapeutics. For instance, nanoparticles and exosomes are improving blood-brain barrier penetration and enabling efficient transport of epigenetic therapy to the CNS. CRISPR/Cas9 epigenome editing is providing new avenues for precise and sustained regulation of epigenetic targets, and stem cell therapy is providing hope for neuronal replacement and regeneration (Rafieezadeh, 2025). The combination of selective inhibitors, targeted degraders, and antibody-proteolysis-targeting chimeric conjugates, and advancements in drug delivery and gene modulation, provides significant promise for the treatment of neurodegenerative diseases, including AD.

It can be concluded that epigenetic modulators have emerged as a compelling and transformative strategy for neurodegenerative diseases, especially AD. Specifically targeting reversible modifications that regulate gene expression is a fundamental approach underlying protein accumulation, synaptic dysregulation, and neuroinflammation. Advances in understanding histone acetylation, DNA methylation, and non-coding RNA networks have revealed critical epigenetic roles associated with the onset and progression of neurodegenerative diseases (Li et al., 2025; Lund et al., 2014). It can offer multiple intervention points beyond conventional symptomatic treatments. Among these, HDAC inhibitors have shown significant promise in restoring cognitive dysregulation and synaptic plasticity in preclinical models, while DNMT modulators and RNA-based strategies are expanding the therapeutic landscape. The emergence of targeted protein degradation – PROTACs enhances the precision and selectivity of epigenetic interventions, potentially by overcoming limitations associated with traditional inhibitors. However, several challenges –

such as isoform selectivity, long-term safety, off-target effects, and efficient delivery across the BBB are major limitations that require combined efforts of medicinal chemistry, molecular biology, and drug delivery platforms. The dynamic and reversible nature of epigenetic regulation provides a significant multifactorial opportunity to modify disease trajectories rather than merely attenuating symptoms. With the continuation of interdisciplinary research and clinical validation, epigenetic modulators hold immense promise for redefining the therapeutic paradigm and improving outcomes in AD and other neurodegenerative disorders.

Acknowledgements

The authors are thankful to the Department of Pharmacy, BITS Pilani-Hyderabad Campus. The authors also thank BioRender.com.

Declaration of generative AI and AI-assisted technologies in the writing process

During the preparation of this work, the author(s) used ChatGPT and Gemini for the formatting process of the write-up of the article according to the journal in order to make the article suitable for the journal. Also, ChatGPT was used to generate Fig. 3, using a specific prompt to present the assay protocol and its functions. After using this tool/service, the author(s) reviewed and edited the content as needed and take(s) full responsibility for the content of the publication.

Conflict of interest

The authors declare that there is no conflict of interest.

References

Alkhammash, A., & Alotaibi, G. (2025). Epigenetic reprogramming as a therapeutic strategy for neurodegenerative diseases: A complex and novel approach. *European Journal of Pharmacology, 1002*, 177853. https://doi.org/10.1016/j.ejphar.2025.177853.

Asadi, K., Radha, K., Satarker, S., & Gurram, P. C. (2025). Histone deacetylase 2 in Alzheimer's disease: A comprehensive molecular blueprint for therapeutic targeting. *Molecular Neurobiology, 62*, 16489–16511. https://doi.org/10.1007/s12035-025-05264-y.

Ateya, N. H., Al-Taie, S. F., Jasim, S. A., Uthirapathy, S., Chaudhary, K., Rani, P., ... Ahmed, J. K. (2025). Histone deacetylation in Alzheimer's diseases (AD); hope or hype. *Cell Biochemistry and Biophysics, 83*, 1537–1553. https://doi.org/10.1007/s12013-025-01670-0.

Behl, T., Kyada, A., Roopashree, R., Nathiya, D., Arya, R., Kumar, M. R., ... Gasmi, A. (2024). Epigenetic biomarkers in Alzheimer's disease: Diagnostic and prognostic relevance. *Ageing Research Reviews, 102*, 102556. https://doi.org/10.1016/j.arr.2024.102556.

Biella, G., Fusco, F., Nardo, E., Bernocchi, O., Colombo, A., Lichtenthaler, S. F., ... Albani, D. (2016). Sirtuin 2 inhibition improves cognitive performance and acts on amyloid-β protein precursor processing in two Alzheimer's disease mouse models. *Journal of Alzheimer's Disease, 53*(3), 1193–1207. https://doi.org/10.3233/JAD-151135.

Bonomi, R. E., Riordan, W., & Gelovani, J. G. (2024). The structures, functions, and roles of class III HDACs (Sirtuins) in neuropsychiatric diseases. *Cells, 13*, 1644. https://doi.org/10.3390/cells13191644.

Bose, C., Hindle, A., Smith, S. C., Strickland, J., Zhang, C., Guzman, I., ... Lawrence, J. J. (2025). Low-dose dietary vorinostat increases brain histone acetylation levels and reduces oxidative stress in an Alzheimer's disease mouse model. *Journal of Alzheimer's Disease, 106*(4), 1360–1382. https://doi.org/10.1177/13872877251352107.

Chen, X., Wales, P., Quinti, L., Zuo, F., Moniot, S., Herisson, F., ... Kazantsev, A. G. (2015). The sirtuin-2 inhibitor AK7 is neuroprotective in models of Parkinson's disease but not amyotrophic lateral sclerosis and cerebral ischemia. *PLoS One, 10*(1), e0116919. https://doi.org/10.1371/journal.pone.0116919.

Chouliaras, L., van den Hove, D., Kenis, G., Draanen, M., Hof, P., Os, J., ... Rutten, B. (2013). Histone deacetylase 2 in the Mouse Hippocampus: Attenuation of age- related increase by caloric restriction. *Current Alzheimer Research, 10*, 868–876. https://doi.org/10.2174/15672050113100080009.

Coneys, R., & Wood, I. C. (2020). Alzheimer's disease: The potential of epigenetic treatments and current clinical candidates. *Neurodegenerative Disease Management, 10*, 543–558. https://doi.org/10.2217/nmt-2019-0034.

Constantin, G., Terrabuio, E., Pietronigro, E., Bani, A., Bianca, V. Della, Laudanna, C., ... Vitulo, N. (2024). CD103–CD8+ T cells promote neurotoxic inflammation in Alzheimer's disease via granzyme K–PAR-1 signaling. https://doi.org/10.21203/rs.3.rs-5203345/v1.

Dai, X.-J., Ji, S.-K., Fu, M.-J., Liu, G.-Z., Liu, H.-M., Wang, S.-P., ... Chen, X.-B. (2024). Degraders in epigenetic therapy: PROTACs and beyond. *Theranostics, 14*, 1464–1499. https://doi.org/10.7150/thno.92526.

Darst, R. P., Pardo, C. E., Ai, L., Brown, K. D., & Kladde, M. P. (2010). Bisulfite sequencing of DNA. *Current Protocols in Molecular Biology, 91*(1), https://doi.org/10.1002/0471142727.mb0709s91.

Das, V., Miller, J. H., Alladi, C. G., Annadurai, N., De Sanctis, J. B., Hrubá, L., & Hajdúch, M. (2024). Antineoplastics for treating Alzheimer's disease and dementia: Evidence from preclinical and observational studies. *Medicinal Research Reviews, 44*(5), 2078–2111. https://doi.org/10.1002/med.22033.

Davis, N., Taylor, B., Abelleira-Hervas, L., Karimian-Marnani, N., Aleksynas, R., Syed, N., ... Sastre, M. (2024). Histone deacetylase-3 regulates the expression of the amyloid precursor protein and its inhibition promotes neuroregenerative pathways in Alzheimer's disease models. *The FASEB Journal, 38*(10), https://doi.org/10.1096/fj.202301762RR.

DNA methylation inhibitors: Retrospective and perspective view (2021) pp. 205 223. https://doi.org/10.1016/bs.acr.2021.03.007.

Fan, S.-J., Huang, F.-I., Liou, J. P., & Yang, C.-R. (2018). The novel histone de acetylase 6 inhibitor, MPT0G211, ameliorates tau phosphorylation and cognitive deficits in an Alzheimer's disease model. *Cell Death & Disease, 9*(6), 655. https://doi.org/10.1038/s41419-018-0688-5.

Fass, D. M., Reis, S. A., Ghosh, B., Hennig, K. M., Joseph, N. F., Zhao, W.-N., ... Haggarty, S. J. (2013). Crebinostat: A novel cognitive enhancer that inhibits histone deacetylase activity and modulates chromatin-mediated neuroplasticity. *Neuropharmacology, 64*, 81–96. https://doi.org/10.1016/j.neuropharm.2012.06.043.

Gao, X., Chen, Q., Yao, H., Tan, J., Liu, Z., Zhou, Y., & Zou, Z. (2022). Epigenetics in Alzheimer's disease. *Frontiers in Aging Neuroscience, 14*. https://doi.org/10.3389/fnagi.2022.911635.

Geng, H., Chen, H., Wang, H., & Wang, L. (2021). The histone modifications of neuronal plasticity. *Neural Plasticity, 2021*, 1–7. https://doi.org/10.1155/2021/6690523.

Gomes, B. A. Q., Silva, J. P. B., Romeiro, C. F. R., dos Santos, S. M., Rodrigues, C. A., Gonçalves, P. R., ... Monteiro, M. C. (2018). Neuroprotective mechanisms of resveratrol in Alzheimer's disease: Role of SIRT1. *Oxidative Medicine and Cellular Longevity, 2018*(1), https://doi.org/10.1155/2018/8152373.

Gräff, J., Kim, D., Dobbin, M. M., & Tsai, L.-H. (2011). Epigenetic regulation of gene expression in physiological and pathological brain processes. *Physiological Reviews, 91*, 603–649. https://doi.org/10.1152/physrev.00012.2010.

Gräff, J., Rei, D., Guan, J.-S., Wang, W.-Y., Seo, J., Hennig, K. M., ... Tsai, L.-H. (2012). An epigenetic blockade of cognitive functions in the neurodegenerating brain. *Nature, 483*, 222–226. https://doi.org/10.1038/nature10849.

Green, K. N., Steffan, J. S., Martinez-Coria, H., Sun, X., Schreiber, S. S., Thompson, L. M., & LaFerla, F. M. (2008). Nicotinamide restores cognition in Alzheimer's disease transgenic mice via a mechanism involving sirtuin inhibition and selective reduction of Thr231-phosphotau. *The Journal of Neuroscience, 28*(45), 11500–11510. https://doi.org/10.1523/JNEUROSCI.3203-08.2008.

Gu, H.-P., Wu, X.-F., Gong, Y.-T., Mu-Yao Wu, Shi, M.-Y., Sun, Y., ... Chen, G. (2023). RGFP966 exerts neuroprotective effect via HDAC3/Nrf2 pathway after surgical brain injury in rats. *Heliyon, 9*(7), e18160. https://doi.org/10.1016/j.heliyon.2023.e18160.

Guan, J.-S., Haggarty, S. J., Giacometti, E., Dannenberg, J.-H., Joseph, N., Gao, J., ... Tsai, L.-H. (2009). HDAC2 negatively regulates memory formation and synaptic plasticity. *Nature, 459*, 55–60. https://doi.org/10.1038/nature07925.

Halder, D., Das, S., Jeyaprakash, R. S., & Joseph, A. (2023). Role of multi-targeted bioactive natural molecules and their derivatives in the treatment of Alzheimer's disease: an insight into structure-activity relationship. *Journal of Biomolecular Structure & Dynamics, 41*, 11286–11323. https://doi.org/10.1080/07391102.2022.2158136.

Hawash, M. (2025). Next-generation HDAC inhibitors: Advancing zinc-binding group design for enhanced cancer therapy. *Cells, 14*, 1997. https://doi.org/10.3390/cells14241997.

He, Y., Guo, Y., Li, X., Wu, J., Feng, Z., Wang, G., & Zhang, M. (2025). Sevoflurane-induced cognitive dysfunction in aged mice mediated by HDAC8-dependent suppression of adult hippocampal neurogenesis via the pCREB/BDNF pathway. *Brain Research Bulletin, 230*, 111497. https://doi.org/10.1016/j.brainresbull.2025.111497.

Herman, J. G., Graff, J. R., Myöhänen, S., Nelkin, B. D., & Baylin, S. B. (1996). Methylation-specific PCR: A novel PCR assay for methylation status of CpG islands. *Proceedings of the National Academy of Sciences, 93*(18), 9821–9826. https://doi.org/10.1073/pnas.93.18.9821.

Himaja, A., Halder, D., Banerjee, S., Biswas, S., Adhikari, N., & Ghosh, B. (2025). PROTAC-based HDAC degradation: A paradigm shift in targeted epigenetic therapies. *ChemMedChem, 20*. https://doi.org/10.1002/cmdc.202500203.

Honer, M. A., Ferman, B. I., Gray, Z. H., Bondarenko, E. A., & Whetstine, J. R. (2024). Epigenetic modulators provide a path to understanding disease and therapeutic opportunity. *Genes & Development, 38*, 473–503. https://doi.org/10.1101/gad.351444.123.

https://clinicaltrials.gov/ [WWW Document] (2026).

https://www.who.int/ (2026). World Health Organization.

Huang, P.-Y., Shih, I.-A., Liao, Y.-C., You, H.-L., & Lee, M.-J. (2023). FT895 impairs mitochondrial function in malignant peripheral nerve sheath tumor cells. *International Journal of Molecular Sciences, 25*(1), 277. https://doi.org/10.3390/ijms25010277.

Huang, H., Lin, S., Garcia, B. A., & Zhao, Y. (2015). Quantitative proteomic analysis of histone modifications. *Chemical Reviews, 115*(6), 2376–2418. https://doi.org/10.1021/cr500491u.

Ji, J., Xia, M., Xia, W., & Miao, G. (2025). Structural classification of HDAC inhibitors and cancer treatment prospects: A review. *Letters in Drug Design & Discovery, 22*, 100165. https://doi.org/10.1016/j.lddd.2025.100165.

Jovic, D., Liang, X., Zeng, H., Lin, L., Xu, F., & Luo, Y. (2022). Single-cell RNA sequencing technologies and applications: A brief overview. *Clinical and Translational Medicine, 12*(3), https://doi.org/10.1002/ctm2.694.

Kaur, R., Pandey, S., Gupta, S., & Singh, J. (2025). Harnessing the potential of long non-coding RNAs in the pathophysiology of Alzheimer's disease. *Experimental Neurology, 385*, 115134. https://doi.org/10.1016/j.expneurol.2024.115134.

Khan, Z. U., Carretero-Rey, M., de León-López, C. A. M., & Navarro-Lobato, I. (2025). Memory-associated immediate early genes: Roles in synaptic function, memory processes, and neurological diseases. *Molecular Neurobiology, 62*, 15885–15915. https://doi.org/10.1007/s12035-025-05203-x.

Kim, M., & Costello, J. (2017). DNA methylation: An epigenetic mark of cellular memory. *Experimental & Molecular Medicine, 49*, e322. https://doi.org/10.1038/emm.2017.10.

Kim, B. S., Lee, C.-H., Chang, G.-E., Cheong, E., & Shin, I. (2016). A potent and selective small molecule inhibitor of sirtuin 1 promotes differentiation of pluripotent P19 cells into functional neurons. *Scientific Reports, 6*(1), 34324. https://doi.org/10.1038/srep34324.

Kitazawa, R., Haraguchi, R., & Kitazawa, S. (2023). Histone modification in histochemistry and cytochemistry. *Acta Histochemica et Cytochemica, 56*. https://doi.org/10.1267/ahc.23-00014.

Lee, H.-Y., Fan, S.-J., Huang, F.-I., Chao, H.-Y., Hsu, K.-C., Lin, T. E., ... Liou, J.-P. (2018). 5-Aroylindoles act as selective histone deacetylase 6 inhibitors ameliorating Alzheimer's disease phenotypes. *Journal of Medicinal Chemistry, 61*(16), 7087–7102. https://doi.org/10.1021/acs.jmedchem.8b00151.

Li, Y., & Tollefsbol, T.O. (2011). Combined chromatin immunoprecipitation and bisulfite methylation sequencing analysis (pp. 239–251). https://doi.org/10.1007/978–1-61779–316-5_18.

Li, F., Wang, Y., Yang, H., Xu, Y., Zhou, X., Zhang, X., ... Bi, J. (2019). The effect of BACE1-AS on β-amyloid generation by regulating BACE1 mRNA expression. *BMC Molecular Biology, 20*, 23. https://doi.org/10.1186/s12867-019-0140-0.

Li, K., Qiu, X., Yang, Q., & Wang, Z. (2025). Histone modifications: Key players in Alzheimer's disease. *Mechanisms of Ageing and Development, 228*, 112102. https://doi.org/10.1016/j.mad.2025.112102.

Lin, F.-L., Yen, J.-L., Kuo, Y. C., Kang, J.-J., Cheng, Y.-W., Huang, W.-J., & Hsiao, G. (2019). HDAC8 inhibitor WK2-16 therapeutically targets lipopolysaccharide-induced mouse model of neuroinflammation and microglial activation. *International Journal of Molecular Sciences, 20*(2), 410. https://doi.org/10.3390/ijms20020410.

Liu, L., Yang, C., & Candelario-Jalil, E. (2021). Role of BET proteins in inflammation and CNS diseases. *Frontiers in Molecular Biosciences, 8*. https://doi.org/10.3389/fmolb.2021.748449.

Lu, W., Kawatani, K., Ren, Y., Nambara, T., Jia, L., Jeevaratnam, S., ... Li, Y. (2026). CI-994 is a dual modulator of class I HDACs and Wnt/β-catenin signaling for the treatment of Alzheimer's disease. *Alzheimer's Research & Therapy, 18*(1), 60. https://doi.org/10.1186/s13195-026-01982-0.

Lund, K., Cole, J. J., VanderKraats, N. D., McBryan, T., Pchelintsev, N. A., Clark, W., ... Adams, P. D. (2014). DNMT inhibitors reverse a specific signature of aberrant promoter DNA methylation and associated gene silencing in AML. *Genome Biology, 15*, 406. https://doi.org/10.1186/s13059-014-0406-2.

Magistri, M., Velmeshev, D., Makhmutova, M., Patel, P., C. Sartor, G., Volmar, C.-H., ... Ali Faghihi, M. (2016). The BET-Bromodomain inhibitor JQ1 reduces inflammation and tau phosphorylation at Ser396 in the brain of the 3xTg model of Alzheimer's disease. *Current Alzheimer Research, 13*(9), 985–995. https://doi.org/10.2174/1567205013666160427101832.

Majid, T., Griffin, D., Criss, Z., Jarpe, M., & Pautler, R. G. (2015). Pharmocologic treatment with histone deacetylase 6 inhibitor (ACY-738) recovers Alzheimer's disease phenotype in amyloid precursor protein/presenilin 1 (APP/PS1) mice. *Alzheimer's & Dementia: Translational Research & Clinical Interventions, 1*(3), 170–181. https://doi.org/10.1016/j.trci.2015.08.001.

Mazzocchi, M., Goulding, S. R., Morales-Prieto, N., Foley, T., Collins, L. M., Sullivan, A. M., & O'Keeffe, G. W. (2022). Peripheral administration of the Class-IIa HDAC inhibitor MC1568 partially protects against nigrostriatal neurodegeneration in the striatal 6-OHDA rat model of Parkinson's disease. *Brain, Behavior, and Immunity, 102*, 151–160. https://doi.org/10.1016/j.bbi.2022.02.025.

Mazzocchi, M., Goulding, S. R., Wyatt, S. L., Collins, L. M., Sullivan, A. M., & O'Keeffe, G. W. (2021). LMK235, a small molecule inhibitor of HDAC4/5, protects dopaminergic neurons against neurotoxin- and α-synuclein-induced degeneration in cellular models of Parkinson's disease. *Molecular and Cellular Neuroscience, 115*, 103642. https://doi.org/10.1016/j.mcn.2021.103642.

McAlpin, B. R., Mahalingam, R., Singh, A. K., Dharmaraj, S., Chrisikos, T. T., Boukelmoune, N., ... Heijnen, C. J. (2022). HDAC6 inhibition reverses long-term doxorubicin-induced cognitive dysfunction by restoring microglia homeostasis and synaptic integrity. *Theranostics, 12*(2), 603–619. https://doi.org/10.7150/thno.67410.

Metzendorf, N. G., Sehlin, D., & Hultqvist, G. (2025). Key considerations for ELISA-based quantification of diverse amyloid beta forms in murine brain homogenates. *Frontiers in Neuroscience, 19*, 1645952. https://doi.org/10.3389/fnins.2025.1645952.

Milne, T.A., Zhao, K., & Hess, J.L. (2009). Chromatin immunoprecipitation (ChIP) for analysis of histone modifications and chromatin-associated proteins (pp. 409–423). https://doi.org/10.1007/978–1-59745–418-6_21.

Miskimen, K. L. S., Chan, E. R., & Haines, J. L. (2017). Assay for transposase-accessible chromatin using sequencing (ATAC-seq) data analysis. *Current Protocols in Human Genetics, 92*(1), https://doi.org/10.1002/cphg.32.

Mitsuishi, Y., Nakano, M., Kojima, H., Okabe, T., & Nishimura, M. (2024). Reduction of amyloid-β production without inhibiting secretase activity by MS-275. *ACS Chemical Neuroscience, 15*(6), 1234–1241. https://doi.org/10.1021/acschemneuro.3c00848.

Nakagawa-Saito, Y., Saitoh, S., Mitobe, Y., Sugai, A., Togashi, K., Suzuki, S., ... Okada, M. (2022). HDAC class I inhibitor domatinostat preferentially targets glioma stem cells over their differentiated progeny. *International Journal of Molecular Sciences, 23*(15), 8084. https://doi.org/10.3390/ijms23158084.

Ndukwe, K., Serrano, P. A., Rockwell, P., Xie, L., & Figueiredo-Pereira, M. E. (2025). Brain-penetrant histone deacetylase inhibitor RG2833 improves spatial memory in females of an Alzheimer's disease rat model. *Journal of Alzheimer's Disease, 104*(1), 173–190. https://doi.org/10.1177/13872877251314777.

Nikseresht, S., Khodagholi, F., & Ahmadiani, A. (2019). Protective effects of ex-527 on cerebral ischemia–reperfusion injury through necroptosis signaling pathway attenuation. *Journal of Cellular Physiology, 234*(2), 1816–1826. https://doi.org/10.1002/jcp.27055.

Ng, B., Avey, D.R., Sultan, F., de Paiva Lopes, K., Fujita, M., Saunders, D., … Wang, Y. (2024). Integrated spatial and single-nuclei transcriptomic analysis of long non-coding RNAs in Alzheimer's disease. https://doi.org/10.1101/2024.10.27.620550.

Olawade, D. B., Rashad, I., Egbon, E., Teke, J., Ovsepian, S. V., & Boussios, S. (2025). Reversing epigenetic dysregulation in neurodegenerative diseases: Mechanistic and therapeutic considerations. *International Journal of Molecular Sciences, 26*, 4929. https://doi.org/10.3390/ijms26104929.

Olufunmilayo, E. O., & Holsinger, R. M. D. (2023). Roles of non-coding RNA in Alzheimer's disease pathophysiology. *International Journal of Molecular Sciences, 24*, 12498. https://doi.org/10.3390/ijms241512498.

Ostaiza-Cardenas, J., Tobar, A. C., Costa, S. C., Calero, D. S., López-Carrera, A., Bermúdez, F. G., & Orellana-Manzano, A. (2025). Epigenetic modulation by life–style: Advances in diet, exercise, and mindfulness for disease prevention and health optimization. *Front-of-Pack Nutrition, 12*. https://doi.org/10.3389/fnut.2025.1632999.

Payne, C. J. (Ed.). (2014). *Epigenetics and epigenomics*InTech. https://doi.org/10.5772/57037.

Park, J., Lee, K., Kim, K., & Yi, S.-J. (2022). The role of histone modifications: From neurodevelopment to neurodiseases. *Signal Transduction and Targeted Therapy, 7*, 217. https://doi.org/10.1038/s41392-022-01078-9.

Peña, C. J. (2026). Epigenetic regulation of brain development, plasticity, and response to early-life stress. *Neuropsychopharmacology: Official Publication of the American College of Neuropsychopharmacology, 51*, 5–15. https://doi.org/10.1038/s41386-025-02179-z.

Piaz, F. D., Vassallo, A., Rubio, O. C., Castellano, S., Sbardella, G., & De Tommasi, N. (2011). Chemical biology of histone acetyltransferase natural compounds modulators. *Molecular Diversity, 15*, 401–416. https://doi.org/10.1007/s11030-010-9299-5.

Prasanth, M. I., Sivamaruthi, B. S., Cheong, C. S. Y., Verma, K., Tencomnao, T., Brimson, J. M., & Prasansuklab, A. (2024). Role of epigenetic modulation in neurodegenerative diseases: Implications of phytochemical interventions. *Antioxidants, 13*, 606. https://doi.org/10.3390/antiox13050606.

Pulya, S., Mahale, A., Bobde, Y., Routholla, G., Patel, T., Swati, ... Ghosh, B. (2021). PT3: A novel benzamide class histone deacetylase 3 inhibitor improves learning and memory in novel object recognition mouse model. *ACS Chemical Neuroscience, 12*, 883–892. https://doi.org/10.1021/acschemneuro.0c00721.

Puzzo, D., Lee, L., Palmeri, A., Calabrese, G., & Arancio, O. (2014). Behavioral assays with mouse models of Alzheimer's disease: Practical considerations and guidelines. *Biochemical Pharmacology, 88*(4), 450–467. https://doi.org/10.1016/j.bcp.2014.01.011.

Qin, Y., Liu, Q., Wang, S., Wang, Q., Du, Y., Yao, J., ... Yang, L. (2024). Santacruzamate A alleviates pain and pain-related adverse emotions through the inhibition of microglial activation in the anterior cingulate cortex. *ACS Pharmacology & Translational Science, 7*(4), 1002–1012. https://doi.org/10.1021/acsptsci.3c00282.

Qin, H., Liu, J., Fang, C., Deng, Y., & Zhang, Y. (2023). DNA methylation: The epigenetic mechanism of Alzheimer's disease. *Ibrain, 9*, 463–472. https://doi.org/10.1002/ibra.12121.

Rafieezadeh, D. (2025). Advances in nanotechnology for targeted drug delivery in neurodegenerative diseases. *American Journal of Neurodegenerative Disease, 14*, 51–57. https://doi.org/10.62347/QHVI3317.

Ramakrishnan, A., Piehl, N., Simonton, B., Parikh, M., Zhang, Z., Teregulova, V., ... Gate, D. (2024). Epigenetic dysregulation in Alzheimer's disease peripheral immunity. *Neuron, 112*, 1235–1248.e5. https://doi.org/10.1016/j.neuron.2024.01.013.

Razick, D. I., Akhtar, M., Wen, J., Alam, M., Dean, N., Karabala, M., ... Siddiqui, S. (2023). The role of sirtuin 1 (SIRT1) in neurodegeneration. *Cureus*. https://doi.org/10.7759/cureus.40463.

Rumbaugh, G., Daws, S. E., Ozkan, E. D., Rojas, C. S., Hubbs, C. R., Aceti, M., ... Miller, C. A. (2015). Pharmacological selectivity within class I histone deacetylases predicts effects on synaptic function and memory rescue. *Neuropsychopharmacology: Official Publication of the American College of Neuropsychopharmacology, 40*, 2307–2316. https://doi.org/10.1038/npp.2015.93.

Saleem, K., Xiao, Z., Zhu, B., Ren, Y., Yan, Z., & Feng, J. (2026). Elevated SGK1 increases Tau phosphorylation and microtubule instability in Alzheimer's patient-derived cortical neurons. *Molecular Psychiatry, 31*, 332–342. https://doi.org/10.1038/s41380-025-03225-4.

Santa-Maria, I., Alaniz, M. E., Renwick, N., Cela, C., Fulga, T. A., Van Vactor, D., ... Crary, J. F. (2015). Dysregulation of microRNA-219 promotes neurodegeneration through post-transcriptional regulation of tau. *Journal of Clinical Investigation, 125*, 681–686. https://doi.org/10.1172/JCI78421.

Selenica, M.-L., Benner, L., Housley, S. B., Manchec, B., Lee, D. C., Nash, K. R., ... Morgan, D. (2014). Histone deacetylase 6 inhibition improves memory and reduces total tau levels in a mouse model of tau deposition. *Alzheimer's Research & Therapy, 6*(1), 12. https://doi.org/10.1186/alzrt241.

Sharma, V. K., Mehta, V., & Singh, T. G. (2020). Alzheimer's disorder: Epigenetic connection and associated risk factors. *Current Neuropharmacology, 18*, 740–753. https://doi.org/10.2174/1570159X18666200128125641.

Siddiqui, S., & Livák, F. (2023). Principles of advanced flow cytometry: A practical guide (pp. 89–114). https://doi.org/10.1007/978-1-0716-2740-2_5.

Singh, M. B., & Sartor, G. C. (2020). BET bromodomains as novel epigenetic targets for brain health and disease. *Neuropharmacology, 181*, 108306. https://doi.org/10.1016/j.neuropharm.2020.108306.

Smith, M. R., Syed, A., Lukacsovich, T., Purcell, J., Barbaro, B. A., Worthge, S. A., ... Marsh, J. L. (2014). A potent and selective Sirtuin 1 inhibitor alleviates pathology in multiple animal and cell models of Huntington's disease. *Human Molecular Genetics, 23*(11), 2995–3007. https://doi.org/10.1093/hmg/ddu010.

Sokpor, G., Xie, Y., Rosenbusch, J., & Tuoc, T. (2017). Chromatin remodeling BAF (SWI/SNF) complexes in neural development and disorders. *Frontiers in Molecular Neuroscience, 10*. https://doi.org/10.3389/fnmol.2017.00243.

Son, S. I., Cao, J., Zhu, C.-L., Miller, S. P., & Lin, H. (2019). Activity-guided design of HDAC11-specific inhibitors. *ACS Chemical Biology, 14*(7), 1393–1397. https://doi.org/10.1021/acschembio.9b00292.

Spires-Jones, T. L., Fox, L. M., Rozkalne, A., Pitstick, R., Carlson, G. A., & Kazantsev, A. G. (2012). Inhibition of sirtuin 2 with sulfobenzoic acid derivative AK1 is non-toxic and potentially neuroprotective in a mouse model of frontotemporal dementia. *Frontiers in Pharmacology, 3*. https://doi.org/10.3389/fphar.2012.00042.

Su, L., Liang, D., Kuang, S.-Y., Dong, Q., Han, X., & Wang, Z. (2020). Neuroprotective mechanism of TMP269, a selective class IIA histone deacetylase inhibitor, after cerebral ischemia/reperfusion injury. *Neural Regeneration Research, 15*(2), 277. https://doi.org/10.4103/1673-5374.265562.

Sun, W., Zhang, N., Liu, B., Yang, J., Loers, G., Siebert, H.-C., ... Zhang, R. (2022). HDAC3 inhibitor RGFP966 ameliorated neuroinflammation in the cuprizone-induced demyelinating mouse model and LPS-stimulated BV2 cells by downregulating the P2X7R/STAT3/NF-κB65/NLRP3 activation. *ACS Chemical Neuroscience, 13*(17), 2579–2598. https://doi.org/10.1021/acschemneuro.1c00826.

Tammaro, A., Daniels, E. G., Hu, I. M., 't Hart, K. C., Reid, K., Juni, R. P., ... Janssens, G. E. (2024). HDAC1/2 inhibitor therapy improves multiple organ systems in aged mice. *IScience, 27*(1), 108681. https://doi.org/10.1016/j.isci.2023.108681.

Tang, L., Nogales, E., & Ciferri, C. (2010). Structure and function of SWI/SNF chromatin remodeling complexes and mechanistic implications for transcription. *Progress in Biophysics and Molecular Biology, 102*, 122–128. https://doi.org/10.1016/j.pbiomolbio.2010.05.001.

Tiwari, P., Dwivedi, R., Kaushik, M., Tripathi, M., & Dada, R. (2025). Genetics and epigenetics of Alzheimer's disease: Understanding pathogenesis and exploring therapeutic potential. *Journal of Molecular Neuroscience, 75*, 72. https://doi.org/10.1007/s12031-025-02363-2.

Thapa, R., Moglad, E., Afzal, M., Gupta, G., Bhat, A. A., Hassan almalki, W., ... Ali, H. (2024). The role of sirtuin 1 in ageing and neurodegenerative disease: A molecular perspective. *Ageing Research Reviews, 102*, 102545. https://doi.org/10.1016/j.arr.2024.102545.

Vargas-Landin, D.B., Pflüger, J., Nguyen, T.V., & Lister, R. (2024). Generation of whole-genome bisulfite sequencing libraries for comprehensive DNA methylome analysis (pp. 383–390). https://doi.org/10.1007/978-1-0716-4051-7_19.

Wagner, F. F., Zhang, Y.-L., Fass, D. M., Joseph, N., Gale, J. P., Weïwer, M., ... Holson, E. B. (2015). Kinetically selective inhibitors of histone deacetylase 2 (HDAC2) as cognition enhancers. *Chemical Science, 6*(1), 804–815. https://doi.org/10.1039/C4SC02130D.

Wang, C., Shen, D., Hu, Y., Chen, J., Liu, J., Huang, Y., ... Ma, H. (2023). Selective targeting of class I HDAC reduces microglial inflammation in the entorhinal cortex of young APP/PS1 mice. *International Journal of Molecular Sciences, 24*(5), 4805. https://doi.org/10.3390/ijms24054805.

Wang, S., Mu, G., Qiu, B., Wang, M., Yu, Z., Wang, W., ... Yang, Y. (2021). The function and related diseases of protein crotonylation. *International Journal of Biological Sciences, 17*, 3441–3455. https://doi.org/10.7150/ijbs.58872.

Woźniak, K., Gardian-Baj, M., Jung, M., Hedesz, P., Jung, M., Żuk-Łapan, A., ... Popczyńska, J. (2024). Alzheimer's disease - A comprehensive review. *Journal of Education, Health and Sport, 56*, 195–209. https://doi.org/10.12775/JEHS.2024.56.013.

Xie, H., Yang, E., Wang, C., Peng, C., & Ji, L. (2023). Targeting HDAC11 activity by FT895 restricts EV71 replication. *Virus Research, 330*, 199108. https://doi.org/10.1016/j.virusres.2023.199108.

Xu, H., & Jia, J. (2021). Single-cell RNA sequencing of peripheral blood reveals immune cell signatures in Alzheimer's disease. *Frontiers in Immunology, 12*. https://doi.org/10.3389/fimmu.2021.645666.

Xu, M., Xiong, L., Qin, Z., Yu, X., Chen, T., Zhang, X., ... Zou, Z. (2026). Epigenetic changes in Alzheimer's disease and interventions for therapy. *Neuropsychiatric Disease and Treatment, 22*, 1–17. https://doi.org/10.2147/NDT.S576404.

Ye, J., Zhong, S., Wan, H., Guo, X., Yao, X., Liu, Q., ... Xiao, S. (2025). Upregulated astrocyte HDAC7 induces Alzheimer-like tau pathologies via deacetylating transcription factor-EB and inhibiting lysosome biogenesis. *Molecular Neurodegeneration, 20*(1), 5. https://doi.org/10.1186/s13024-025-00796-2.

Zhang, L., Liu, C., Wu, J., Tao, J., Sui, X., Yao, Z., ... Qin, C. (2014). Tubastatin A/ACY-1215 improves cognition in Alzheimer's disease transgenic mice. *Journal of Alzheimer's Disease, 41*, 1193–1205. https://doi.org/10.3233/JAD-140066.

Zhang, X., Wang, X., Shivashankar, G. V., & Uhler, C. (2022). Graph-based autoencoder integrates spatial transcriptomics with chromatin images and identifies joint biomarkers for Alzheimer's disease. *Nature Communications, 13*(1), 7480. https://doi.org/10.1038/s41467-022-35233-1.

Zhang, Y., Zhao, Y., Ao, X., Yu, W., Zhang, L., Wang, Y., & Chang, W. (2021). The role of non-coding RNAs in Alzheimer's disease: From regulated mechanism to therapeutic targets and diagnostic biomarkers. *Frontiers in Aging Neuroscience, 13*. https://doi.org/10.3389/fnagi.2021.654978.

Zhang, M., Wang, W., Ye, Q., Fu, Y., Li, X., Yang, K., ... Li, W.-D. (2024). Histone deacetylase inhibitors VPA and WT161 ameliorate the pathological features and cognitive impairments of the APP/PS1 Alzheimer's disease mouse model by regulating the expression of APP secretases. *Alzheimer's Research & Therapy, 16*(1), 15. https://doi.org/10.1186/s13195-024-01384-0.

Zhang, Z.-Y., & Schluesener, H. J. (2013). Oral administration of histone deacetylase inhibitor MS-275 ameliorates neuroinflammation and cerebral amyloidosis and improves behavior in a mouse model. *Journal of Neuropathology & Experimental Neurology, 72*(3), 178–185. https://doi.org/10.1097/NEN.0b013e318283114a.

Zheng, Z., Zeng, S., Liu, C., Li, W., Zhao, L., Cai, C., ... He, Y. (2021). The DNA methylation inhibitor RG108 protects against noise-induced hearing loss. *Cell Biology and Toxicology, 37*(5), 751–771. https://doi.org/10.1007/s10565-021-09596-y.

Zhao, W., Zhang, J., Chen, K., Yuan, J., Zhai, L., & Tan, M. (2025). Mass spectrometry-based characterization of histone post-translational modification. *Current Opinion in Chemical Biology, 88*, 102622. https://doi.org/10.1016/j.cbpa.2025.102622.

Zhou, Y., Ge, Y., Liu, Q., Li, Y.-X., Chao, X., Guan, J.-J., ... Zhang, Q. (2021). LncRNA BACE1-AS promotes autophagy-mediated neuronal damage through the miR-214-3p/ATG5 signalling axis in Alzheimer's disease. *Neuroscience, 455*, 52–64. https://doi.org/10.1016/j.neuroscience.2020.10.028.

CHAPTER NINE

Integrated SDH, COX histochemistry, and MHC fibre typing in skeletal muscles: Methods and interpretation

Aastha Munjal[a,1], Ridhi Singh[a,1], Ashwini Bhelave[a,1], Shwetha Sekar[b], Ninitha Asirvatham-Jeyaraj[b], and Nagalingam R. Sundaresan[a,*]

[a]Department of Microbiology and Cell Biology, Indian Institute of Science, Bengaluru, India
[b]Department of Biotechnology, Bhupat and Jyoti Mehta School of Biosciences, Indian Institute of Technology Madras, Chennai, India
*Corresponding author. e-mail address: rsundaresan@iisc.ac.in

Contents

1. Introduction 224
2. Sirtuins and skeletal muscle metabolism: Applications of SDH/COX and MHC fibre typing in metabolic research 225
   2.1 Exercise training 225
   2.2 Obesity and insulin resistance 227
   2.3 Ageing and sarcopenia 227
   2.4 Genetic manipulation of metabolic regulators 229
3. Biochemical principles underlying SDH/COX histochemistry and MHC fibre typing in skeletal muscles 229
   3.1 Succinate dehydrogenase (SDH) activity assay 229
   3.2 Cytochrome c oxidase (COX) histochemistry 230
   3.3 Muscle fibre typing using MHC immunohistochemistry 230
4. Methodology 231
   4.1 Tissue sectioning 231
   4.2 SDH staining 233
   4.3 COX staining 234
   4.4 MHC fibre typing 235
5. Quantitative image analysis 238
   5.1 Imaging acquisition standards 238
   5.2 Statistical considerations 239
6. Interpretation of results 239
   6.1 SDH/COX staining 239
   6.2 MHC fibre typing 244

[1] Equal authorship

Methods in Enzymology, Volume 733
ISSN 0076-6879, https://doi.org/10.1016/bs.mie.2026.05.046

7. Strengths and limitations of SDH/COX and MHC fibre typing histochemical approaches 244
7.1 Strengths 244
7.2 Limitations 244
8. Troubleshooting 245
8.1 SDH Staining 245
8.2 COX staining 245
8.3 MHC immunostaining 245
9. Future directions 248
References 248

Abstract

Skeletal muscle plays a vital role in metabolic homeostasis, accounting for the majority of glucose uptake, lipid oxidation, and adaptive thermogenesis. Its plasticity enables rapid, controlled remodelling in response to exercise, nutrients, hormonal changes, ageing, and disease. This metabolic plasticity is due to fibre-type heterogeneity. Each muscle fibre has distinct contractile and bioenergetic properties. Sirtuins are known critical regulators of skeletal muscle mitochondrial content and oxidative metabolism. Sirtuins, are NAD^+-dependent acetylases and deacetylases that regulate mitochondrial biogenesis, redox balance, and cellular response to stress. Thus, studying the role of sirtuins in muscle physiology requires assays that can identify metabolic and contractile phenotypes. In this chapter, we provide a comprehensive histochemistry protocol for succinate dehydrogenase (SDH) and cytochrome c oxidase (COX) to assess mitochondrial oxidative capacity, and Myosin Heavy Chain (MHC) immunohistochemistry to assess fibre-type classification. Additionally, we have discussed detailed guidance for troubleshooting the critical steps of the protocol, including cryoinjury, tissue sectioning, staining optimisation, and imaging.

1. Introduction

Skeletal muscle accounts for 30–40 % of body weight and exhibits highly heterogeneous tissue composition and metabolic profiles (Wong et al., 2025; Talbot & Maves, 2016). Based on distinct enzyme profiles, contractile kinetics, and metabolic preferences, muscle fibres are classified as slow-oxidative type I muscle fibres, fast-oxidative/glycolytic type IIA, and fast-glycolytic type IIX/IIB (Talbot & Maves, 2016). Oxidative fibres depend on mitochondrial respiration for energy generation, whereas glycolytic fibres generate ATP anaerobically (Bourdeau Julien et al., 2018). Muscles also switch between glycolytic and oxidative fibres (Pereyra et al., 2022). This heterogeneity in fibre types and the ability to switch between fibre types enables the muscle to generate force rapidly, exhibit endurance, and maintain systemic metabolic homeostasis in response to changes in nutrition, energy demand (such as during physical activity or disuse), and with age (Wong et al., 2025; Pette & Staron, 2000).

At the molecular level, skeletal muscle homeostasis and fibre type switching are mainly regulated by a family of post-translational modifying enzymes, Sirtuins. Sirtuins are NAD^+-dependent histone deacetylases that regulate various metabolic functions in the muscles, including glucose and lipid homeostasis, mitochondrial biogenesis, and various cellular stress responses (Smith et al., 2000; Schwer & Verdin, 2008). They are known to coordinate metabolic pathways to maintain homeostasis under stress conditions such as caloric restriction, metabolic disorders, or nutrient stress (Yu & Auwerx, 2009).

Histological assays that help characterise metabolic and contractile phenotypes in muscle fibres are essential to study the role of Sirtuins in skeletal muscle metabolic homeostasis. Staining for mitochondrial enzymes SDH/COX helps visualise mitochondrial density and activity, and enables phenotypic assessment of oxidative metabolism at the fibre level in skeletal muscle (Ross, 2011). Myosin Heavy Chain (MHC) immunohistochemistry helps distinguish slow-oxidative, fast-oxidative/glycolytic, and fast-glycolytic fibres based on MHC protein isoform expression (Sokoloff et al., 2007; Bloemberg & Quadrilatero, 2012). Altogether, integrating SDH, COX, and MHC fibre typing assays provides a powerful framework for studying mitochondrial content and oxidation to understand the regulatory role of Sirtuins in skeletal muscle pathophysiology.

2. Sirtuins and skeletal muscle metabolism: Applications of SDH/COX and MHC fibre typing in metabolic research

2.1 Exercise training

Exercise training can result in enhanced oxidative capacity, fibre-type transitions, or both. SDH/COX and MHC fibre typing characterise exercise-induced adaptations and also help distinguish contractile remodelling (MHC) from metabolic remodelling (SDH/COX). The main types of exercise adaptations, along with their staining and interpretation, are listed in Table 1.

Sirtuins are critical for metabolic rewiring during exercise training adaptations. They regulate key mitochondrial enzymes to coordinate

Table 1 SDH/COX and MHC fibre typing across different training adaptations.

Training type	MHC change	SDH	COX	Interpretation	References
Endurance Training (Eg.: treadmill running, swimming)	Fast to slow shift	Increase	Increase	Enhanced oxidative metabolism and fatigue resistance	Lin et al. (2002), Handschin et al. (2007)
Resistance Training (Eg.: weight-loaded jumping, squats, weight-loaded vertical ladder climbing)	Minimal fibre-type shift (often Type IIX/IIB hypertrophy)	Slight increase or unchanged	Slight increase or unchanged	Strength gain with modest metabolic remodelling	Parry et al. (2020)

mitochondrial and contractile adaptations in skeletal muscles (Gibril et al., 2024). Exercise increases NAD^+ levels in skeletal muscle, activating SIRT1 and SIRT3, which deacetylate and activate PGC-1α, promoting mitochondrial biogenesis and oxidative metabolism (Palacios et al., 2009; Huang et al., 2016). This, in turn, enhances SDH and COX activity and supports the shift towards oxidative (Type I/IIa) fibres seen with endurance training, thereby augmenting both oxidative capacity and fatigue resistance. Mechanistically, endurance exercise activates pathways such as AMPK, CaMK, and PGC-1α signalling (Chen et al., 2025). These pathways coordinate mitochondrial biogenesis, oxidative enzyme expression, and slow-twitch fibre gene programs (Chen et al., 2025).

2.2 Obesity and insulin resistance

In obesity and insulin resistance, mitochondrial oxidative enzyme activity is significantly impaired in skeletal muscles. This results in oxidative fibre depletion (Oberbach et al., 2006; Fiorenza et al., 2024). SDH/COX staining is used to visualise these changes. Although Sirtuins are not directly implicated in regulating SDH or COX activity, they are known to regulate the activity of several mitochondrial oxidative enzymes, resulting in impaired mitochondrial oxidation observed in insulin-resistant muscles. For instance, both SIRT1 and SIRT3 are depleted in insulin-resistant muscles (Zhang et al., 2015; Lantier et al., 2015). SIRT1 deacetylates and activates PGC-1α to promote mitochondrial biogenesis under conditions of energy depletion and stress (Gurd, 2011) Depletion of SIRT1 is accompanied by a loss of mitochondrial content in insulin-resistant skeletal muscles, as reflected by reduced SDH/COX staining (Zhang et al., 2015). On the other hand, SIRT3, the primary mitochondrial deacetylase, is known to deacetylate and regulate the activity of several enzymes involved in the TCA cycle, fatty acid oxidation, and Complex I of the electron transport chain (ETC) (Lantier et al., 2015). Thus, SIRT3 depletion in insulin-resistant muscles leads to reduced mitochondrial oxidative activity and reduced SDH/COX staining (Lantier et al., 2015).

2.3 Ageing and sarcopenia

Ageing-associated sarcopenia is characterised by progressive loss of muscle mass, strength, and metabolic flexibility (Larsson et al., 2019). At the molecular level, ageing is associated with decreased NAD^+ levels and reduced sirtuin expression (Poljsak et al., 2023). Reduced SIRT1 and SIRT3 in ageing muscles impair mitochondrial function, increase oxidative

Table 2 Known regulatory targets of Sirtuins in skeletal muscles.

Sirtuin	Known role in skeletal muscle health	Effect on COX/SDH levels	Effect on MHC fibre type transition	References
SIRT1	SIRT1 activates PGC-1α, increases mitochondrial biogenesis, and promotes oxidative remodelling	SIRT1 overexpression: Increased SDH (dark blue fibres) and COX (dark brown fibres)	SIRT1 transgenic mice: Increased Type I oxidative fibres (fast to slow transition)	Chalkiadaki et al. (2014)
SIRT2	SIRT2 alleviates stress-induced atrophy, regulates FOXO1 and FOXO3	SIRT2 knockout and muscle-specific SIRT2-transgenic mice: No significant difference in the SDH and COX in muscles (blue/mixed brown fibres)	SIRT2 knockout mice: Increased Type I and Type IIA muscle fibres	Tamta et al. (2025)
SIRT3	SIRT3 levels increase in caloric-restricted mice, regulate Pyruvate Dehydrogenase (PDH), and increase oxidative capacity in muscles	SIRT3 overexpression: Increased SDH (dark blue/purple fibres) and COX (dark brown fibres)	SIRT3 transgenic mice: Increased type I muscle fibres in the quadriceps muscle	Lin et al. (2014)
SIRT5	SIRT5 slows skeletal muscle ageing, desuccinylates TBK1, and improves fibre diameter	SIRT5 Overexpression: Increased SDH/COX SIRT5 knockdown/knockout: reduced mitochondrial function	Not validated	Zhao et al. (2025), Zhang et al. (2024)
SIRT6	SIRT6 downregulates Sox6 (key repressor of slow fibre-specific genes), protects against glucocorticoid-induced skeletal muscle atrophy, and regulates IGF/PI3K/AKT signalling	SIRT6 knockout mice: Reduced SDH activity (less than 50 % SDH positive) and reduced COX (light brown fibres); SIRT6 transgenic mice: Increased SDH activity (more dark blue fibres)	SIRT6 knockout mice: Increased type-IIB (glycolytic fibres) and decreased oxidative fibres	Mishra et al. (2022), Song et al. (2022)

stress, and reduce muscle regeneration capacity (Li & Cai, 2022). SDH/COX staining and MHC fibre typing reveal both contractile remodelling and mitochondrial dysfunction observed during ageing. A hallmark of ageing muscle is mitochondrial dysfunction, especially impaired ETC activity (Zhang et al., 2025). SDH staining often shows the presence of SDH^+/COX^- fibres and increased mitochondrial DNA deletions (Fogarty et al., 2020). Mechanistically, this occurs due to accumulation of mitochondrial DNA mutations, oxidative stress, impaired mitochondrial quality control, and reduced mitophagy (Herbst et al., 2007). MHC fibre typing also reveals preferential loss of type II fibres due to atrophy, accompanied by a relative increase in proportion of type I fibres observed due to denervation-reinnervation by slow motor neurons (Rowan et al., 2012; Horwath et al., 2025).

2.4 Genetic manipulation of metabolic regulators

Transcriptional coactivators and mitochondrial regulators coordinate contractile and oxidative remodelling in muscle. Sirtuins act as NAD^+-sensitive metabolic regulators linking transcriptional control to mitochondrial enzyme activity (Chalkiadaki et al., 2014; Tamta et al., 2025; Kane & Sinclair, 2018). Various studies utilising genetically engineered animal models of Sirtuins (including SIRT2 knockout and transgenic over-expressing mice) have reported distinct regulatory targets of Sirtuins in skeletal muscles, as listed in Table 2.

3. Biochemical principles underlying SDH/COX histochemistry and MHC fibre typing in skeletal muscles

3.1 Succinate dehydrogenase (SDH) activity assay

Succinate dehydrogenase is an enzyme present in the inner mitochondrial membrane (Huang & Millar, 2013). It participates in both the Krebs cycle and the Electron Transport Chain (Huang & Millar, 2013; Moosavi et al., 2019). In the Krebs cycle, it catalyses the oxidation of succinate to fumarate (Krebs & Eggleston, 1940). The hydrogen atoms released during this reaction are transferred to FAD, reducing it to $FADH_2$. SDH then transfers electrons from $FADH_2$ to ubiquinone, linking oxidative phosphorylation to the ETC in the mitochondria (Krebs & Eggleston, 1940; Ryan & O'Neill, 2020).

Since SDH is a vital, stable, and active enzyme embedded in the mitochondrial membrane, it serves as a marker of active mitochondria in cells. As a result, SDH staining is indicative of the mitochondrial density and activity in tissues (Punsoni et al., 2017). To estimate the mitochondrial content using the SDH activity histochemical assay, unfixed cryopreserved tissue sections with preserved SDH activity are treated with Nitro Blue Tetrazolium (NBT). NBT acts as an electron acceptor that changes colour from colourless to blue upon accepting electrons released during the SDH-catalysed oxidation of succinate to fumarate in the tissue mitochondria (Punsoni et al., 2017; Zogby et al., 2017). As a result, highly oxidative tissues are stained dark blue/purple, whereas tissues with lower oxidative capacity are stained a lighter blue or remain uncoloured. Factors such as substrate concentration, incubation time, and section thickness all influence the SDH staining (Punsoni et al., 2017; Zogby et al., 2017).

3.2 Cytochrome c oxidase (COX) histochemistry

Cytochrome c oxidase is the terminal complex in the electron transport chain that transfers electrons from cytochrome c to molecular oxygen, generating a proton gradient across the inner and outer mitochondrial membranes (Kadenbach & Huttemann, 2015). As a result, it serves as a marker for mitochondrial respiration. To estimate the mitochondrial content using the COX histochemical assay, unfixed cryopreserved tissue sections with preserved COX activity are treated with 3,3′-diaminobenzidine (DAB) (Ross, 2011; Seligman et al., 1968). DAB serves as a chromogenic substrate that, in the presence of cytochrome c, is oxidised by COX to a brown precipitate. In this manner, healthy muscle fibres with functional mitochondria stain brown, whereas COX-deficient muscles remain colourless (Ross, 2011; Seligman et al., 1968).

3.3 Muscle fibre typing using MHC immunohistochemistry

Myosin Heavy Chain protein is a major motor protein that determines muscle contraction, speed, and regulates energy through its ATPase activity (Pette & Staron, 2000). Muscle fibre typing by immunohistochemistry is based on the principle that different muscle fibres express distinct MHC protein isoforms, resulting in heterogeneous fibre bundles in skeletal muscle. Type I fibres express MHC-I, type IIA fibres (oxidative) express MHC-IIA, type IIB fibres (glycolytic) express MHC-IIB, and type IIX (glycolytic) fibres express MHC-IIX protein isoform (Pette & Staron, 2000). Using isoform-specific monoclonal antibodies against MHC proteins provides accurate molecular classification and helps determine shifts in muscle fibre type (Pette & Staron, 2000; Serrano et al., 1996). Studies have shown that fibre transitions from glycolytic to oxidative or from oxidative to glycolytic occur during various metabolic conditions. This transition can be characterised using MHC fibre typing immunohistochemistry (Kelsey et al., 2020).

4. Methodology

4.1 Tissue sectioning

4.1.1 Materials and reagents

OCT (optimal cutting temperature) compound, isopentane, cryomoulds, liquid nitrogen, pre-coated positively charged glass slides, aluminium foil, stainless steel trough, forceps.

4.1.2 Equipment

Cryostat, freezer (−80 °C).

4.1.3 Tissue collection

Once the muscle of interest is harvested, it must be preserved appropriately and with care to avoid cryoinjury.

1. Collect the harvested tissue directly into moulds containing OCT (optimal cutting temperature) compound at room temperature.
2. Pre-cool the isopentane by dipping the trough containing it into liquid nitrogen until white crystals form. This process usually takes five to ten minutes.
3. Once the isopentane crystals begin to form, place the trough at room temperature, and immerse the OCT moulds in isopentane until the OCT compound transitions completely from transparent to white (Ghag et al., 2025). This process usually takes thirty seconds to one minute.
4. Pour out the excess isopentane accumulated in the OCT moulds, and immediately wrap the moulds in aluminium foil, place securely in an airtight box, and store at −80 °C until cryosectioned (Ghag et al., 2025; Battey et al., 2024).

4.1.4 Tissue cryosectioning

1. Prior to beginning cryosectioning of the OCT-embedded tissues, turn on the cryostat and allow it to reach the optimised chamber and head temperatures for your tissue.

2. Maintain a low ambient room temperature (16–18 °C) to keep the cryostat temperature stable.
3. Place 2 separate blades for trimming and sectioning inside the cooling cryostat (Ghag et al., 2025). Additionally, place a set of appropriately labelled, charged or Poly-L-lysine-coated glass slides at room temperature (Huang et al., 1983).
4. Once the desired temperature is reached, load the OCT blocks onto the chuck using a small amount of OCT at room temperature, and place them in the cryostat chamber until the OCT is dry and the block is securely loaded onto the chuck (Ghag et al., 2025).
5. Initially, load the trimming blade and trim 100 μm sections of excess OCT until the tissue becomes visible. Use a gentle-bristle brush to brush off the shavings (Battey et al., 2024).
6. Once visible, cut the tissue into 8–10 μm thick sections. Use a fine brush to carefully lay the section flat on the cryostat stage and collect it onto a charged glass slide by gently pressing the slide onto the section.
7. Place the glass slide containing the section inside the cryostat until ready to be stored (Battey et al., 2024).

4.1.5 Storage conditions

1. Avoid storage and stain the sections immediately to prevent loss of enzymatic activity. If unavoidable, the sections can be stored securely for a short term (1–2 weeks) in a parafilm-wrapped and aluminium-foil-covered glass box at −80 °C until ready for staining.

4.1.6 Notes

1. Care should be taken to avoid the formation of bubbles in and around the tissue once it is

   placed into the OCT-containing mould (Battey et al., 2024). Ideally, the tissue should not be exposed at either end of the block and should be placed in the centre of the mould, completely immersed in the OCT compound.
2. For cryosectioning of skeletal muscles, the recommended cryostat chamber temperature is −21 °C.
3. Ensure that the OCT block is loaded onto the chuck in the desired orientation for cryosectioning. Typically, transverse muscle sectioning is desirable for downstream histological applications (Ghag et al., 2025).
4. Avoid freeze-thawing tissue blocks to prevent tissue from turning brittle and cracking during subsequent sectioning attempts.

4.2 SDH staining

4.2.1 *Materials and reagents*

Phosphate buffer (PBS, pH 7.4), sodium succinate, magnesium chloride, nitro blue tetrazolium, double-distilled water, ethanol, xylene, DPX mounting medium, hydrophobic barrier pen.

4.2.2 *Equipment*

Light microscope with a digital camera and imaging software, chemical fume hood, pH meter, vortex, incubator.

4.2.3 *Reagent preparation*

1. Prepare 0.2 M phosphate buffer using Na_2HPO_4 and NaH_2PO_4, and adjust the pH to 7.4.
2. To this buffer, add 0.2 M sodium succinate, followed by 0.1 M $MgCl_2$ and 2.4 mM NBT (nitro blue tetrazolium).
3. Vortex to mix the solution.

4.2.4 *SDH staining*

1. If staining is being performed using slides stored at −80 °C, allow them to dry at room temperature briefly for one to five minutes, and gently wipe off any condensate forming on the slide.
2. Using a hydrophobic barrier pen (PAP or Dako), outline the section on the slide.
3. Add 100 μL of SDH solution to it.
4. Incubate the slide at 37 °C for 1 h in the dark. If the solution changes from yellow to blue/purple after incubation, the reaction has worked.
5. Tilt and dab the slide to remove excess solution.
6. Wash the section with double-distilled water for 3 min to remove unbound stain particles adhered onto the slide. Repeat the wash if necessary.
7. Subsequently, dehydrate the sample by incubating it in 50 % ethanol for 2 min.
8. Dip the slides into a trough containing xylene for 3 min, then air-dry them for 10 min at room temperature.
9. Once dry, add DPX mounting medium onto the section to prepare the slide for imaging (Fig. 1).

4.2.5 *Notes*

1. Take care to pipette the staining solution gently along the drawn boundary and not directly onto the section.

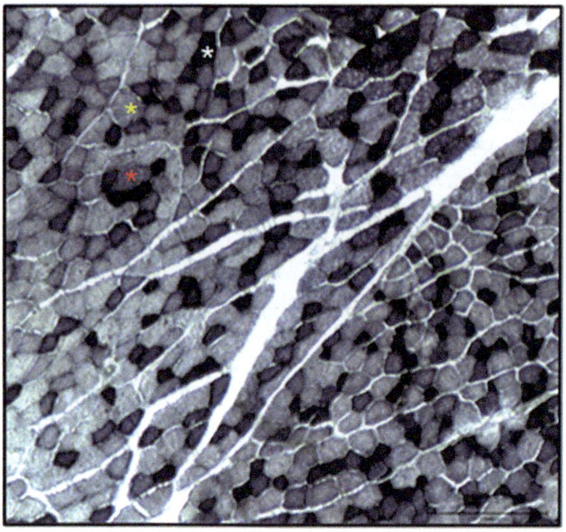

Fig. 1 Histochemical staining for Succinate dehydrogenase in mouse tibialis anterior muscle section. Scale bar = 275 μm (magnification 10x). In the image, the white asterisk represents predominantly oxidative fibres, i.e. Type I fibre (dark blue); the red asterisk represents oxidative and glycolytic fibres, i.e. Type IIA fibres (intermediate blue); and the yellow asterisks represent glycolytic fibres, i.e. Type IIB fibres (light blue).

2. The working incubation medium for SDH staining is to be freshly prepared prior to staining.
3. The solution will appear heterogeneous even after vortexing. Thus, mix the solution immediately before adding it to the slides for staining.

4.3 COX staining

4.3.1 Materials and reagents

Phosphate buffer (PBS, pH 7.4), cytochrome oxidase, bovine catalase, DAB (3,3'-diamino benzidine), double-distilled water, ethanol, xylene, DPX mounting medium, hydrophobic barrier pen.

4.3.2 Equipment

Light microscope with a digital camera and imaging software, chemical fume hood, pH meter, vortex, incubator.

4.3.3 Reagent preparation

The working incubation medium for COX staining is to be freshly prepared prior to staining.

1. Prepare 0.1 M phosphate buffer of pH 7.0.

2. To this buffer, add 100 μM Cytochrome oxidase, 2 μg/mL bovine catalase, and 1X DAB (3,3'-diamino benzidine).
3. Vortex to mix the solution before directly adding onto the slides for staining.

4.3.4 COX staining

1. If staining is being performed using slides stored at −80 °C, allow them to dry at room temperature briefly for 60 min.
2. Gently wipe off any condensate forming on the slide if present.
3. Using a hydrophobic barrier pen (PAP or Dako), outline the section on the slide.
4. Add 100 μL of COX solution to it.
5. Incubate the slide at 37 °C in the dark for 30 min (until the sample appears brown).
6. Tilt the slide and dab off the excess solution, followed by three 1X PBS washes for 10 min each and one final wash with double-distilled water for 3 min at room temperature.
7. Dehydrate the sample by incubating it in 50 % ethanol for 2 min.
8. Dip the slides into a trough containing xylene for 3 min and air-dry them for 10 min at room temperature.
9. Once dry, add DPX mounting medium onto the section to prepare the slide for imaging. Storing the slide overnight before imaging will yield clearer images (Fig. 2).

4.3.5 Notes

1. The working incubation medium for SDH staining is to be freshly prepared prior to staining.
2. The solution will appear heterogeneous even after vortexing. Thus, mix the solution immediately before adding it to the slides for staining.
3. Take care to pipette the staining solution gently along the drawn boundary and not directly onto the section

4.4 MHC fibre typing

4.4.1 Materials and reagents

Phosphate buffer (PBS, pH 7.4), tween 20 (0.1 %) in PBS, bovine serum albumin, acetone, primary antibody: BAD5, 6H1, SC71, and BF-F3, 1, Alexa-Fluor conjugated secondary antibody, 85 % glycerol solution (mounting medium), hydrophobic barrier pen, transparent nail polish.

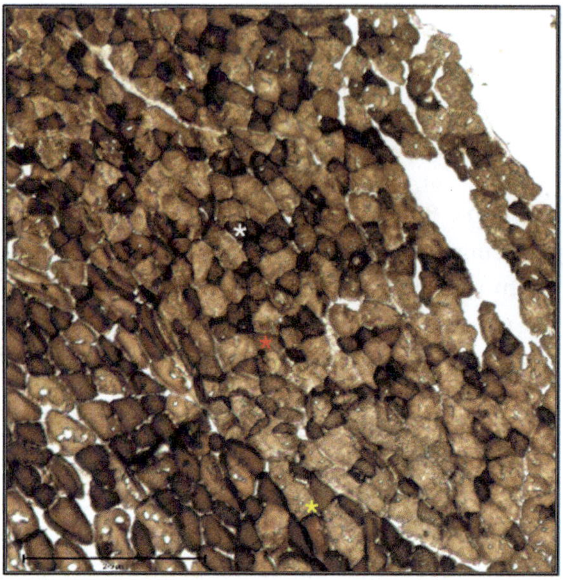

Fig. 2 Histochemical staining for cytochrome c oxidase (COX), mouse tibialis anterior muscle section. Scale bar = 275 μm (magnification 10x). In the image, the white asterisk represents Type I fibres (dark brown), the red asterisk represents COX intermediate fibres, and the yellow asterisks represent COX-deficient, Type II fibres (light brown).

4.4.2 Equipment

Chemical fume hood, pH meter, vortex, humidified chamber.

4.4.3 Primary antibody preparation

1. Prepare 5% BSA in wash buffer (PBS + 0.05 % Tween20).
2. Add BAD5 (MHC-I) and 6H1 (MHC-IIX) primary antibody at a 1:50 dilution.
3. Mix well and use this cocktail mixture for one tissue section.
4. Prepare another antibody cocktail with SC71 (MHC-IIA) and BF-F3 (MHC-IIB) at a 1:50 dilution in 5 % BSA to stain the other tissue section in the same orientation. This ensures that the same section of tissue is stained for all four types of muscle fibres. [Prepare 200 μl of primary antibody to cover two tissue sections].

4.4.4 Secondary antibody preparation

Prepare the secondary antibody by adding Alexa Fluor-conjugated secondary antibody specific to the primary antibody's host species into 1% BSA at a 1:250 dilution. Prepare 200 μl of secondary antibody to cover two tissue sections.

4.4.5 MHC fibre typing

1. If staining is being performed on freshly made slides, store them at −20 °C for 12−16 h before staining.
2. Remove the slides from −20 °C or −80 °C and incubate them at room temperature (RT) for 30 min.
3. Mark the boundary around the OCT-embedded tissue section on the slide using a hydrophobic barrier pen (PAP or Dako pen).
4. Add ice-cold acetone to the tissue section and incubate at −20 °C for 10 min.
5. Gently wash the tissue section with water using a pipette at an angle of 15−20 ° to avoid dislodging the tissue.
6. Remove the water using a cotton swab or a tissue paper, then slowly add 200 μL of 5% BSA (made in 0.1 % Tween 20) to the tissue section and incubate at RT for 30 min.
7. Carefully remove the BSA using a pipette or by gently dabbing the slide onto the absorbent tissue paper.
8. Wash the tissue twice for 1 min per wash using a wash buffer.
9. Place the slides in a humid chamber, mark the boundary again with a PAP pen, and add the primary antibody to the tissue section.
10. Incubate the slides at RT for 90 min or at 4 °C for 12−14 h in a humid chamber.
11. Remove the primary antibody and wash the slide using wash buffer, repeating 3 times for 5 min each at RT. To remove excess wash buffer, use a cotton swab or tissue wick.
12. Add the cocktail of the respective Alexa Fluor-conjugated secondary antibody to the tissue sections (100−200 μl per section).
13. Place the slides in a humid chamber and incubate the slides for 60 min at RT in the dark.
14. After secondary antibody staining, use a tissue wick to remove excess antibody, then wash the tissue with PBS twice for 5 min each at RT.
15. Remove the excess water and mount the tissue using 85 % glycerol solution.
16. Seal the coverslipwith transparent nail polish.
17. Perform confocal microscopy using appropriate lasers (Fig. 3).

4.4.6 Notes

1. Ensure the volume of antibody solutionsis sufficient to cover the surface of the slides.

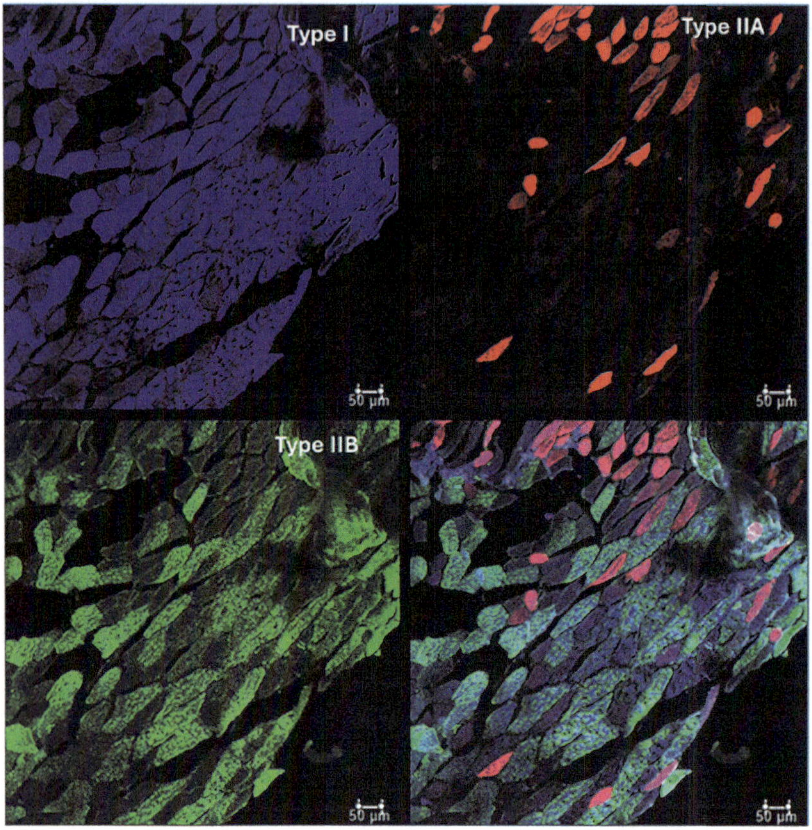

Fig. 3 Muscle fibre typing or MHC staining of a mouse tibialis anterior muscle section. Scale bar = 50 μm (magnification 63x). The image represents Type I fibre in blue; Type IIA fibre in red; Type IIB fibre in green; and the merged image of all fibre types.

2. Use a cotton swab, tissue wick, or a pipette at a 10–20° angle from the slide to add and remove solution during the antibody-washing and probing to avoid dislodging or damaging the tissue sections.

5. Quantitative image analysis

5.1 Imaging acquisition standards

5.1.1 For SDH/COX staining

1. Use a light microscope equipped with a digital camera to capture images initially at 10X objective magnification, then at 20X.

2. Inspect overall muscle morphology at low magnification before capturing detailed staining patterns at higher magnification.
3. Save the images in standard format (.jpeg or.tiff) for publication.

5.1.2 For MHC Fibre typing

1. Use a confocal microscope equipped with 488 nm, 594 nm and 647 nm lasers.
2. Capture images initially at 10X objective magnification, then at 40X or 60X.
3. Inspect overall muscle morphology at low magnification before capturing detailed staining patterns at higher magnification.
4. Save the images in standard format (.jpeg or.tiff) for publication.

5.2 Statistical considerations

1. Typically, 400–600 fibres per animal or human sample should be analysed for SDH/COX and MHC Fibre Typing to achieve statistical significance for larger muscles such as gastrocnemius and quadriceps.
2. For smaller muscles (e.g. soleus, biceps) and atrophied muscles, analyse the whole muscle tissue section.
3. A minimum of 3 biological replicates is recommended to reliably infer results.
4. Perform all statistical analyses using GraphPad Prism.
5. Import data from the Microsoft.xls file into a new GraphPad Prism file and assess the dataset for normal distribution by running normality tests (D'Agostino-Pearson and Shapiro-Wilk).
6. For datasets with a Gaussian distribution, perform parametric tests. For datasets that do not follow a Gaussian distribution, use nonparametric tests.
7. Plot the data as a scatter plot with mean and standard deviation for each experimental group.

6. Interpretation of results

6.1 SDH/COX staining

SDH staining in skeletal muscles is a marker of the oxidative capacity of the fibre (Punsoni et al., 2017; Sher & Cardasis, 1976). Staining patterns of different muscle types in rodents are listed in Table 3 (Punsoni et al., 2017; Kalmar et al., 2012). Importantly, SDH staining is an enzyme activity assay,

Table 3 SDH staining patterns in rodent skeletal muscles.

Muscle	Predominant fibre type composition	Metabolic profile	SDH staining pattern	References
Quadriceps	Mixed (Type IIB/X predominant) (1) Vastus lateralis, medialis, rectus femoris (mainly Type IIA/IIX/IIB fibres) (2) Vastus intermedius predominantly Type I fibres	Predominantly glycolytic with oxidative regions	Weak to moderate (region dependent)	Burkholder et al. (1994)
Gastrocnemius (White region)	Type IIB/X fibres mostly	Glycolytic	Weak/Light	Cornachione et al. (2011)
Gastrocnemius (Red region)	Type IIA fibres and some Type I fibres	Oxidative-glycolytic	Moderate to strong	Cornachione et al. (2011)
Tibialis Anterior	Mostly Type IIB/X fibres, some Type IIA fibres	Primarily glycolytic	Weak to moderate (regional variation)	Giacomello et al. (2020)
Soleus	Mostly Type I fibres ($96.1 \pm 2.9\,\%$), the remaining are Type IIA fibres	Highly oxidative and fatigue resistant	Dark in the majority of fibres	Soukup et al. (2002)
EDL (Extensor Digitorum Longus)	Mostly Type IIB fibres ($75.7 \pm 2.2\,\%$), Type IIA fibres ($18.8 \pm 1.7\,\%$), and remaining Type I fibres	Primarily glycolytic	Weak to moderate	Soukup et al. (2002)

not a structural marker of mitochondria. While staining intensity generally correlates with mitochondrial content, it more directly reflects functional SDH enzymatic activity (Ustunel & Demir, 1997). Thus, two muscle fibres with similar mitochondrial numbers could show different SDH intensities if their mitochondrial function differs. Therefore, SDH intensity should be interpreted as a measure of functional oxidative capacity that correlates with, but does not directly equate to, total mitochondrial content in the muscle (Punsoni et al., 2017; Zogby et al., 2017; Mahadev Bhat & Sieck, 2024).

Similar to SDH, COX staining intensity indicates oxidative metabolic capacity and is typically stronger in oxidative fibres. However, while SDH is a nuclear-encoded mitochondrial enzyme that remains functional even when mitochondrial DNA is damaged, COX is a mitochondrial genome-encoded mitochondrial enzyme that is dysfunctional in mitochondrial myopathies (Murphy et al., 2012). Thus, dual staining with SDH and COX helps distinguish mitochondrial quantity from respiratory integrity. Each skeletal muscle type has a distinct SDH/COX dual-staining pattern, determined by mitochondrial content, activity, and metabolic profile (Ross, 2011; Murphy et al., 2012). When interpreting results from dual SDH/COX staining, it is important to distinguish between mitochondrial quantity and quality. While an increase in mitochondrial quantity generally suggests enhanced oxidative capacity (as observed in endurance-trained or type I-rich muscles), it is better estimated more directly using structural or abundance-based markers such as electron microscopy, mitochondrial DNA copy number, citrate synthase activity, or outer membrane proteins like TOM20 or VDAC (Blakely et al., 2013). In contrast, mitochondrial quality depends on intact electron transport chain complexes, preserved cristae structure, balanced fusion-fission dynamics, and minimal oxidative damage (Blakely et al., 2013). Thus, a muscle fibre may be characterised with low SDH/COX staining despite having greater mitochondrial abundance if the mitochondrial oxidative function is impaired. This distinction becomes particularly relevant in pathological conditions (Blakely et al., 2013; Wattez et al., 2023; Phadke, 2017). SDH/COX dual staining patterns observed under different physiological and pathological conditions are listed in Table 4.

Accurate interpretation of mitochondrial status in a tissue sample should ideally integrate structural markers of mitochondrial content with enzymatic or respiratory assays of mitochondrial function to determine whether changes reflect a loss of mitochondria, impaired mitochondrial performance, or both.

Table 4 Interpretation of SDH/COX staining under different conditions.

Condition	SDH activity	COX activity	Appearance	References
Healthy	Normal	Normal	Deep blue and brown mosaic	Kalmar et al. (2012)
Ageing	Increased SDH^{++}	Reduced (increased proportion of COX^{-} fibres)	Blue	Kimoloi et al. (2022)
Glucocorticoid induced-atrophy	Increased SDH^{+}	Increased COX^{-}	Overall reduced colour	Tamta et al. (2025)
Mitochondrial myopathy	Increased (compensatory)	Negative	Blue	Murphy et al. (2012), Murgia et al. (2019)
Intermediate/transitioning to pathology	Normal	Intermediate	Violet-grey mosaic	Murphy et al. (2012)
Nuclear DNA defects (mitochondrial biogenesis unaffected)	Reduced	Normal	Light intensity brown-blue mosaic	Hedberg-Oldfors et al. (2022)
Nuclear DNA defects (mitochondrial biogenesis affected)	Reduced	Reduced	Light intensity brown-blue mosaic	Hedberg-Oldfors et al. (2022)

Table 5 MHC fibre type transition observed in different pathophysiological conditions.

Condition	MHC transition	Functional interpretation	Metabolic change	References
Huntington's Disease	Type I to Type IIA	Slow to fast contraction	Slightly reduced oxidative metabolism	Chaturvedi et al. (2009)
Spinal Cord Injuries	Type IIA to Type IIX	Fast, more glycolytic	Reduced mitochondrial content	Kelsey et al. (2020)
Weightlifting and Sprinting	Type I to Type IIB/Type IIX	Maximum speed, fatigue-prone	Reduced oxidative capacity	Plotkin et al. (2021)
Long-Distance Running and Cycling	Type II to Type I	Slow, fatigue-resistant	Increased oxidative capacity	Lin et al. (2002), Plotkin et al. (2021)
Aging rats	Type IIB to IIX	Fast to slow, less fatigue	Reduced oxidative capacity	Larsson et al. (1993), Zhang et al. (2010)

6.2 MHC fibre typing

Muscle MHC isoforms determine contraction speed, force production, and fatigue resistance (Pette & Staron, 2000). MHC isoform shift reflects a change in the contractile phenotype of a muscle fibre (Murach et al., 2019). Fibre type transition may be attributed to adaptive physiological processes (such as endurance training) or pathological conditions (such as disuse, glucocorticoids, or denervation). These fibre type transitions reflect the underlying metabolic changes in skeletal muscle, as listed in Table 5.

7. Strengths and limitations of SDH/COX and MHC fibre typing histochemical approaches

7.1 Strengths

Key strengths of SDH/COX and MHC fibre typing-based histological approaches are:

1. Preservation of muscle tissue architecture allows for high spatial resolution and visualisation of metabolic heterogeneity within and between individual fibres. Additionally, regional adaptations (such as red versus white muscle compartments) can be directly compared within the same section.
2. Combining SDH/COX staining with MHC fibre typing enables interpreting the correlation between single-fibre metabolic and contractile muscle function.
3. Staining protocols are suitable for large experimental cohorts.
4. Staining protocols are cost-effective and scalable due to minimal requirement for advanced instrumentation.

7.2 Limitations

Despite their utility, SDH/COX and MHC fibre typing-based histological methods are characterised by some major limitations:

1. They require technical precision and tedious optimisation of various parameters, including incubation time, substrate concentration, temperature, and section thickness.
2. SDH/COX staining only enables measuring enzyme activity at the time of staining, rather than estimation of real-time mitochondrial respiration or ATP flux - thus yielding no information regarding the dynamic bioenergetic performance of the muscle.

8. Troubleshooting

Some general notes of caution for each histology method are listed below.

8.1 SDH Staining

1. Prepare fresh succinate and NBT solutions immediately before staining.
2. Ensure that NBT has dissolved thoroughly.
3. Maintain consistent $MgCl_2$ concentration for all tissue types.
4. Avoid tissue drying before incubation with the SDH staining buffer.
5. Standardise incubation time for each type of tissue individually.
6. Maintain homogeneous incubation temperature during staining.
7. Mix the SDH buffer properly before adding it to the tissue sections.
8. Use uniform section thickness (8–10 μm).
9. Capture images with identical exposure settings.

8.2 COX staining

1. Prepare fresh DAB prior to staining.
2. Carefully set the pH of staining solution (~7.0).
3. Avoid over-incubation with the staining buffer to prevent non-specific or background staining.
4. Include a strong oxidative control (e.g., cardiac muscle).
5. Avoid section folding and use of sections with uneven thickness.

8.3 MHC immunostaining

1. Use freshly cryosectioned tissue for staining.
2. Do not thaw the slides for over 60 min.
3. If using fixatives such as 4 % PFA or acetone, wash the sample to remove excess fixative prior to staining.
4. Optimise blocking conditions, including blocking buffer composition, and incubation time (e.g., 1−5 % BSA for 30 min to 1 h).
5. Avoid excessive detergent when staining the surface isoforms.
6. Validate antibody specificity.
7. Set up controls with no primary antibody to ensure that the secondary antibody is not emitting signal non-specifically.
8. Minimize freeze–thaw cycles for sectioned tissue.
9. Pipette gently throughout the protocol to avoid damaging the tissue section.

A guide to troubleshooting common errors in cryosectioning and staining is outlined in Table 6.

Table 6 Common observations and troubleshooting suggestions for SDH/COX staining and MHC fibre typing.

Sr. No.	Observation	Potential errors	Solutions
1.	Cryoinjury in the tissue section. 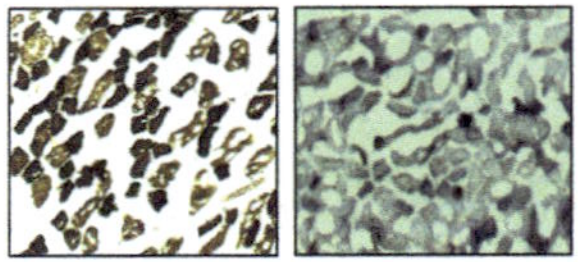	Liquid nitrogen has come into direct contact with the tissue during embedding, causing cryoinjury.	(1) Submerge the tissue OCT moulds into the pre-cooled isopentane trough in ambient conditions. (2) Do not allow the moulds to come in contact with liquid nitrogen or vapours.
2.	Cloudy deposit over the tissue section during SDH staining. 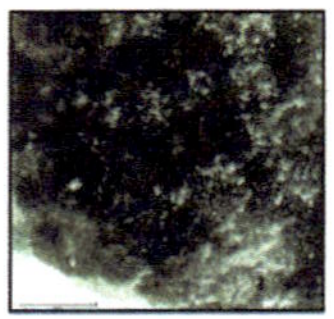	The concentration of $MgCl_2$ or phosphate buffer may be too high. [$MgCl_2$ forms aggregates with phosphate buffer, forming magnesium phosphate precipitates].	(1) Decrease $MgCl_2$ concentration to as low as 10 mM. (2) Wash with 1X PBS after incubation in SDH buffer.
3.	Cracks in the tissue section. 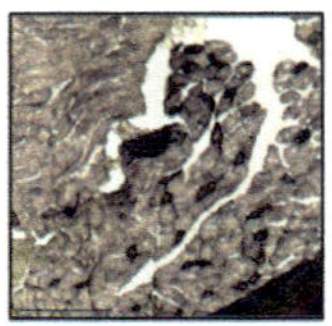	The tissue is drying out.	(1) Do not incubate the tissue in 50 % ethanol for longer than the recommended time. (2) Observe the tissue section at each step of the protocol; do not over-incubate or air-dry between two steps. [Drying time can vary with tissue section thickness and the ambient temperature]. 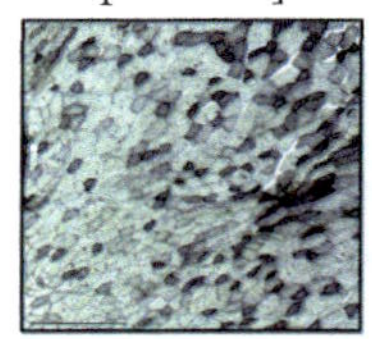

4.	Edge effect and uneven staining. 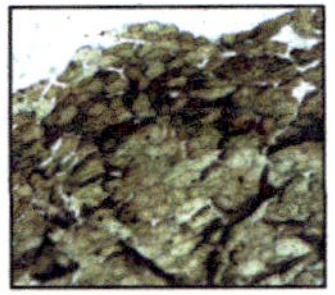 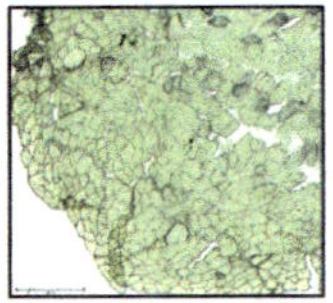	(1) Accumulation of precipitate on the edges or, (2) Lack of homogeneity in the SDH/COX staining buffer or, (3) Folding of tissue edge during cryosectioning or staining.	(1) Use a tissue wick to dry the tissue instead of dabbing the glass slides onto a tissue. (2) Coat the glass slides with Poly-L-lysine to prevent the tissue section from peeling off the slides. (3) For SDH staining: optimise $MgCl_2$ concentration. (4) For SDH staining: mix the SDH buffer before adding it to each tissue section.
5.	A few stained fibres on the edges. 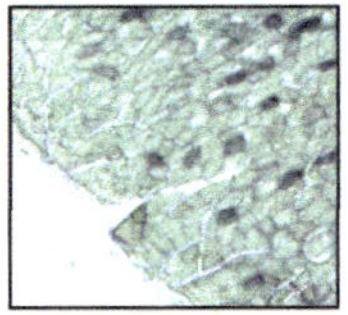	(1) OCT on the tissue or, (2) Drying out of the tissue or, (3) Insufficient time given for washing.	(1) Properly remove the OCT. (2) Do not over-dry the tissue. (3) Provide sufficient incubation time for washing the section. 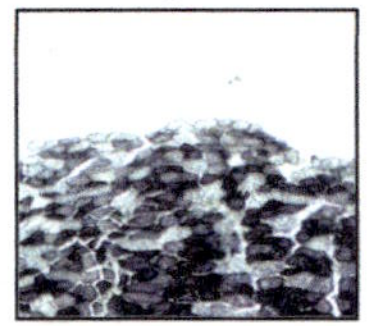
6.	Shrunken muscle fibres and an aberrant staining pattern with high background and weak signal during MHC fibre typing.	After acetone fixation, any residual acetone that is not washed off properly can dehydrate the tissue, reducing antigen accessibility and primary antibody binding.	Wash the tissue sections three times in 1X PBS for 5 min each.

9. Future directions

Advancements aimed at making SDH, COX, and MHC fibre typing histochemical techniques more user-friendly and less prone to error and misinterpretation include the automation of whole-tissue section scanning and the quantification of enzyme activity across thousands of muscle fibres. Automated sectioning of the tissue sample could reduce its susceptibility to cryo-injury and user-induced errors. This will help avoid bias in results and save time. Additionally, muscle histochemistry can be paired with transcriptomics and genomics using intact tissue to understand the impact at the genetic level, as well as metabolic states driven by molecules such as Sirtuins. Tagging this methodology with artificial intelligence (AI) can make it even more informative, helping us predict metabolism from morphology and detect early pathology.

References

Battey, E., et al. (2024). Protocol for muscle fiber type and cross-sectional area analysis in cryosections of whole lower mouse hindlimbs. *STAR Protocols, 5*(4), 103424.

Blakely, E. L., et al. (2013). Pathogenic mitochondrial tRNA point mutations: Nine novel mutations affirm their importance as a cause of mitochondrial disease. *Human Mutation, 34*(9), 1260–1268.

Bloemberg, D., & Quadrilatero, J. (2012). Rapid determination of myosin heavy chain expression in rat, mouse, and human skeletal muscle using multicolor immunofluorescence analysis. *PLoS One,* 7(4), e35273.

Bourdeau Julien, I., Sephton, C. F., & Dutchak, P. A. (2018). Metabolic networks influencing skeletal muscle fiber composition. *Frontiers in Cell and Developmental Biology, 6*, 125.

Burkholder, T. J., et al. (1994). Relationship between muscle fiber types and sizes and muscle architectural properties in the mouse hindlimb. *Journal of Morphology, 221*(2), 177–190.

Chalkiadaki, A., et al. (2014). Muscle-specific SIRT1 gain-of-function increases slow-twitch fibers and ameliorates pathophysiology in a mouse model of duchenne muscular dystrophy. *PLoS Genetics, 10*(7), e1004490.

Chaturvedi, R. K., et al. (2009). Impaired PGC-1alpha function in muscle in Huntington's disease. *Human Molecular Genetics, 18*(16), 3048–3065.

Chen, J., et al. (2025). AMPK/SIRT1/PGC-1alpha signaling pathway: Molecular mechanisms and targeted strategies from energy homeostasis regulation to disease therapy. *CNS Neuroscience & Therapeutics, 31*(11), e70657.

Cornachione, A. S., et al. (2011). Characterization of fiber types in different muscles of the hindlimb in female weanling and adult Wistar rats. *Acta Histochemica et Cytochemica, 44*(2), 43–50.

Fiorenza, M., et al. (2024). Reducing the mitochondrial oxidative burden alleviates lipid-induced muscle insulin resistance in humans. *Science Advances, 10*(44), eadq4461.

Fogarty, M. J., et al. (2020). Aging reduces succinate dehydrogenase activity in rat type IIx/IIb diaphragm muscle fibers. *Journal of Applied Physiology (1985), 128*(1), 70–77.

Ghag, N., et al. (2025). Cryopreservation method for preventing freeze-fracture of small muscle samples. *Bio-Protocol, 15*(1), e5145.

Giacomello, E., et al. (2020). Age dependent modification of the metabolic profile of the tibialis anterior muscle fibers in C57BL/6J mice. *International Journal of Molecular Sciences, 21*(11).
Gibril, B. A. A., Xiong, X., Chai, X., Xu, Q., Gong, J., & Xu, J. (2024). Unlocking the nexus of sirtuins: A comprehensive review of their role in skeletal muscle metabolism, development, and disorders. *International Journal of Biological Sciences, 20*(8), 3219–3235. https://doi.org/10.7150/ijbs.96885.
Gurd, B. J. (2011). Deacetylation of PGC-1alpha by SIRT1: Importance for skeletal muscle function and exercise-induced mitochondrial biogenesis. *Applied Physiology, Nutrition, and Metabolism = Physiologie Appliquee, Nutrition et Metabolisme, 36*(5), 589–597.
Handschin, C., et al. (2007). Skeletal muscle fiber-type switching, exercise intolerance, and myopathy in PGC-1alpha muscle-specific knock-out animals. *The Journal of Biological Chemistry, 282*(41), 30014–30021.
Hedberg-Oldfors, C., et al. (2022). Respiratory chain dysfunction in perifascicular muscle fibres in patients with dermatomyositis is associated with mitochondrial DNA depletion. *Neuropathology and Applied Neurobiology, 48*(7), e12841.
Herbst, A., et al. (2007). Accumulation of mitochondrial DNA deletion mutations in aged muscle fibers: Evidence for a causal role in muscle fiber loss. *The Journals of Gerontology. Series A, Biological Sciences and Medical Sciences, 62*(3), 235–245.
Horwath, O., et al. (2025). Ageing leads to selective type II myofibre deterioration and denervation independent of reinnervative capacity in human skeletal muscle. *Experimental Physiology, 110*(2), 277–292.
Huang, C. C., et al. (2016). Effect of exercise training on skeletal muscle SIRT1 and PGC-1alpha expression levels in rats of different age. *International Journal of Medical Sciences, 13*(4), 260–270.
Huang, S., & Millar, A. H. (2013). Succinate dehydrogenase: The complex roles of a simple enzyme. *Current Opinion in Plant Biology, 16*(3), 344–349.
Huang, W. M., et al. (1983). Improved section adhesion for immunocytochemistry using high molecular weight polymers of L-lysine as a slide coating. *Histochemistry, 77*(2), 275–279.
Kadenbach, B., & Huttemann, M. (2015). The subunit composition and function of mammalian cytochrome c oxidase. *Mitochondrion, 24*, 64–76.
Kalmar, B., Blanco, G., & Greensmith, L. (2012). Determination of muscle fiber type in rodents. *Current Protocols in Mouse Biology, 2*(3), 231–243.
Kane, A. E., & Sinclair, D. A. (2018). Sirtuins and NAD(+) in the development and treatment of metabolic and cardiovascular diseases. *Circulation Research, 123*(7), 868–885.
Kelsey, P. D., et al. (2020). Pathophysiology, biomarkers, and therapeutic modalities associated with skeletal muscle loss following spinal cord injury. *Brain Sciences, 10*(12).
Kimoloi, S., et al. (2022). Combined fibre atrophy and decreased muscle regeneration capacity driven by mitochondrial DNA alterations underlie the development of sarcopenia. *Journal of Cachexia, Sarcopenia and Muscle, 13*(4), 2132–2145.
Krebs, H. A., & Eggleston, L. V. (1940). The oxidation of pyruvate in pigeon breast muscle. *The Biochemical Journal, 34*(3), 442–459.
Lantier, L., et al. (2015). SIRT3 is crucial for maintaining skeletal muscle insulin action and protects against severe insulin resistance in high-fat-fed mice. *Diabetes, 64*(9), 3081–3092.
Larsson, L., et al. (1993). An age-related type IIB to IIX myosin heavy chain switching in rat skeletal muscle. *Acta Physiologica Scandinavica, 147*(2), 227–234.
Larsson, L., et al. (2019). Sarcopenia: Aging-related loss of muscle mass and function. *Physiological Reviews, 99*(1), 427–511.
Li, H., & Cai, Z. (2022). SIRT3 regulates mitochondrial biogenesis in aging-related diseases. *Journal of Biomedical Research, 37*(2), 77–88.

Lin, J., et al. (2002). Transcriptional co-activator PGC-1 alpha drives the formation of slow-twitch muscle fibres. *Nature, 418*(6899), 797–801.
Lin, L., et al. (2014). Regulation of skeletal muscle oxidative capacity and muscle mass by SIRT3. *PLoS One, 9*(1), e85636.
Mahadev Bhat, S., & Sieck, G. C. (2024). Heterogeneous distribution of mitochondria and succinate dehydrogenase activity in human airway smooth muscle cells. *FASEB BioAdvances, 6*(6), 159–176.
Mishra, S., et al. (2022). Sirtuin 6 inhibition protects against glucocorticoid-induced skeletal muscle atrophy by regulating IGF/PI3K/AKT signaling. *Nature Communications, 13*(1), 5415.
Moosavi, B., et al. (2019). The assembly of succinate dehydrogenase: A key enzyme in bioenergetics. *Cellular and Molecular Life Sciences: CMLS, 76*(20), 4023–4042.
Murach, K. A., et al. (2019). Fiber typing human skeletal muscle with fluorescent immunohistochemistry. *Journal of Applied Physiology (1985), 127*(6), 1632–1639.
Murgia, M., et al. (2019). Proteomics of cytochrome c oxidase-negative versus -positive muscle fiber sections in mitochondrial myopathy. *Cell Reports, 29*(12), 3825–3834 e4.
Murphy, J. L., et al. (2012). Cytochrome c oxidase-intermediate fibres: Importance in understanding the pathogenesis and treatment of mitochondrial myopathy. *Neuromuscular Disorders: NMD, 22*(8), 690–698.
Oberbach, A., et al. (2006). Altered fiber distribution and fiber-specific glycolytic and oxidative enzyme activity in skeletal muscle of patients with type 2 diabetes. *Diabetes Care, 29*(4), 895–900.
Palacios, O. M., et al. (2009). Diet and exercise signals regulate SIRT3 and activate AMPK and PGC-1alpha in skeletal muscle. *Aging (Albany NY), 1*(9), 771–783.
Parry, H. A., Roberts, M. D., & Kavazis, A. N. (2020). Human skeletal muscle mitochondrial adaptations following resistance exercise training. *International Journal of Sports Medicine, 41*(6), 349–359.
Pereyra, A. S., et al. (2022). Skeletal muscle undergoes fiber type metabolic switch without myosin heavy chain switch in response to defective fatty acid oxidation. *Molecular Metabolism, 59*, 101456.
Pette, D., & Staron, R. S. (2000). Myosin isoforms, muscle fiber types, and transitions. *Microscopy Research and Technique, 50*(6), 500–509.
Phadke, R. (2017). Myopathology of adult and paediatric mitochondrial diseases. *Journal of Clinical Medicine, 6*(7).
Plotkin, D. L., et al. (2021). Muscle fiber type transitions with exercise training: Shifting perspectives. *Sports (Basel), 9*(9).
Poljsak, B., et al. (2023). The central role of the NAD+ molecule in the development of aging and the prevention of chronic age-related diseases: Strategies for NAD+ modulation. *International Journal of Molecular Sciences, 24*(3).
Punsoni, M., et al. (2017). Succinate dehydrogenase B (SDHB) immunohistochemistry for the evaluation of muscle biopsies. *Applied Immunohistochemistry & Molecular Morphology: AIMM / Official Publication of the Society for Applied Immunohistochemistry, 25*(9), 645–650.
Ross, J. M. (2011). Visualization of mitochondrial respiratory function using cytochrome c oxidase/succinate dehydrogenase (COX/SDH) double-labeling histochemistry. *Journal of Visualized Experiments : JoVE*(57), e3266.
Rowan, S. L., et al. (2012). Denervation causes fiber atrophy and myosin heavy chain co-expression in senescent skeletal muscle. *PLoS One,* 7(1), e29082.
Ryan, D. G., & O'Neill, L. A. J. (2020). Krebs cycle reborn in macrophage immunometabolism. *Annual Review of Immunology, 38*, 289–313.
Schwer, B., & Verdin, E. (2008). Conserved metabolic regulatory functions of sirtuins. *Cell Metabolism,* 7(2), 104–112.

Seligman, A. M., et al. (1968). Nondroplet ultrastructural demonstration of cytochrome oxidase activity with a polymerizing osmiophilic reagent, diaminobenzidine (DAB). *The Journal of Cell Biology, 38*(1), 1–14.
Serrano, A. L., et al. (1996). Myosin isoforms and muscle fiber characteristics in equine gluteus medius muscle. *The Anatomical Record, 244*(4), 444–451.
Sher, J., & Cardasis, C. (1976). Skeletal muscle fiber types in the adult mouse. *Acta Neurologica Scandinavica, 54*(1), 45–56.
Smith, J. S., et al. (2000). A phylogenetically conserved NAD+-dependent protein deacetylase activity in the Sir2 protein family. *Proceedings of the National Academy of Sciences of the United States of America, 97*(12), 6658–6663.
Sokoloff, A. J., et al. (2007). Immunohistochemical characterization of slow and fast myosin heavy chain composition of muscle fibres in the styloglossus muscle of the human and macaque (Macaca rhesus). *Archives of Oral Biology, 52*(6), 533–543.
Song, M. Y., et al. (2022). Sirt6 reprograms myofibers to oxidative type through CREB-dependent Sox6 suppression. *Nature Communications, 13*(1), 1808.
Soukup, T., Zacharova, G., & Smerdu, V. (2002). Fibre type composition of soleus and extensor digitorum longus muscles in normal female inbred Lewis rats. *Acta Histochemica, 104*(4), 399–405.
Talbot, J., & Maves, L. (2016). Skeletal muscle fiber type: Using insights from muscle developmental biology to dissect targets for susceptibility and resistance to muscle disease. *Wiley Interdisciplinary Reviews. Developmental Biology, 5*(4), 518–534.
Tamta, A. K., et al. (2025). SIRT2 attenuates stress-induced skeletal muscle atrophy by inhibiting glucocorticoid receptor signalling. *bioRxiv*.
Ustunel, I., & Demir, R. (1997). A histochemical, morphometric and ultrastructural study of gastrocnemius and soleus muscle fiber type composition in male and female rats. *Acta Anat (Basel), 158*(4), 279–286.
Wattez, J. S., et al. (2023). Loss of skeletal muscle estrogen-related receptors leads to severe exercise intolerance. *Molecular Metabolism, 68*, 101670.
Wong, J. P. H., et al. (2025). Skeletal muscle proteomics: Considerations and opportunities. *npj Metabolic Health and Disease, 3*(1), 30.
Yu, J., & Auwerx, J. (2009). The role of sirtuins in the control of metabolic homeostasis. *Annals of the New York Academy of Sciences, 1173*(0 1), E10-9.
Zhang, F. M., et al. (2024). Transcriptome profiling of fast/glycolytic and slow/oxidative muscle fibers in aging and obesity. *Cell Death & Disease, 15*(6), 459.
Zhang, H. H., et al. (2015). SIRT1 attenuates high glucose-induced insulin resistance via reducing mitochondrial dysfunction in skeletal muscle cells. *Experimental Biology and Medicine (Maywood, NJ), 240*(5), 557–565.
Zhang, M. Y., Zhang, W. J., & Medler, S. (2010). The continuum of hybrid IIX/IIB fibers in normal mouse muscles: MHC isoform proportions and spatial distribution within single fibers. *American Journal of Physiology. Regulatory, Integrative and Comparative Physiology, 299*(6), R1582–R1591.
Zhang, X., et al. (2024). Desuccinylation of TBK1 by SIRT5 regulates inflammatory response of macrophages in sepsis. *Cell Reports, 43*(12), 115060.
Zhang, X., et al. (2025). Mitochondrial dysfunction in the regulation of aging and aging-related diseases. *Cell Communication and Signaling: CCS, 23*(1), 290.
Zhao, Q., et al. (2025). SIRT5 safeguards against primate skeletal muscle ageing via desuccinylation of TBK1. *Nature Metabolism*, 7(3), 556–573.
Zogby, A. M., et al. (2017). Skeletal muscle fiber-type specific succinate dehydrogenase activity in cerebral palsy. *Muscle & Nerve, 55*(1), 122–124.

CHAPTER TEN

Analysis of copper toxicity by western blotting of secretory proteins translocated via the Sec61 channel

Smriti Anand, Prateeksha Thakur, Vandana Anjana, Santoshi Acharjee, Rajshree Pal, Sugandh Sharma, and Raghuvir Singh Tomar*

Department of Biological Sciences, Indian Institute of Science Education and Research, Bhopal, Madhya Pradesh, India

*Corresponding author. e-mail address: rst@iiserb.ac.in

Contents

1. Introduction 254
2. Rational 259
3. Methodology 260
   3.1 Transformation 260
   3.2 Yeast growth condition 265
   3.3 Whole cell extract preparation 266
   3.4 Western blotting 268
4. Data analysis and interpretation 273
5. Conclusion 274
Acknowledgments 275
Author's contribution 275
Data availability statement 275
Conflict of interest 275
References 275

Abstract

As a trace element, copper plays a vital role in regulating cellular homeostasis by serving as a cofactor for various enzymes. Therefore, maintaining an adequate copper concentration inside the cell is crucial. Even moderate dysregulation of homeostasis can lead to cytotoxicity, protein aggregation, resulting in cell death. One of the significant cellular consequences in *Saccharomyces cerevisiae* is the inhibition of Sec61-mediated protein translocation in the secretory pathway, resulting in the accumulation of the cytosolic form of precursor secretory proteins, such as Gas1 and CPY, in the cytosol. The accumulation of unprocessed precursor proteins in the cytosol provides a method to read out translocon dysfunction. This chapter describes experimental

Methods in Enzymology, Volume 733
ISSN 0076-6879, https://doi.org/10.1016/bs.mie.2026.05.028

approaches for analysing copper-induced translocation defects in yeast, with a primary focus on Western blot–based detection of precursor and mature secretory protein species. Given the high conservation of copper homeostasis and the secretory pathway across eukaryotes, these methods offer a robust framework for investigating the molecular mechanisms underlying copper toxicity and its impact on protein folding and maturation. This biochemical assay further enables the identification and characterization of chemical or genetic modulators that mitigate copper-induced translocation defects. Compounds that restore Sec61 function or improve ER targeting and maturation reduce the accumulation of precursor forms and promote the recovery of mature protein species. Thus, this approach provides a robust platform for assessing copper toxicity and for screening chemical agents that rescue ER translocation and protein maturation under metal stress conditions.

1. Introduction

There are many trace elements that play critical roles in our cellular processes. Maintaining their levels is important, and cells regulate homeostasis through various transporters and sequestration mechanisms to control their concentrations. Copper, for example, is one of these trace elements that play several key roles in the human body. As a redox-active metal, it acts as a cofactor for enzymes such as cytochrome c oxidase (CcO), copper/zinc superoxide dismutase (Cu/Zn SOD or SOD1), and Fet3 (Ferro-O_2-oxidoreductase), which are involved in mitochondrial respiration, defence against oxidative stress, iron uptake, DNA damage response, and many other processes (Acharjee et al., 2025; Festa & Thiele, 2011). Copper inside the cell exists in two oxidation states: Cu^+ (cuprous ion) and Cu^{2+} (cupric ion), allowing it to act as either an electron donor or acceptor (Falcone & Hureau, 2023; Festa & Thiele, 2011). Within the cell, copper is never found in its free metal ion form; instead, it exists either bound to proteins or weakly attached to cytosolic ligands like glutathione in the form Cu-GSH complex, which helps in maintaining the subcellular labile copper pool the bioavailable or exchangeable pool (Kim et al., 2023; Maryon et al., 2013; Rae et al., 1999). Cu^+ has a higher affinity to bind with the thiol or thioether group containing amino acid, like cysteine and methionine, while Cu^{2+} shows more affinity towards the secondary carboxyl group or imidazole, as in the case of aspartic/glutamic acid and histidine, respectively (Falcone & Hureau, 2023; Rubino & Franz, 2012). Thus, the balance of the labile copper pool is essential, as dysregulation can lead to protein misfolding, oxidative stress, and cuproptosis. An excess of copper can promote the production of reactive oxygen species and non-specifically attach to protein thiol groups, thus impairing protein structure and function. Imbalances in copper levels are associated with various health issues. These include Wilson's disease, Menkes' disease, neurodegenerative disorders, and cancer (Chen et al., 2022). Dysregulation of copper homeostasis within the cell also leads to cell death through a process known as cuproptosis (Tsvetkov et al., 2022). Cuproptosis is a form of regulated cell death triggered by the aggregation of Cu-bound lipoylated mitochondrial proteins, and the subsequent reduction in Fe–S (iron–sulfur) clusters induces proteotoxic stress, ultimately leading to cell death due to excessive Cu^{2+} levels (Tsvetkov et al., 2022). Thus, understanding the process by which Cu accumulation causes cellular toxicity is an important first step toward developing new, effective therapies.

Cells manage metal homeostasis through their specific or common transporters and chelators. The highly conserved copper transporter Ctr1 is responsible for the majority of copper uptake into cells (Nevitt et al., 2012). In *Saccharomyces cerevisiae,* Ctr1 is involved in the uptake of Cu^{+}, following the Cu^{2+} reduction to Cu^{+} by the cell surface metalloreductases Fre1and Fre2 (Georgatsou et al., 1997). Chaperons play an important role in transporting the copper to its active site in the protein. Metallothionein plays a crucial role in regulating intracellular copper levels, like Cup1, which specifically chelates copper. The expression of these transporters and chelators is regulated by the Mac1 and Ace1 proteins. When the cell senses a copper deficiency, it leads to the expression of Mac1, which then increases copper uptake by upregulating the expression of *CTR1* and *FRE1*. On the other hand, high copper conditions trigger the degradation of Ctr1. When copper levels are increased, Ace1 activates *CUP1*, facilitating the sequestration of copper (Thiele, 1988). Additional buffering is provided by amino acids and small molecules, such as cysteine, histidine, and glutathione, which can bind to copper ions and reduce their reactivity. This coordinated regulation ensures that copper concentration is maintained within the cell (Anjana et al., 2025).

A recently discovered consequence of copper dysregulation is the inhibition of the Sec61 translocation channel (Acharjee et al., 2025; Saha & Tomar, 2022). The conserved heterotrimeric Sec61 complex, located in the endoplasmic reticulum (ER) membrane, is responsible for both co-translational and post-translational translocation of secretory and membrane proteins through the ER for glycosylation and maturation. It also facilitates the retrograde transport of ER-misfolded proteins for degradation, thereby maintaining cellular protein homeostasis (Schafer & Wolf, 2009). Additionally, the disruption of Sec61-mediated translocation leads to several disease conditions, including Autosomal Dominant Tubulointerstitial Kidney Disease (ADTKD) and Immunodeficiency (CVID/plasma cell defects), among others (Bolar et al., 2016; Lang et al., 2017; Schubert et al., 2018; Sicking et al., 2021). The proteins that mature through the secretory pathway are first synthesized in the cytosol in their inactive form and are then translocated to the ER through post-translational translocation, as seen with Gas1 or CPY, as shown in Fig. 1, or through co-translational translocation,

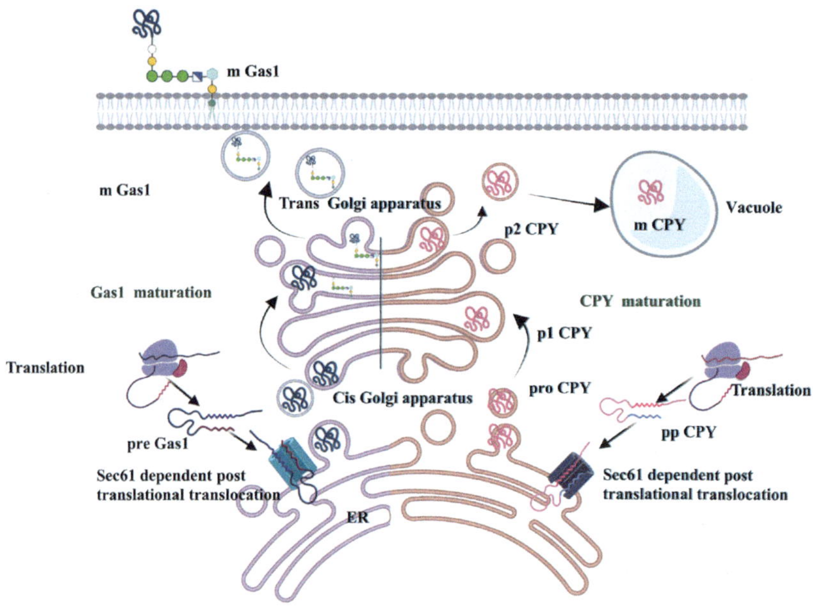

Fig. 1 Maturation and trafficking of secretory reporter proteins Gas1 and CPY in *Saccharomyces cerevisiae*. This schematic illustrates the major steps of translation, ER translocation, post-translational modification, and organelle-specific maturation of two classical secretory pathway markers, Gas1 and CPY. Both proteins are synthesized in the cytosol and targeted to the ER by an N-terminal signal peptide. After post-translational translocation into the ER lumen, they undergo initial N-glycosylation, folding, and quality control. The Gas1 precursor (pre-Gas1) inserts into the ER membrane through Sec61, is N-glycosylated, and covalently attached to a glycosylphosphatidylinositol (GPI) anchor. Properly folded Gas1 is transported to the Golgi, where it receives additional glycan remodeling before exiting through secretory vesicles to the plasma membrane. Mature Gas1 (mGas1), a GPI-anchored β-1,3-glucanosyltransferase, functions in cell wall remodeling and integrity. Prepro-CPY (ppCPY) enters the ER lumen, where its signal peptide is removed, and the pro-enzyme receives core N-glycans. In the Golgi, CPY matures from p1 to p2 CPY as glycans are extended and processed. p2 CPY is sent through the late Golgi to the vacuole using the VPS sorting pathway. In the vacuole, the pro-peptide is removed, creating fully mature CPY (mCPY), which is an active vacuolar hydrolase.

as observed with DPAPB (Jiang et al., 2008; Yi et al., 2021). In the endoplasmic reticulum, proteins are folded and post-translationally modified, which are then targeted to the Golgi, where further maturation occurs, and subsequently transported to their active sites. For example, a β-(1, 3)-glucan elongase protein glycophospholipid-anchored surface protein (Gas1) is post-translationally modified through the attachment of a glycolipid moiety called glycosylphosphatidylinositol (GPI) at its C-terminal end, which is important for its attachment to the outer leaflet of the plasma membrane (Nuoffer et al., 1991). Similarly, carboxypeptidase Y (CPY) is synthesized as a pre-proenzyme, which enters the ER lumen through post-translational translocation, where the signal sequence is cleaved, resulting in the formation of procarboxypeptidase Y (proCPY). After this, the enzyme is folded through disulfide bond formation and glycosylation to form the p1CPY structure; the enzyme is then transported to the Golgi, where outer mannose residues are added, resulting in the p2CPY form. Finally, inside the vacuole, proteinase B cleaves the p2CPY to form the mature CPY (Pilon et al., 1998). It was found that when there is an increase in copper inside the cell, copper leads to the dysregulation of secretory protein translocation through Sec61, resulting in the cytosolic accumulation of immature, non-glycosylated forms of secretory proteins (Fig. 2). In the case of Gas1 and CPY, there is an accumulation of pre-Gas1 form and ppCPY form, respectively. Similarly, in dipeptidyl aminopeptidase B (DPAPB) and invertase (Suc2), there is a copper-induced accumulation of the pre-pro forms of DPAPB and Suc2 (Acharjee et al., 2025). Importantly, this defect is highly specific to copper, as other transition metals tested, including zinc, iron, and cadmium, do not elicit the same response.

The secretory pathway is necessary for maintaining protein balance, ensuring their correct folding, modification, and targeting to their functional destination (Barlowe & Miller, 2013). Disruptions in this pathway can cause endoplasmic reticulum (ER) stress, leading to the accumulation of misfolded proteins. These aggregates activate the unfolded protein response (UPR) pathway and spread to the cytosolic and mitochondrial areas, affecting overall protein quality control (Hamdan et al., 2017; Hetz & Papa, 2018). In higher eukaryotes, ER stress and protein aggregation are closely linked to neurodegenerative disorders like Alzheimer's, Parkinson's, and Huntington's diseases (Hetz & Saxena, 2017; Scheper & Hoozemans, 2015). Notably, irregular copper levels have been found in these conditions, where copper binds incorrectly to proteins prone to aggregation, such as amyloid-β, α-synuclein, and tau. This binding promotes their

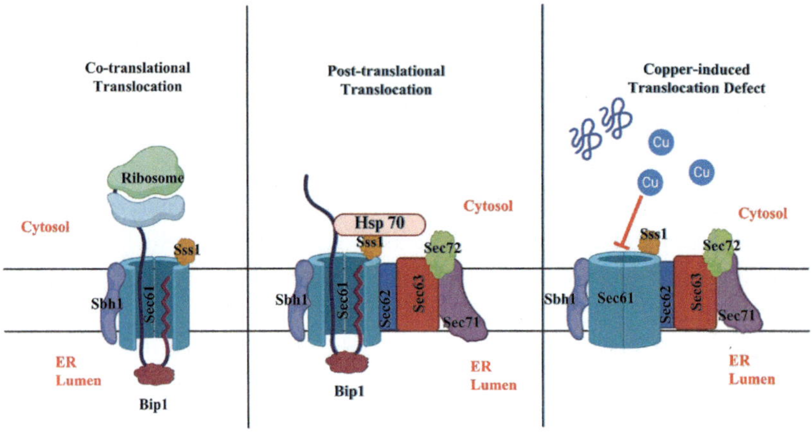

Fig. 2 Co-translational and post-translational ER translocation pathways and copper-induced translocation defects in *Saccharomyces cerevisiae*. Co-translational and post-translational ER translocation pathways and copper-induced translocation defects in *Saccharomyces cerevisiae*. This schematic illustrates the translocation process of newly synthesized secretory and membrane proteins from the cytoplasm to the endoplasmic reticulum (ER), encompassing both co-translational and post-translational translocation. (Left) Co-translational translocation: Ribosome-bound nascent polypeptides are targeted to the Sec61 translocon, composed of Sec61, Sbh1, and Sss1. Along with translation, the growing polypeptide is directed to the ER through the sec61 translocon, where the chaperone BiP (Kar2) assists in initial folding. This mode is typical for membrane proteins and large secretory proteins requiring coupled translation–translocation. (Middle) Post-translational translocation: Precursor Polypeptides that are fully synthesized in the cytoplasm are maintained in an unfolded state by cytosolic Hsp70 and translocated to the Sec62–Sec63 complex, which associates with Sec61 to drive channel opening. Auxiliary components Sec71 and Sec72 stabilize substrate engagement with the translocon. BiP in the ER lumen provides the pulling force necessary for vectorial translocation. This pathway is the primary mode for small or signal-sequence–bearing secretory proteins in yeast and is crucial for the maturation of the precursor form of polypeptides. (Right) Copper-induced translocation defect: Increased copper level leads to inhibition of the sec61 function, resulting in impairment of both co- and post-translational translocation. As a result, precursor proteins accumulate in the cytosol.

misfolding and oxidative crosslinking. These aggregates can disturb cellular metabolism, mitochondrial function, and gene regulation (Ahmadi et al., 2019; Bagheri et al., 2018; Kim et al., 2013). Recent findings suggest that histone deacetylases (HDACs) play a crucial role in regulating stress responses in these situations. HDACs control the transcription programs related to metal detoxification, ER stress, and protein balance. Several HDAC inhibitors have shown potential in reducing protein aggregation and nerve cell toxicity (Chuang et al., 2009). Therefore, understanding how copper disrupts the movement of secretory proteins and promotes protein aggregation can offer valuable insights into the link between metal-related protein imbalance and HDAC-related neurodegeneration.

We have demonstrated a western blotting-based technique to detect copper-induced translocation defects in the concentration of precursor forms accumulation relative to mature forms (Acharjee et al., 2025; Saha & Tomar, 2022). As in the case of Gas1, the precursor, unglycosylated cytosolic form, is smaller than the mature form and thus migrates faster than the mature form. Therefore, through western blot, we can verify the accumulation of the cytosolic precursor Gas1 protein upon copper treatment, indicating that copper suppresses the maturation of the Gas1 protein. The phenotype can be alleviated by sequestration of copper. Treatment with the copper chelator bathocuproine sulfonate (BCS), supplementation with thiol group-containing amino acids such as cysteine, histidine, or reduced glutathione, addition of zinc, or overexpression of *CUP1* can alleviate copper accumulation, resulting in a decrease in the accumulation of the precursor form of the protein. And some genetic dysregulation can alleviate copper toxicity. Together, these observations underscore that copper acts directly on the translocation machinery and that its toxic effects can be modulated by both genetic and metabolic factors.

In this chapter, we describe methodological approaches for analysing copper toxicity by western blotting of secretory proteins translocated via the Sec61 channel using *S. cerevisiae* as a model system. Through western blotting, we distinguish precursor and mature forms of secretory proteins as a sensitive readout of Sec61 activity.

2. Rational

Cells maintain protein trafficking and maturation through a series of regulated processes. Sec61-mediated translocation in the ER provides an important step in maintaining secretory protein homeostasis. Any disturbance at this early entry point into the ER rapidly propagates into protein misfolding, impaired maturation, and ultimately protein aggregation. Therefore, a proper evaluation of protein trafficking and maturation is crucial for understanding proteotoxicity caused by protein aggregation. In *Saccharomyces* cerevisiae, copper is known to affect the early entry of secretory proteins through the Sec61 complex, leading to the accumulation of the precursor form of the secretory protein. This presents a valuable opportunity to investigate the molecular mechanisms underlying metal-induced protein aggregation. Monitoring this process biochemically provides important insight into the early events that weaken protein quality control.

The yeast *Saccharomyces cerevisiae* is a valuable system for these studies, thanks to its ease of genetic manipulation, conserved ER translocation pathway to that of other organisms, and the availability of well-studied secretory reporter proteins. By examining the ratio of precursor and mature forms of secretory proteins, such as Gas1 and CPY, researchers can assess how copper influences the efficiency of ER translocation and folding. Using western blot detection to track these forms provides a simple, sensitive, and reliable method for measuring the degree of copper-induced disruption.

This method not only helps identify conditions that worsen or improve translocation issues but also allows testing of possible protective mechanisms, such as the effects of chelators, antioxidants, or genetic changes. The approach can also be used to study broader cellular responses associated with ER stress and protein aggregation. Overall, the described assay gives a foundation for exploring how changes in metal balance lead to challenges in protein processing and trafficking.

3. Methodology

3.1 Transformation

3.1.1 Equipment

1. Sterilized glass spreaders
2. Syringe
3. PES Syringe filter(0.45 μm)
4. PCR machine
5. Agarose gel electrophoresis
6. Chemi Doc Bio-Rad
7. Laminar flow hood
8. 30 °C shaker incubator
9. Heating block

3.1.2 Reagents

1. YPD Medium
2. 1X TE(Tris-EDTA) buffer, pH 7.5, sterile

3. 10X lithium acetate stock solution: 1 M lithium acetate, pH 7.5 (adjust pH with dilute acetic acid), filter sterilized
4. 50 % (w/v) PEG (polyethylene glycol)
5. Salmon sperm single-stranded carrier DNA (ssDNA)
6. Transforming DNA (template)
7. Selection media plate

3.1.3 *Procedure*

1. **Plasmid Constructs and epitope Tagging**
   a. **Amplicon-mediated Myc tagging of the CPY**
   Amplicon preparation for epitope (Myc) tagging in *Saccharomyces cerevisiae* begins with designing two long primers that will direct the integration event. Each primer carries about 40 bp of homology to the genomic region surrounding the target gene's (CPY) stop codon, one upstream and one downstream, followed by a universal priming sequence that anneals to a plasmid containing the epitope tag and a selectable marker (Janke et al., 2004; Longtine et al., 1998) Using these primers, the plasmid is amplified by PCR to generate a linear tagging cassette in which the gene-specific homology arms flank the epitope tag and the marker. The resulting PCR product is then purified to remove excess primers, nucleotides, and any residual template plasmid so that only the linear cassette enters the yeast cells. This purified amplicon is introduced into *Saccharomyces cerevisiae* using a standard LiOAc/PEG transformation method in the presence of carrier DNA. Once inside the cell, the yeast homologous recombination machinery recognizes the 40-bp flanking regions and carries out a precise double-crossover event at the native gene locus, inserting the epitope tag and selectable marker immediately downstream of the coding sequence.
   b. **Plasmid-mediated expression of GFP-tagged Gas1**
   For experiments requiring expression of a Gas1-GFP fusion protein from a plasmid. In this case, the complete plasmid harbouring the Gas1 coding sequence fused in-frame to GFP, together with its selectable marker, is directly introduced into *Saccharomyces cerevisiae* using the standard LiOAc/PEG transformation procedure (Ha et al., 2014) and medium containing the appropriate selection to ensure retention of the GAS1-GFP plasmid. Only transformants that successfully acquire the plasmid grow under these conditions, allowing for immediate downstream analysis of Gas1-GFP expression by western blot or fluorescence microscopy.

2. Yeast competent cells preparation

a. Desired Yeast strain is primary inoculated in 5 ml YPD in a sterile culture tube and incubated overnight, ~12–14 h at 30 °C, 200 rpm (Fig. 3).

b. A 0.2 OD600 secondary culture is established in 10 ml of media using the overnight primary culture and grown at 30 °C, 200 rpm, until it reaches the mid-log phase (OD 0.8−1).

c. At $OD_{600} \approx 1.0$, harvest cells by centrifugation at 3000 rpm for 5 min at 25 °C.

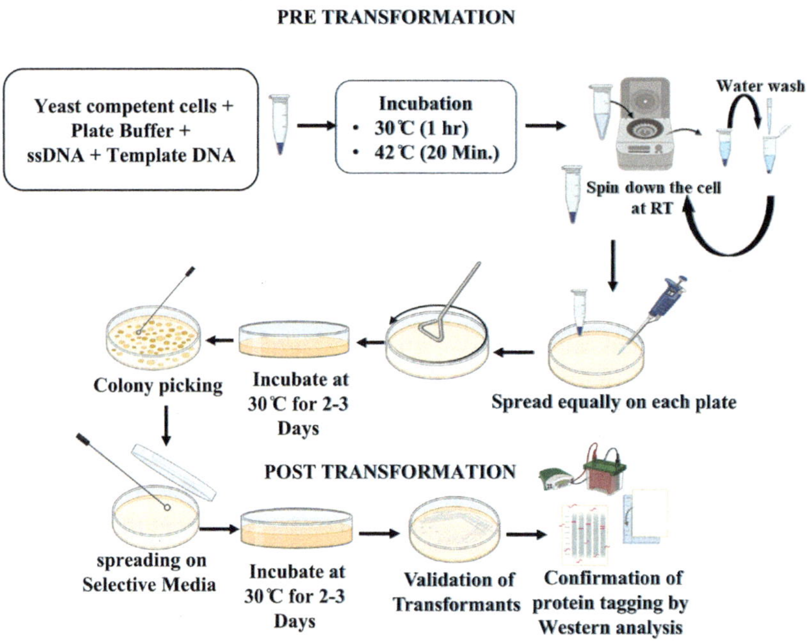

Fig. 3 **PCR-based epitope tagging and homologous recombination-mediated genomic integration in *Saccharomyces cerevisiae*.** Long primers containing ~40 bp of homology to the genomic region flanking the target gene's stop codon are used to amplify an epitope-tag–selection-marker cassette from a plasmid template. The PCR amplicon produced includes the coding sequence junction linked to the epitope tag and selectable marker, surrounded by homology arms on both upstream and downstream sides. When transforming into yeast, the cassette integrates at the endogenous site via homologous recombination, substituting the native 3′ end with the tagged variant while maintaining the upstream and downstream genomic environment. Successful integrants include the properly inserted epitope tag and marker at the target site after the recombination-mediated exchange of the native and altered segments.

d. Wash pellets with sterile 1 ml Milli-Q water and transfer to a 1.5 ml microcentrifuge tube, and pellet down by centrifugation at 3000 rpm. Repeat this step twice.
e. Remove water and resuspend the pellet in 1x LiOAc-TE buffer and centrifuge at 3000 rpm at 25 °C. Repeat this step twice.
f. Resuspend the final pellet in 300 μL 1x LiOAc-TE buffer. Mix gently by tapping.
g. Keep the competent cell suspension 20 min on room temperature on ice, or at 4 °C until you proceed to the transformation step.

Notes:-

1. The entire process should be done in highly sterilized conditions to avoid contamination.
2. In the laminar hood, discard the supernatant by decanting without disturbing the pellet.
3. Gently mix the pellet by tapping, avoiding vortexing, which can shear cells and reduce viability.
4. All steps should be carried out at room temperature.

3. **Transformation of yeast cells**
   a. Prepare a sterilized work area. All required materials (pipettes, sterile tubes, syringe filters, spreaders, etc.) are placed inside a laminar flow hood and UV-sterilized for 20 min prior to use (Fig. 4).
   b. Salmon sperm single-stranded carrier DNA (ssDNA) was loosened by incubating at 42 °C.
   c. Prepare fresh Plate buffer by mixing PEG (50% w/v), 100 mM TE, and 100 mM LiOAc, followed by filtration through a 0.2 μm filter.
   d. Prepare the reaction setup in Microcentrifuge Tubes. In each setup, add 20 μL of ssDNA, 300 μL of plate buffer, 50 μL of competent cells, and the plasmid. Prepare the control setup separately without the plasmid. The mixture is gently mixed by tapping.
   e. All transformation reactions setups were incubated at 30 °C for 45 min, followed by a heat shock at 42 °C for 15 min.
   f. Pellet down Cells 3000 rpm for 5 min, remove supernatant, wash pellets with sterile Milli-Q water, and resuspend the pellet in approx. 100 μL Milli-Q water.
   g. The remaining transformation mixture is spread onto selective agar plates using a sterilized spreader.
   h. Incubate Plates at 30 °C for 2–3 days until colonies appear.

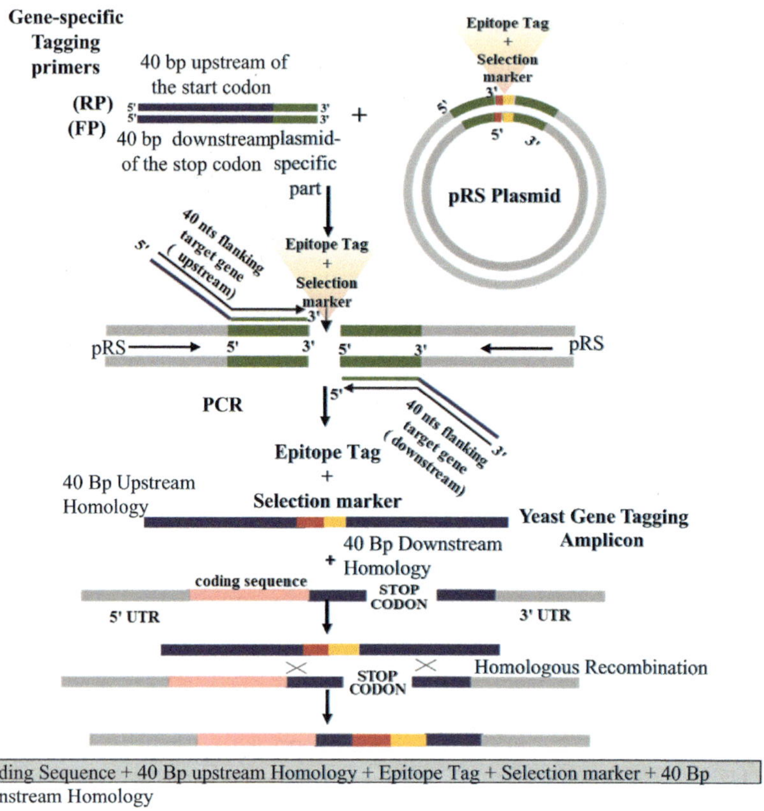

Fig. 4 Diagrammatic representation of the lithium acetate (LiOAc)-based transformation method used for introducing DNA into *Saccharomyces cerevisiae*. For the transformant reaction setup, denatured ssDNA, plate buffer, competent cell, and plasmid/amplicon were added. The reaction mixture was gently mixed and incubated at 30 °C for 45 min, followed by a heat shock at 42 °C for 15 min. The reaction mixture was pelleted down by centrifugation at 3000 rpm for 5 min at room temperature and then washed with water. The pellet is then resuspended in 100 μL of water, and the entire amount is spread on a selective media plate and incubated at 30 °C to allow the growth of transformants. Positive colonies were confirmed through western blotting.

i. Restreak individual colonies onto fresh selective plates to obtain clonal isolates. Confirm correct plasmid uptake or epitope tagging by western blot analysis. Table 1

Notes:-

1. Prepare Transformation mixture for positive control and Spread mixture on SC agar plate.

Table 1 Plasmid used for assay.

Plasmid name	References
pFA-6a-13myc	Acharjee et al. (2025)
pRS415-Gas1-GFP	Ha et al. (2014)
pRS413-Gas1-GFP	Ha et al. (2014)
pRS426GPD-DAP2-HA	Junne et al. (2015)
pRS426GPD-Suc2-HA	Junne et al. (2015)
pRS426GPD-H1-HA	Junne et al. (2015)

3.2 Yeast growth condition

3.2.1 Equipment

1. 30 °C shaker incubator
2. Culture flask
3. Culture tube

3.2.2 Reagents

1. Amino acids (Sigma Aldrich)
2. YNB without ammonium sulphate (HiMedia G090)
3. Ammonium sulphate (Sigma Aldrich)
4. D-glucose (Sigma Aldrich-158968)
5. Bacto agar (BD-214010)
6. Copper chloride dihydrate ($CuCl_2{\cdot}2\,H_2O$) (Sigma Aldrich-307483)
7. TCA(TriChloric Acid)

3.2.2.1 Media composition

Yeast was grown in synthetic complete (SC) medium containing 1.81 g/L amino acid mix: adenine, arginine, glutamate, glycine, aspartate, isoleucine, lysine, histidine, methionine, phenylalanine, leucine, proline, serine, threonine, tryptophan, tyrosine, uracil, and valine. For selective plasmid maintenance, dropout media was prepared separately, containing all amino acids except Leu, His, Ura, and Trp. 0.18 % Complete SC or 0.1 % dropout media mixed with 0.17 % of YNB without ammonium sulphate, 0.5 % of ammonium sulphate, and 2 % of D-glucose. All media were sterilized by autoclaving at 121°C for 20 min.

3.2.3 Procedure

1. **Culture conditions**
   a. Inoculate freshly restreaked transformed yeast cells in 5 ml of selection medium and grow overnight at 30 °C in a shaker incubator.
   b. Set 0.2 OD_{600} secondary culture in 10 ml media using the overnight primary culture and grow them at 30 °C with shaking (200 rpm) until they reach mid-log phase (OD 0.8−1).
2. **Copper treatment**
   a. Add the $CuCl_2$ solution to the cultures at the required concentration, using healthy mid-log phase cells (OD_{600} = 0.8−1), and incubate for 2 h at 30 °C and 200 rpm. The concentration of $CuCl_2$ is determined on the basis of growth phenotype in the presence of $CuCl_2$.
3. **Cell harvesting and washing**
   a. Harvest cells by centrifugation at 3000 rpm for 5 min at 4 °C.
   b. Wash pellets once with ice-cold distilled water, followed by 20 % (w/v) trichloroacetic acid (TCA).
   c. Harvested cells were used for downstream assays.

Notes:-
1. Always prepare fresh $CuCl_2$ solution to avoid oxidation.
2. Ensure equivalent cell density and growth phase before copper addition to allow valid comparisons.

3.3 Whole cell extract preparation

Glass Bead lysis,(Conzelmann et al., 1988) LiAc/NaOH protein extraction method (Zhang et al., 2011), Sonication (Liu et al., 2013), NaOH b-mercaptoethanol method (Riezman et al., 1983) etc., are frequent procedures known for the preparation of yeast protein extracts. In this chapter, we have elaborated on the glass bead lysis method for preparing whole-cell protein extracts (Fig. 5).

3.3.1 Equipment

1. 1 mm Glass beads:- Cat. no. Z250473 (SIGMA-Aldrich)
2. Automated vortex machine
3. Centrifuge
4. Sterile needle
5. Spirit burner
6. Heating block

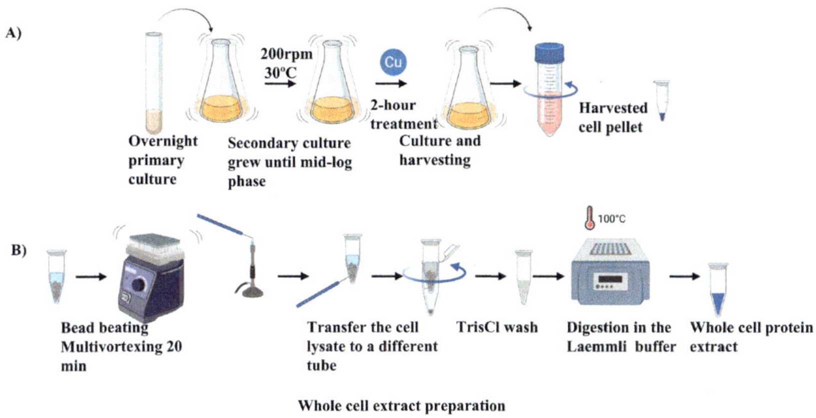

Fig. 5 Cell culture and whole cell protein extraction. (A) Yeast cells, such as those expressing Gas1 on plasmids or CPY-tagged constructs, are first selected on an appropriate media plate. Cells were inoculated from a plate in primary culture medium. A secondary culture was established at an optical density (OD_{600}) of 0.2 from the primary culture and grown to the mid-log phase at 30 °C in a 200 rpm shaker incubator. The cells were then treated with copper for 2 h. After treatment, cells are harvested for protein extraction. (B) Whole-cell protein extraction was done as depicted in the figure. The whole cell extract was lysed with the help of 1 mm glass beads beading in 200 µL of 20 % TCA. The lysate was transferred to fresh MCTs by puncturing, followed by centrifugation. Then the lysate was washed in 0.5 M Tris-Cl, and digested in Laemmli sample buffer and 0.5 M Tris-Cl at 100 °C.

3.3.2 Reagents

1. 20 % (w/v) trichloroacetic acid (TCA):- Cat. No. 1.94971.0521(SIGMA-Aldrich)
2. Tris-Cl (0.5 M, pH 7.5)
3. Laemmli sample buffer (0.06 M Tris ± HCl, pH 6.8, 5 % glycerol, 2 % SDS, 4 % β-mercaptoethanol, 0.0025 % bromophenol blue) (Laemmli, 1970)

3.3.3 Procedure

Glass bead lysis (Conzelmann et al., 1988) was used with a few optimizations. Laemmli sample buffer was prepared as described by Horvath and Riezman. SDS ± PAGE was performed according to Laemmli (Horvath & Riezman, 1994; Laemmli, 1970).

1. Cell pellet resuspended in 200 µL of 20 % TCA and broken with 250 µL of 1 mm glass beads: suspension vortexed for 15 mins on an automated vortex machine at 4 °C.

2. Bottom of the MCTs pierced with a red-hot sterile needle, and the entire content transferred to new MCTs by centrifugation at 6000 rpm for 10 min at 0.5 M Tris-HCl (pH 6.8), 4 °C.
3. Remove the supernatant and resuspend the pellet in 800 μL of 0.5 M Tris-Cl, pH 7.5. Then, centrifuge at high speed at 4 °C.
4. Suspend the pellet in Laemmli sample buffer and 0.5 M Tris-Cl, pH 7.5, at a ratio of 3:1, respectively.
5. Boil the suspension at 100 °C for 10 min to fully denature and linearize the proteins.
6. Transfer the sample to ice and centrifuge at high speed to remove cell debris from soluble protein.

Notes:-

1. If the colour of the protein suspension becomes yellow upon the addition of Laemmli sample buffer or after boiling at 100 °C. Please wash the samples with 0.5 M Tris-HCl (pH 7.5) again to neutralize the acidic pH caused by TCA.

3.4 Western blotting

3.4.1 Equipment

3.4.1.1 For polyacrylamide gel casting

1. Spacer Plates with 1.0 mm Integrated Spacers
2. Short Plates
3. Comb, 15-well, 1.0 mm, 26 μL
4. Tetra Cell Casting Stand with Clamp Kit for 2 Core System

3.4.1.2 For polyacrylamide gel run and western blotting

1. Tetra Vertical Electrophoresis Cell
2. Blotter With Wire Electrodes
3. Nitrocellulose Membrane, Roll, 0.45 μm, 30 cm × 3.5 m- Cat. no. 1620115 (Bio Rad)

3.4.1.3 For antibody probing and visualization

1. Rocker shaker
2. Washing Box
3. LI-COR infrared imaging machine

3.4.2 Reagents

3.4.2.1 For polyacrylamide gel casting

1. Stacking Buffer:- 0.5 M Tris-HCl (pH 6.8)

2. Resolving Buffer:- 1.5 M Tris-HCl (pH 8·8)
3. SDS solution (Sodium Dodecyl Sulfate):- Cat.no. L3771 (SIGMA-Aldrich): 10 % (w/v)
4. APS (Ammonium Persulfate):- Cat. no. A3678 (SIGMA-Aldrich): 10 % (w/v)
5. TEMED (Tetramethylethylenediamine):- Cat. no. MB026 (HiMedia)
6. 2-Propanol:- Cat. no. 1945240521 (Emplura)

3.4.2.2 For polyacrylamide gel run and western blotting

1. Electrode running buffer:- 0.192 M glycine, 0.025 M Tris, and 0.1 % SDS.
2. Transfer buffer:- 0.192 M glycine, 0.025 M Tris, 10 % methanol, and 0.03 % SDS.
3. 1.5X SDS Loading dye.
4. Pre-stained protein ladder- Cat. no. 161-0374 (Bio-rad).

3.4.2.3 For antibody probing and visualization

1. 2.5 % BSA (Bovine Serum Albumin)- Cat. no. MB083 (HiMedia).
2. TBST (Tris-Buffered Saline with Tween-20)- 20 mM Tris-HCl (pH 7.5), 150 mM NaCl, 0.1 % (v/v) Tween-20:- Cat. no. 8.22184.0521 Merck.
3. Primary antibodies and secondary antibodies used are listed in Table 2.

Table 2 Antibodies used.

Primary antibody	Secondary antibody
α-GFP (catalog no.: G1544; Sigma)	Goat anti-rabbit IgG secondary antibody (catalog no.: A32734; Invitrogen)
α-Myc monoclonal antibody (catalog no.: MA1-980; Invitrogen)	Goat anti-mouse IgG secondary antibody (catalog no.: 926–32210; Odyssey).
α-HA (catalog no.: 26183; Invitrogen)	Goat anti-mouse IgG secondary antibody (catalog no.: 926–32210; Odyssey).
α-TBP polyclonal antiserum raised in a rabbit	Goat anti-rabbit IgG secondary antibody (catalog no.: A32734; Invitrogen)

3.4.3 Procedure

1. **Polyacrylamide gel casting:** Dual percentage/Gradient gels were prepared to better resolve the precursor and mature forms of Gas1 and CPY proteins while retaining αTBP (Tata binding protein- the protein loading control) in the same gel. For separating CPY forms, an 8 % to 12 % gel is prepared, and for Gas1, a 10 % to 15 % gel is prepared, as described below (Fig. 6).

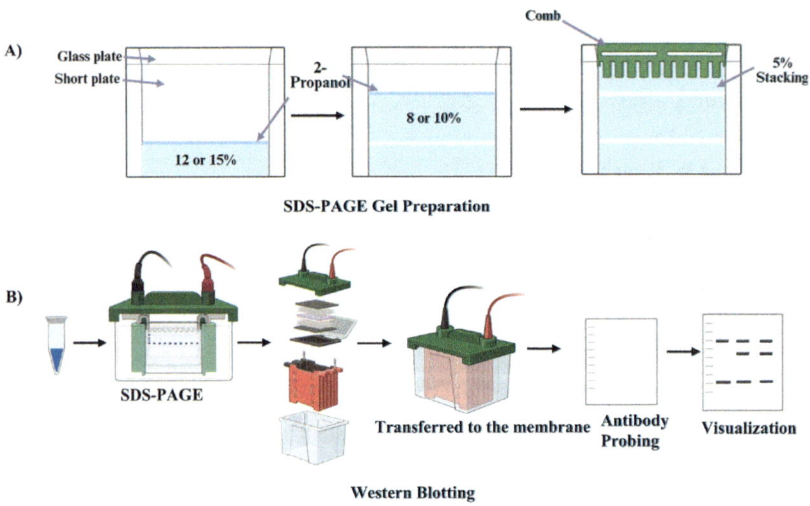

Fig. 6 **Workflow for detecting copper-induced protein translocation defects.** (A) To retain protein loading control αTBP (TATA-binding protein) within the same gel, a gradient gel was prepared. Gas1 and CPY were resolved at 10 % and 8 % gels, followed by higher percent gels at 15 and 12, respectively, to slow down the loading control run. This will ensure that the protein of interest and the loading control can be transferred to the same nitrocellulose membrane, thereby decreasing the error associated with protein loading. A higher percentage gel was poured first, followed by a lower percentage gel. A resolving gel (12 % or 15 %) is poured between glass plates, followed by an overlay of 2-propanol (or isopropanol) to flatten the surface and ensure polymerization. After solidification and removal of 2-propanol, the resolving gel (8 % or 10 %) is poured, followed by the addition of 2-propanol. After solidification and removal of 2-propanol, the stacking gel (5 %) is poured, and the comb is inserted to form wells. (B) Equal amounts of total protein are loaded onto an SDS–PAGE gel. After electrophoresis, proteins are transferred to a membrane and probed with specific antibodies for Gas1 or CPY. Western blot analysis reveals precursor accumulation (pre-Gas1, ppCPY/p1CPY), which indicates impaired ER translocation due to copper. Mature forms act as internal controls for normal trafficking and processing. This assay allows for sensitive detection of copper-induced defects in the secretory translocation pathway.

	Separating gel: 1.5 M Tris-HCl (pH 8·8) (10 ml)				Stacking: 0.5 M Tris-HCl (pH 6.8) (5 ml)
For 2 gels	8 %	10 %	12 %	15 %	5 %
Acrylamide	2.66 ml	3.33 ml	4 ml	5 ml	0.833 ml
Tris-HCl Buffer	2.5 ml	2.5 ml	2.5 ml	2.5 ml	1.25 ml
0.1 % Ammonium persulfate	100 µl	100 µl	100 µl	100 µl	50 µl
0·1% SDS	100 µl	100 µl	100 µl	100 µl	100 µl
0.1 % (v/v) tetramethylethylenediamine (TEMED)	8 µl	8 µl	10 µl	10 µl	4 µl
Autoclaved MilliQ	4.73 ml	4.06 ml	3.39 ml	2.39 ml	2.861 ml

a. Gels containing 8 % and 12 % or 10 % and 15 % separating gel with 5 % stacking gel were prepared using 30 % stock solution of acrylamide and 0·8 % of N, N'-Methylenebisacrylamide.

b. Cast 10 cm wide, 6 cm long, and 1 mm thick separating gel in spacer plates with 1.0 mm integrated spacer and short plates in the tetra cell casting module.

c. Final concentration of reagents in separating gels was as follows: 1.5 M Tris-HCl (pH 8·8), 0.1 % ammonium persulfate, and 0·1% SDS. The gels were polymerized by the addition of 0.1 % (v/v) tetramethylethylenediamine (TEMED).

d. To prepare a gradient of two percent, pour nearly 1.5 ml of the separating solution containing 12 % acrylamide for an 8/12 % gel and 15 % acrylamide for a 10/15 % gel. 2-Propanol is added to the surface of the separating solution.

e. After solidification, remove 2-propanol followed by water rinse and pour nearly 2.5 ml of the separating solution containing 8 % acrylamide for an 8/12 % gel and 10 % acrylamide for a 10/15 % gel. 2-Propanol is added to the surface of the separating solution.

f. After solidification of separating gel, remove 2-propanol and cast 1 cm of stacking gel on the same with 15 wells, 1.5 mm, comb.

g. The stacking gels of 5 % acrylamide have the following final concentrations of reagents: 0.5 M Tris-HCl (pH 6.8), 0.1 % ammonium persulfate, and 0.1 % SDS. The gels were polymerized in the same manner as the separating gel, i.e., by the addition of 0.1 % (v/v) TEMED.

Notes:-

1. Remove the 2-propanol completely from the gel, as it can lead to bubble formation in the gel.

2. Polyacrylamide gel electrophoresis and western blotting

a. The electrode buffer, also known as the electrode running buffer, with a pH of 8.3, was used to resolve the protein via electrophoresis.
b. The gel is completely immersed in the tetra vertical electrophoresis cell, containing the running buffer. Ladder/ Marker is loaded at the end.
c. Gels were run at a constant 90 V until the target protein resolved clearly, generally when it migrated to the mid-gel region while retaining TBP within the gel matrix (2–3 h).
d. The separated proteins were transferred to a nitrocellulose membrane from the gel. Prepare a transfer sandwich with sponge, filter papers, a nitrocellulose membrane, and filter papers, and place the sandwich in the transfer apparatus containing chilled transfer buffer.
e. Transfer of protein to nitrocellulose membrane was carried out at 90 volts and 300 mA of current for 90 min at 4 °C.

Notes:

1. Protein samples were loaded with 1.5X SDS loading dye to normalize the volume.

3. Antibody probing and visualization

a. Block the nitrocellulose membrane with 2.5 % BSA in TBST for 45 min at 4 °C.
b. Probe with primary antibody prepared in 2.5 % BSA and incubate for 90−120 min at 4 °C on a rocker.
c. Wash the membrane 3 times with 1X TBST for 7 min with continuous shaking on the rocker at RT.
d. Probe with respective secondary antibody prepared in 2.5 % BSA and incubate for 45−60 min at 4 °C on a rocker.
e. Wash the membrane 3 times with 1X TBST for 7 min with continuous shaking on the rocker at RT and keep for drying in the dark.

f. Visualize the dried blot in the LI-COR infrared imaging machine at the required intensity.

Notes:-

1. Incubate and wash the secondary antibody in the dark to avoid photobleaching.

4. Data analysis and interpretation

Immunoblotting of Gas1 has two bands: a precursor Gas1 band (Pre Gas1 immature, unglycosylated forms of secretory Gas1 proteins of molecular weight 60 kDa) and one band of the mature Gas1 (m-Gas1), mature glycosylated form of molecular weight 125 kD (Fig. 7) (Liu & Chang, 2008; Nuoffer et al., 1991; Yi et al., 2021). According to the literature, it is well established that when a Sec61-dependent translocation defect occurs through the ER, there is an accumulation of pre-Gas1 in the cytosol. Similarly, when cells are treated with increasing concentrations of

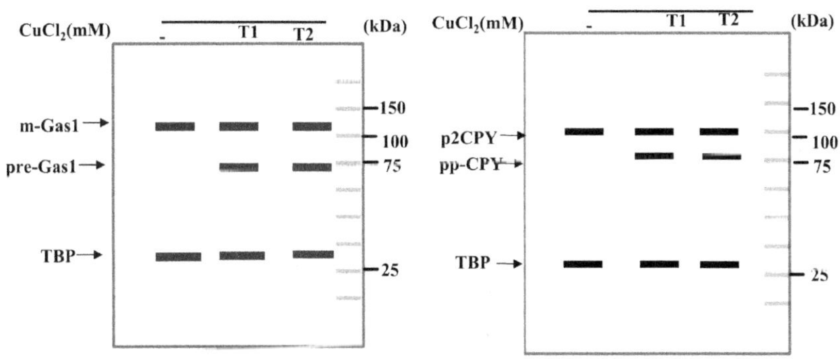

Fig. 7 Representative schematic of western blot patterns showing precursor and mature forms of Gas1 and CPY under copper-induced translocation stress. This diagram illustrates the expected migration pattern of Gas1 and CPY during western blot analysis. A copper-induced sec61 translocation defect result in the accumulation of precursor forms, which migrate at different molecular weights compared to their fully processed mature species. For Gas1, the precursor pre-Gas1 is faster than mature Gas1 (mGas1) because of the presence of a signal peptide and incomplete GPI-anchor/Golgi glycan modifications in pre-Gas1, while mature Gas1 (mGas1) is properly glycosylated and GPI-anchored. For CPY, Golgi-modified precursor (p2CPY) migrates higher than the early ER precursor (p1CPY) under copper-induced translocation defect. This representative blot illustrates that the accumulation of precursor form directly reflects defects in translocation, serving as a reliable biochemical marker of copper-induced translocation failure.

copper, we observe an increased accumulation of precursor Gas1, indicating a copper-induced Sec61 translocation defect. For the quantification of copper-induced protein accumulation, we measure the band signals of the pre-Gas1 and mature Gas1 using ImageJ software and then normalize these signals to the TBP control. To determine the total protein level of Gas1, the normalized intensities were summed. The ratio of Gas1 immature to mature forms was determined by calculating the percentage of immature or mature protein in relation to the total protein level, using the formula (normalized intensity of immature or mature protein/intensity of total protein) $\times$ 100.

Similarly, immunoblotting of CPY (carboxypeptidase Y) shows different forms of CPY, including p2 CPY, pp CPY, and mCPY (Fig. 7). In the case of CPY, when there is a translocation defect, then there is an accumulation of ppCPY, which is the cytosolic preproenzyme form (Pilon et al., 1998; Yi et al., 2021). Similarly, when copper treatment is present, there is an increased accumulation of ppCPY, indicating that copper induces cytosolic protein aggregation. The accumulation of ppCPY is calculated in the same way as above.

5. Conclusion

In this chapter, we describe an experimental design to investigate protein aggregation resulting from a copper-induced defect in Sec61 translocation. We will track the maturation and processing of secretory proteins that rely on the ER-Golgi secretory pathway. This method utilizes secretory pathway proteins, such as Gas1 and CPY, as traditional markers of maturation. It provides a sensitive and quantitative way to assess how copper impacts both co-translational and post-translational translocation pathways.

Logarithmically growing yeast cells treated with copper, followed by subsequent culture harvest, protein isolation, and application of the western blot technique, allow us to observe the accumulation of precursor forms of proteins. Pre-Gas1, ppCPY, and p1/p2 CPY, precursors of Gas1 and CPY, directly indicate disruption of Sec61-dependent maturation of these proteins. Each stage of Gas1 and CPY maturation results in different shifts in electrophoretic mobility. This allows us to clearly distinguish between proteins that successfully enter the ER and undergo further maturation and those that do not due to the inhibition of the Sec61 translocon by toxic copper.

The assays we describe show that copper disrupts Sec61-mediated translocation, regardless of the pathway involved. This suggests that copper interferes with a fundamental function of the Sec61 channel or its regulatory components. The method is highly reproducible and only needs basic yeast culture and western blotting tools, making it easy to adapt for testing genetic or chemical factors that affect secretory pathway function through copper toxicity.

Overall, this approach provides a robust foundation for understanding the molecular basis of copper toxicity and how the Sec61 machinery contributes to maintaining proteostasis under metal stress. Combining yeast genetics with well-defined reporter proteins and simple biochemical analysis makes this method highly effective for studying ER translocation and its disruption by metal stress.

Acknowledgments

We acknowledge all the laboratory members for their assistance and suggestions throughout this study. V.A., S.A., P.T., S.S., R.P., and S.A. acknowledge ICMR, CSIR, UGC, and IISER Bhopal, Government of India, for providing fellowship support. All the figures were created using https://BioRender.com.

Author's contribution

S.A., P.T., V.A., and S.A. prepared the figures. S.A., V.A., and P.T. wrote the original draft. R.S.T. supervision and conceptualization. R.S.T., S.A., P.T., V.A., S.A., R.P., and S.S., writing – review & editing. R.S.T. project administration and funding acquisition. This work was supported by the funds from IISER Bhopal to R.S.T.

Data availability statement

Data will be made available upon request.

Conflict of interest

The authors declare that they have no conflicts of interest with the contents of this article.

References

Acharjee, S., Pal, R., Anand, S., Thakur, P., Anjana, V., Singh, R., ... Tomar, R. S. (2025). Mutations in histones dysregulate copper homeostasis leading to defect in Sec61-dependent protein translocation mechanism in Saccharomyces cerevisiae. *Journal of Biological Chemistry, 301*(2), https://doi.org/10.1016/j.jbc.2025.108163.

Ahmadi, S., Zhu, S. L., Sharma, R., Wu, B., Soong, R., Majumdar, R. D., ... Kraatz, H. B. (2019). Aggregation of microtubule binding repeats of tau protein is promoted by Cu. *Acs Omega, 4*(3), 5356–5366. https://doi.org/10.1021/acsomega.8b03595.

Anjana, V., Anand, S., Thakur, P., Pal, R., Acharjee, S., Sharma, S., ... Tomar, R. S. (2025). Metabolic regulation of copper homeostasis governs the Sec61-dependent protein translocation process in. *Molecular and Cellular Biology*. https://doi.org/10.1080/10985549.2025.2569577.

Bagheri, S., Squitti, R., Haertlé, T., Siotto, M., & Saboury, A. A. (2018). Role of copper in the onset of Alzheimer's disease compared to other metals. *Frontiers in Aging Neuroscience, 9*. https://doi.org/10.3389/fnagi.2017.00446.

Barlowe, C. K., & Miller, E. A. (2013). Secretory protein biogenesis and traffic in the early secretory pathway. *Genetics, 193*(2), 383–410. https://doi.org/10.1534/genetics.112.142810.

Bolar, N. A., Golzio, C., Zivna, M., Hayot, G., Van Hemelrijk, C., Schepers, D., ... Loeys, B. L. (2016). Heterozygous loss-of-function SEC61A1 mutations cause autosomal-dominant tubulo-interstitial and glomerulocystic kidney disease with anemia. *American Journal of Human Genetics, 99*(1), 174–187. https://doi.org/10.1016/j.ajhg.2016.05.028.

Chen, L., Min, J., & Wang, F. (2022). Copper homeostasis and cuproptosis in health and disease. *Signal Transduction and Targeted Therapy, 7*(1), 378. https://doi.org/10.1038/s41392-022-01229-y.

Chuang, D. M., Leng, Y., Marinova, Z., Kim, H. J., & Chiu, C. T. (2009). Multiple roles of HDAC inhibition in neurodegenerative conditions. *Trends in Neurosciences, 32*(11), 591–601. https://doi.org/10.1016/j.tins.2009.06.002.

Conzelmann, A., Riezman, H., Desponds, C., & Bron, C. (1988). A major 125-kd membrane glycoprotein of Saccharomyces cerevisiae is attached to the lipid bilayer through an inositol-containing phospholipid. *The EMBO Journal, 7*(7), 2233–2240. https://doi.org/10.1002/j.1460-2075.1988.tb03063.x.

Falcone, E., & Hureau, C. (2023). Redox processes in Cu-binding proteins: The "in-between" states in intrinsically disordered peptides. *Chemical Society Reviews, 52*(19), 6595–6600. https://doi.org/10.1039/d3cs00443k.

Festa, R. A., & Thiele, D. J. (2011). Copper: An essential metal in biology. *Current Biology: CB, 21*(21), R877–R883. https://doi.org/10.1016/j.cub.2011.09.040.

Georgatsou, E., Mavrogiannis, L. A., Fragiadakis, G. S., & Alexandraki, D. (1997). The yeast Fre1p/Fre2p cupric reductases facilitate copper uptake and are regulated by the copper-modulated Mac1p activator. *Journal of Biological Chemistry, 272*(21), 13786–13792. https://doi.org/10.1074/jbc.272.21.13786.

Ha, C. W., Kim, K., Chang, Y. J., Kim, B., & Huh, W. K. (2014). The β-1,3-glucanosyltransferase Gas1 regulates Sir2-mediated rDNA stability in. *Nucleic Acids Research, 42*(13), 8486–8499. https://doi.org/10.1093/nar/gku570.

Hamdan, N., Kritsiligkou, P., & Grant, C. M. (2017). ER stress causes widespread protein aggregation and prion formation. *Journal of Cell Biology, 216*(8), 2295–2304. https://doi.org/10.1083/jcb.201612165.

Hetz, C., & Papa, F. R. (2018). The unfolded protein response and cell fate control. *Molecular Cell, 69*(2), 169–181. https://doi.org/10.1016/j.molcel.2017.06.017.

Hetz, C., & Saxena, S. (2017). ER stress and the unfolded protein response in neurodegeneration. *Nature Reviews Neurology, 13*(8), 477–491. https://doi.org/10.1038/nrneurol.2017.99.

Horvath, A., & Riezman, H. (1994). Rapid protein extraction from Saccharomyces cerevisiae. *Yeast (Chichester, England), 10*(10), 1305–1310. https://doi.org/10.1002/yea.320101007.

Janke, C., Magiera, M. M., Rathfelder, N., Taxis, C., Reber, S., Maekawa, H., ... Knop, M. (2004). A versatile toolbox for PCR-based tagging of yeast genes: New fluorescent proteins, more markers and promoter substitution cassettes. *Yeast (Chichester, England), 21*(11), 947–962. https://doi.org/10.1002/yea.1142.

Jiang, Y., Cheng, Z. L., Mandon, E. C., & Gilmore, R. (2008). An interaction between the SRP receptor and the translocon is critical during cotranslational protein translocation. *Journal of Cell Biology, 180*(6), 1149–1161. https://doi.org/10.1083/jcb.200707196.

Junne, T., Wong, J., Studer, C., Aust, T., Bauer, B. W., Beibel, M., ... Hoepfner, D. (2015). Decatransin, a new natural product inhibiting protein translocation at the Sec61/SecYEG translocon. *Journal of Cell Science, 128*(6), 1217–1229. https://doi.org/10.1242/jcs.165746.

Kim, D. K., Song, J. W., Park, J. D., & Choi, B. S. (2013). Copper induces the accumulation of amyloid-beta in the brain. *Molecular & Cellular Toxicology, 9*(1), 57–66. https://doi.org/10.1007/s13273-013-0009-0.

Kim, J. E., Jeon, S., & Lindahl, P. A. (2023). Discovery of an unusual copper homeostatic mechanism in yeast cells respiring on minimal medium and an unexpectedly diverse labile copper pool. *Journal of Biological Chemistry, 299*(12), https://doi.org/10.1016/j.jbc.2023.105435.

Laemmli, U. K. (1970). Cleavage of structural proteins during the assembly of the head of bacteriophage T4. *Nature, 227*(5259), 680–685. https://doi.org/10.1038/227680a0.

Lang, S., Pfeffer, S., Lee, P. H., Cavalie, A., Helms, V., Forster, F., & Zimmermann, R. (2017). An update on Sec61 channel functions, mechanisms, and related diseases. *Frontiers in Physiology, 8*, 887. https://doi.org/10.3389/fphys.2017.00887.

Liu, D., Zeng, X. A., Sun, D. W., & Han, Z. (2013). Disruption and protein release by ultrasonication of yeast cells. *Innovative Food Science & Emerging Technologies, 18*, 132–137. https://doi.org/10.1016/j.ifset.2013.02.006.

Liu, Y., & Chang, A. (2008). Heat shock response relieves ER stress. *The EMBO Journal, 27*(7), 1049–1059. https://doi.org/10.1038/emboj.2008.42.

Longtine, M. S., McKenzie, A. I. I. I., Demarini, D. J., Shah, N. G., Wach, A., Brachat, A., ... Pringle, J. R. (1998). Additional modules for versatile and economical PCR-based gene deletion and modification in. *Yeast (Chichester, England), 14*(10), 953–961. https://doi.org/10.1002/(Sici)1097-0061(199807)14:103.3.Co;2-L.

Maryon, E. B., Molloy, S. A., & Kaplan, J. H. (2013). Cellular glutathione plays a key role in copper uptake mediated by human copper transporter 1. *American Journal of Physiology-Cell Physiology, 304*(8), C768–C779. https://doi.org/10.1152/ajpcell.00417.2012.

Nevitt, T., Ohrvik, H., & Thiele, D. J. (2012). Charting the travels of copper in eukaryotes from yeast to mammals. *Biochimica et Biophysica Acta, 1823*(9), 1580–1593. https://doi.org/10.1016/j.bbamcr.2012.02.011.

Nuoffer, C., Jeno, P., Conzelmann, A., & Riezman, H. (1991). Determinants for glycophospholipid anchoring of the Saccharomyces cerevisiae GAS1 protein to the plasma membrane. *Molecular and Cellular Biology, 11*(1), 27–37. https://doi.org/10.1128/mcb.11.1.27-37.1991.

Pilon, M., Römisch, K., Quach, D., & Schekman, R. (1998). Sec61p serves multiple roles in secretory precursor binding and translocation into the endoplasmic reticulum membrane. *Molecular Biology of the Cell, 9*(12), 3455–3473. https://doi.org/10.1091/mbc.9.12.3455.

Rae, T. D., Schmidt, P. J., Pufahl, R. A., Culotta, V. C., & O'Halloran, T. V. (1999). Undetectable intracellular free copper: The requirement of a copper chaperone for superoxide dismutase. *Science (New York, N. Y.), 284*(5415), 805–808. https://doi.org/10.1126/science.284.5415.805.

Riezman, H., Hase, T., Vanloon, A. P. G. M., Grivell, L. A., Suda, K., & Schatz, G. (1983). Import of proteins into mitochondria - a 70 kilodalton outer-membrane protein with a large carboxy-terminal deletion is still transported to the outer-membrane. *Embo Journal, 2*(12), 2161–2168. https://doi.org/10.1002/j.1460-2075.1983.tb01717.x.

Rubino, J. T., & Franz, K. J. (2012). Coordination chemistry of copper proteins: How nature handles a toxic cargo for essential function. *Journal of Inorganic Biochemistry, 107*(1), 129–143. https://doi.org/10.1016/j.jinorgbio.2011.11.024.

Saha, N., & Tomar, R. S. (2022). Copper inhibits protein maturation in the secretory pathway by targeting the Sec61 translocon in Saccharomyces cerevisiae. *Journal of Biological Chemistry, 298*(8), 102170. https://doi.org/10.1016/j.jbc.2022.102170.

Schafer, A., & Wolf, D. H. (2009). Sec61p is part of the endoplasmic reticulum-associated degradation machinery. *The EMBO Journal, 28*(19), 2874–2884. https://doi.org/10.1038/emboj.2009.231.

Scheper, W., & Hoozemans, J. J. M. (2015). The unfolded protein response in neurodegenerative diseases: A neuropathological perspective. *Acta Neuropathologica, 130*(3), 315–331. https://doi.org/10.1007/s00401-015-1462-8.

Schubert, D., Klein, M. C., Hassdenteufel, S., Caballero-Oteyza, A., Yang, L., Proietti, M., ... Grimbacher, B. (2018). Plasma cell deficiency in human subjects with heterozygous mutations in Sec61 translocon alpha 1 subunit (SEC61A1). *The Journal of Allergy and Clinical Immunology, 141*(4), 1427–1438. https://doi.org/10.1016/j.jaci.2017.06.042.

Sicking, M., Lang, S., Bochen, F., Roos, A., Drenth, J. P. H., Zakaria, M., ... Linxweiler, M. (2021). Complexity and specificity of Sec61-channelopathies: Human diseases affecting gating of the Sec61 complex. *Cells, 10*(5), https://doi.org/10.3390/cells10051036.

Thiele, D. J. (1988). ACE1 regulates expression of the Saccharomyces cerevisiae metallothionein gene. *Molecular and Cellular Biology, 8*(7), 2745–2752. https://doi.org/10.1128/mcb.8.7.2745-2752.1988.

Tsvetkov, P., Coy, S., Petrova, B., Dreishpoon, M., Verma, A., Abdusamad, M., ... Golub, T. R. (2022). Copper induces cell death by targeting lipoylated TCA cycle proteins. *Science (New York, N. Y.), 375*(6586), 1254–1261. https://doi.org/10.1126/science.abf0529.

Yi, J. K., Fujita, H., Mandon, E. C., & Gilmore, R. (2021). Rapid inactivation of the yeast Sec complex selectively blocks transport of post-translationally translocated proteins. *Journal of Biological Chemistry, 297*(4), https://doi.org/10.1016/j.jbc.2021.101171.

Zhang, T., Lei, J., Yang, H., Xu, K., Wang, R., & Zhang, Z. (2011). An improved method for whole protein extraction from yeast Saccharomyces cerevisiae. *Yeast (Chichester, England), 28*(11), 795–798. https://doi.org/10.1002/yea.1905.

CHAPTER ELEVEN

Histone deacetylase as a potential therapeutic target for memory dysfunction during aging and neurodegenerative disorders

Padmanabh Singh and Mahendra Kumar Thakur*
Biochemistry and Molecular Biology Laboratory, Department of Zoology, Institute of Science, Centre of Advanced Study, Banaras Hindu University, Varanasi, India
*Corresponding author. e-mail address: mktian2007@gmail.com

Contents

1. Introduction 280
   1.1 Histone acetylation 281
   1.2 Histone deacetylases 282
   1.3 Class I HDACs 282
   1.4 Histone deacetylase 2 282
   1.5 Therapeutic potential of HDAC inhibitors 284
2. Expression of HDAC2 upregulated in the hippocampus of old mice during age associated memory decline 284
   2.1 Animal 284
   2.2 Equipment 285
   2.3 Reagents 285
   2.4 Procedure 285
3. Histone acetylation level declined in the hippocampus of old mice during age associated memory decline 287
   3.1 Animal 287
   3.2 Equipment 287
   3.3 Reagents 288
   3.4 Procedure 288
4. HDAC2 inhibition improved learning and memory in old mice 289
   4.1 Animal 289
   4.2 Equipment 289
   4.3 Reagents 289
   4.4 Procedure 289
5. HDAC2 inhibition increased histone acetylation level in the hippocampus of old mice 290
   5.1 Animal 290

Methods in Enzymology, Volume 733
ISSN 0076-6879, https://doi.org/10.1016/bs.mie.2026.05.037

5.2 Equipment 290
5.3 Reagents 290
5.4 Procedure 291
6. HDAC2 inhibition increased histone acetylation level at the promoter of synaptic plasticity genes in the hippocampus of old mice 292
6.1 Animal 292
6.2 Equipment 292
6.3 Reagents 292
6.4 Procedure 292
7. Conclusion 293
Acknowledgment 294
References 294

Abstract

Brain aging is a natural biological process characterized by gradual structural, molecular, and functional changes in the brain over time. Aging is the major risk factor for neurodegenerative diseases characterized by progressive neuronal loss, cognitive decline, and impaired motor function. In recent years, growing evidence has highlighted the importance of epigenetic regulation in the development and progression of these disorders. Among these epigenetic regulators, histone deacetylases (HDACs) have emerged as critical modulators of gene expression and neuronal function. Dysregulation of HDAC activity has been increasingly implicated in the pathogenesis of several neurodegenerative diseases. Abnormal HDAC expression or activity can alter transcription of genes involved in neuronal survival and synaptic integrity, which leads to neuronal dysfunction and degeneration. Pharmacological inhibition of HDAC activity has been shown to restore histone acetylation levels, enhance the expression of neuroprotective genes, and improve synaptic plasticity in various experimental models.

1. Introduction

Memory is a fundamental cognitive function enabling the encoding, storage, and retrieval of previously acquired experiences and information. This acquired information is stabilized in the form of long-term memory by strengthening the synaptic connections and is closely associated with the transcription and translation of synaptic plasticity-related genes (Kandel, 2001). Synaptic plasticity genes play a pivotal role in the regulation of learning, memory formation, and cognitive stability. These genes regulate molecular mechanisms that strengthen synaptic connections and structural remodeling, which are critical for maintaining memory functions. The expression of these synaptic plasticity genes is altered during normal aging and age-associated neurodegenerative diseases. These alterations in the expression and function of synaptic plasticity genes are closely associated with the decline of learning and memory during aging and neurodegenerative diseases (Navakkode & Kennedy, 2024).

Previous research revealed that epigenetic mechanisms play an important role in the regulation of these synaptic plasticity genes, and its dysregulation can lead to memory impairment. Histone proteins (H2A, H2B, H3, and H4) that are part of the chromatin complex play an important role in regulating gene expression through epigenetic mechanisms. The covalent modifications of histone proteins, such as acetylation, phosphorylation, and methylation, regulate the condensation and relaxation of chromatin and thus influence gene expression. Acetylation on the N-terminal tail of histones positively regulates gene expression. Histone acetylation is tightly regulated by histone acetyltransferases (HATs), which add an acetyl group, and histone deacetylases (HDACs), which remove acetyl groups from the histones. The balance between HATs and HDACs is important for maintaining normal homeostasis, and this balance is affected during normal aging and age-associated neurodegenerative diseases (Peserico & Simone, 2011). Based on dysregulation of chromatin modifications and gene expression, several hypotheses have been proposed related to impaired epigenetic mechanisms, nuclear signaling, and transcriptional dysregulation is associated with neurodegeneration (Pereira et al., 2024; Robakis, 2003).

1.1 Histone acetylation

Histone proteins undergo several post-translational modifications such as acetylation, phosphorylation, methylation, etc. Among these, the histone acetylation is correlated with active gene expression, while deacetylation leads to gene repression. Histone acetylation and deacetylation are catalyzed by histone acetyltransferases (HATs), which add an acetyl group ($-COCH_3$) from acetyl CoA, and histone deacetylases (HDACs), which remove the acetyl group on the lysine residues from N terminal tail of histones. Acetylation on histone reduces its interaction with DNA, that results in a relaxed or open chromatin which allows the binding of transcription factors on the promoter region of a gene and thus active gene transcription. On the contrary, removal of the acetyl group from histones increases the interaction between histone and DNA, resulting in a closed or condensed chromatin that does not allow the binding of transcription factors and thus gene repression.

1.2 Histone deacetylases

HDACs are grouped into four classes based on their sequence, cellular localization, and cofactor requirements. Class I HDACs have sequence similar to yeast Rpd3 protein and comprises of HDAC1, HDAC2, HDAC3 and HDAC8, and primarily located in the nucleus and responsible for chromatin remodeling. Class II HDACs have sequence similar to yeast Hda1 protein and comprise of HDAC4, HDAC5, HDAC6, HDAC7, HDAC9 and HDAC10. They mostly shuttle between the nucleus and cytoplasm, regulating deacetylation of both histones and non-histone proteins and modulating cell signaling pathways. Both Class I and II HDACs use Zn^{2+} as cofactor for enzyme activities. Class III HDACs or sirtuins (SIRT1–SIRT7) require NAD+ as cofactor for activity and regulate metabolic and stress response pathways. Sirtuins have sequence similar to Sir2 protein of yeast. Class IV HDACs comprise of HDAC11, mostly localized to nucleus and have both deacetylase and deacylase activity.

1.3 Class I HDACs

Class I HDACs are ubiquitously expressed and predominantly localized in the nucleus and are mostly involved in the regulation of transcription, cell cycle progression, and maintaining genomic stability. HDAC1 and HDAC2 are part of transcriptional repressor complexes such as those involving Sin3, NuRD, and CoREST. They play an important role in the central nervous system, mainly associated with learning and memory. Similarly, HDAC3 is part of nuclear receptor co-repressor complexes (NCoR/SMRT) and plays an important role in lipid metabolism and immune response. HDAC8 plays specialized roles in development, such as skull morphogenesis, and is essential for maintaining muscle differentiation pathways. By deacetylating histones, they effectively silence genes involved in differentiation and cellular response to DNA damage, promoting proliferation and survival.

1.4 Histone deacetylase 2

Among class I HDACs, HDAC2 is most widely studied in relation to learning memory, aging, and neurodegeneration. Dysregulation of HDAC2 activity has been strongly associated with age-associated cognitive decline as well as the progression of various neurodegenerative diseases, including AD, PD, and HD. Guan et al. (2009) reported that HDAC2 overexpression decreased spine density, histone H3K14/K12 acetylation, negatively regulates synaptic plasticity gene expression, such as BDNF,

GluR, EGR1, etc., and impaired memory consolidation. Gräff et al. (2012) reported that the expression of HDAC2 increased in brain of AD patients and the hippocampus of CK-p25 mice. Further, knockdown of HDAC2 rescued synaptic plasticity and memory consolidation in CK-p25 mice.

Tau pathology involves the abnormal accumulation of tau protein into neurofibrillary tangles inside neurons, which leads to neuronal dysfunction and death in AD. HDAC2 upregulation is associated with higher production and aggregation of tau phosphorylation and Aβ, resulting in decreased dendritic spine density (Guan et al., 2009; Ishizuka et al., 2014). Liu et al. (2017) first reported that upregulation of HDAC2 expression correlated with tau hyperphosphorylation and aggregation in the AD mouse model. Nakatsuka et al. (2021) reported that HDAC2 knockdown increased dendritic length and density of mushroom like spine in the CA1 region in the APP/PS1 transgenic neurodegenerative mouse model. Further, HDAC2 knockdown increased freezing behavior in the hippocampal-dependent contextual fear conditioning test.

HDAC2 plays a significant role in the pathogenesis of PD through its involvement in epigenetic regulation, neural cell survival, neuroinflammation, and oxidative stress. Recent evidence suggests that dysregulated HDAC2 may aggravate cellular stress and neurodegeneration of dopaminergic neurons in PD pathology. Tan et al. (2018) reported that the expression of HDAC2 was upregulated in the microglia in the substantia nigra of PD patients as compared to control subjects. These results showed that HDAC2 may be involved in microglia-mediated neuroinflammation in the substantia nigra and loss of dopaminergic neurons in PD patients.

Our lab studied the role of HDAC2 in regulation of learning and memory in amnesic mice model and during aging. We observed that the expression of HDAC2 was upregulated in the hippocampus of old as well as scopolamine-induced amnesic mice. This upregulation in HDCA2 expression is associated with decreased histone acetylation at H3K9 and H3K14 position. This upregulation of HDAC2 negatively regulates the expression of synaptic plasticity genes i.e. Arc, EGR1, Homer1 and Narp. Chromatin immunoprecipitation studies showed that the level of HDAC2 increased at the promoter of these genes in aged as well as amnesic mice. Inhibition of HDAC2 by sodium butyrate (NaB) restored hippocampal-dependent memory consolidation both in old and amnesic mice. Further, NaB treated mice showed increased H3K9 and H3K14 histone acetylation at the promoter of synaptic plasticity genes (Singh & Thakur, 2014; Singh et al., 2015; Singh & Thakur, 2018).

1.5 Therapeutic potential of HDAC inhibitors

HDAC inhibitors are being actively investigated for their neuroprotective effects. These small molecules counteract HDAC-mediated repression by promoting histone acetylation, thereby upregulating genes essential for neuronal health (D'Mello, 2009). HDAC inhibitors are classified into several structural groups: Hydroxamic acids (e.g., vorinostat, trichostatin A), Short-chain fatty acids (e.g., sodium butyrate (NaB), valproic acid), Benzamides (e.g., entinostat), Cyclic peptides (e.g., romidepsin) (Ji et al., 2025). Among these classes, Hydroxamic acids and Short-chain fatty acids are widely used in different neurological and neurodegenerative disorders (Zhang et al., 2024). Animal studies showed that HDAC inhibitors such as NaB restored histone acetylation and improved learning and memory (Garcez et al., 2018; Govindarajan et al., 2011; Singh & Thakur, 2018). Specific inhibitors, such as vorinostat (HDAC1, HDAC2) and trichostatin A, have shown efficacy in reducing neuroinflammation and improving synaptic plasticity in animal models (Camelo et al., 2005). While preclinical data are compelling, significant challenges remain in translating these findings into safe and effective clinical therapies.

Here we described in detailed about the methods to study the role of HDAC2 in age-associated hippocampal memory declined in mice. Further, we used HDAC inhibitor NaB to recover age associated memory decline in mice.

2. Expression of HDAC2 upregulated in the hippocampus of old mice during age associated memory decline

2.1 Animal

1. Animal Species – Swiss albino strain, male mice of 25–30 g of weight.
2. Age and groups: Young (10 ± 2 weeks), adult (30 ± 2 weeks) and old (80 ± 2 weeks).
3. Animal Housing – Temperature 24 ± 2 °C, 12/12 h light/dark cycle with *ad libitum* supply of food and water.
4. All animal care and experimental procedures were performed according to the Institutional Animal Ethical Committee (IAEC) of Banaras Hindu University.

2.2 Equipment

Plastic box (40 cm × 50 cm × 50 cm in size), camera, plastic objects, surgical instruments, scissors, forceps, centrifuge tubes, tissue homogenizer, centrifuge, weighing balance, spectrophotometer.

2.3 Reagents

Ethanol, Phosphate-buffered saline (PBS), RIPA lysis buffer, PMSF, Bovine serum abumin, Bradford reagent, Tris.HCl, NaCl, SDS, Triton X-100, Ethylenediaminetetraacetic acid (EDTA), phenylmethylsulfonyl fluoride (PMSF), Laemmli buffer, 10% acrylamide solution (acrylamide: bis-acrylamide, 29:1), ammonium persulfate (APS), TEMED, glycine, polyvinylidene difluoride (PVDF) membrane, Ponceau-S, fat free skimmed milk, primary antibodies (HDAC2 and β-actin), HRP conjugated secondary antibodies, Tween 20, ECL reagents, X-ray film.

2.4 Procedure

1. Novel object recognition (NOR) test was used to check hippocampal dependent learning and memory.
2. The NOR test comprises of habituation phase, training phase and test phase.
3. During habituation, each mouse was allowed to explored the open box for 5 min with out any objects for two consecutive days.
4. During training, each mouse was allowed to interact with two similar objects for 5 min in the box for one day.
5. During test phase (48 hr after training), one familiar object was replaced with a novel object of similar dimension and different shape.
6. After that mice were placed in the box and allowed to interacts with both the objects (familiar and novel objects) for 5 min.
7. To remove odor of mice, after each trial, the box and objects were cleaned with 70 % ethanol.
8. The data were analyzed by ANY-maze software (Stoelting Co., USA) and time spent with novel object (TNovel) and time spent with familiar object (TFamiliar) were calculated.
9. The percentage of time spent and the Discrimination Index (DI) was calculated as TNovel/(TNovel+TFamiliar)× 100 and TNovel-TFamiliar/(TNovel+TFamiliar) for novel object respectively (Fig. 1).
10. After behavioral test, mice were sacrificed, brain was isolated and hippocsmpus was dissected out.

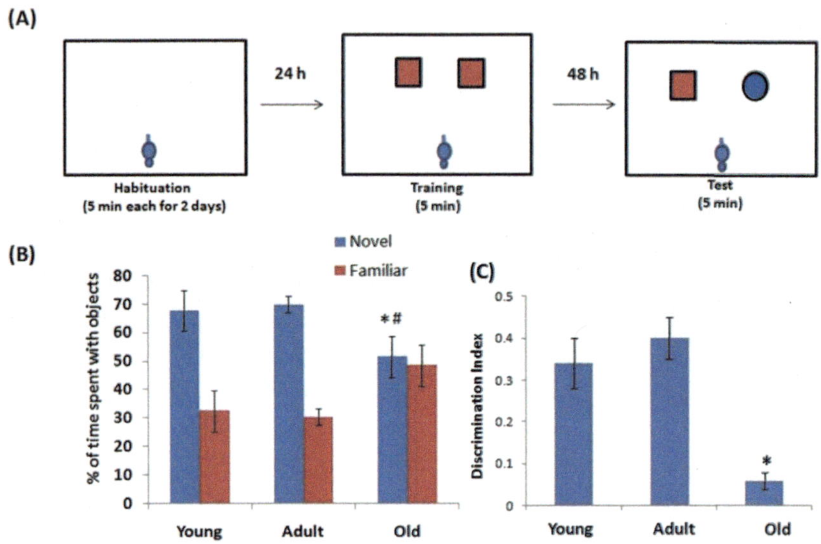

Fig. 1 Novel object recognition test in young, adult and old mice. (A) Experimental set up, (B) Percentage of time spent with novel and familiar objects and (C) Discrimination index. *Image adapted from Singh and Thakur (2014).*

11. Total protein was isolated from the hippocampus using RIPA lysis buffer with protease inhibitor.
12. Tissue homogenate was prepared using a Dounce homogenizer.
13. The tissue homogenate was centrifuged at 12,000 × *g* at 4 °C for 10 min
14. The pellet was discarded and supernatant was collected in separate centrifuge tubes.
15. Protein was estimated by Bradford (1976) methods using a spectrophotometer.
16. Proteins were separated using denaturing SDS-polyacrylamide gel electrophoresis (SDS-PAGE) and transferred onto PVDF membrane.
17. The efficiency of protein transfer was checked by Ponceau-S and the membrane was blocked with 5 % (w/v) skimmed milk for 3 h at room temperature (RT).
18. Thereafter, the membrane was incubated with primary antibody HDAC2 (1:15000, Sigma) at 4 °C overnight.
19. Next day, membrane was washed in PBST (0.1 % Tween 20) three times 5 min each and incubated with HRP conjugated secondary antibody (1:3000, Bangalore Genei) for 3 h at RT.

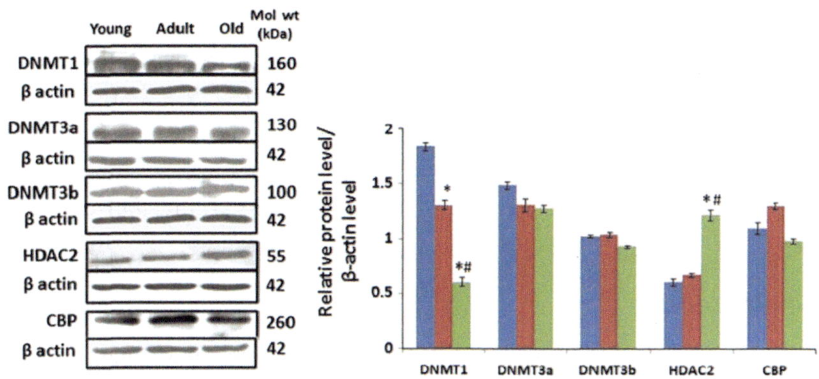

Fig. 2 Expression analysis of chromatin modifying enzymes in the hippocampus of young, adult and old mice. *Image adapted from Singh and Thakur (2014).*

20. Membrane was washed in PBST three times 5 min each and signal was detected using ECL reagents on X-ray film.

21. Further, the membrane was stripped and incubated with primary β-actin antibody (1:10,000, Sigma) followed HRP conjugated secondary antibody (1:3000, Bangalore Genei) for 3 h at room temperature.

22. Membrane was washed in PBST three times 5 min each and signal was detected using ECL reagents on X-ray film.

23. Signal intensities were analyzed in AlphaEaseFC software (Alpha Innotech Corp, USA) (Fig. 2).

3. Histone acetylation level declined in the hippocampus of old mice during age associated memory decline

3.1 Animal

1. Animal Species – Swiss albino strain, male mice of 25–30 g of weight.
2. Age and groups: Young (10 ± 2 weeks), adult (30 ± 2 weeks) and old (80 ± 2 weeks).

3.2 Equipment

Surgical instruments, scissors, forceps, centrifuge tubes, tissue homogenizer, centrifuge, weighing balance, spectrophotometer.

3.3 Reagents

PBS, RIPA lysis buffer, PMSF, Bovine serum abumin, Bradford reagent, Tris.HCl, NaCl, SDS, Triton X-100, EDTA, PMSF, Laemmli buffer, 10 % acrylamide solution (acrylamide: bis-acrylamide, 29:1), APS, TEMED, glycine, PVDF membrane, Ponceau-S, fat free skimmed milk, primary antibodies (H3K9 Ac, H3K14 Ac and H3), HRP conjugated secondary antibodies, Tween 20, ECL reagents, X-ray film.

3.4 Procedure

Histone acetylation was analysed in the hippocampus of young, adult and old mice. After behavioral analysis, hippocampus was isolated and processed for acetylation analysis.

1. For histone acetylation analysis, protein was isolated from the hippocampus using RIPA lysis buffer with protease inhibitor.
2. Tissue homogenate was prepared using a Dounce homogenizer.
3. The tissue homogenate was centrifuged at 12,000 × *g* at 4 °C for 10 min
4. The pellet was discarded and supernatant was collected in separate centrifuge tubes.
5. Protein was estimated by Bradford methods (Bradford, 1976) using a spectrophotometer.
6. Proteins were separated using denaturing SDS-PAGE and transferred onto PVDF membrane.
7. The efficiency of protein transfer was checked by Ponceau-S and the membrane was blocked with 5% (w/v) skimmed milk for 3 h at RT.
8. Thereafter, the membrane was incubated with primary antibody H3K9 Ac (1:1000 Millipore), H3K14 Ac (1:1000, Epigentek) at 4 °C overnight.
9. Next day, membrane was washed in PBST three times 5 min each and incubated with HRP conjugated secondary antibody (1:3000, Bangalore Genei) for 3 h at RT.
10. Membrane was washed in PBST three times 5 min each and signal was detected using ECL reagents on X-ray film.
11. Further, the membrane was stripped and incubated with primary H3 antibody (1:10,000, Sigma) followed HRP conjugated secondary antibody (1:3000, Bangalore Genei) for 3 h at room temperature.
12. Membrane was washed in PBST three times 5 min each and signal was detected using ECL reagents on X-ray film.
13. Signal intensities were analyzed AlphaEaseFC software (Fig. 3).

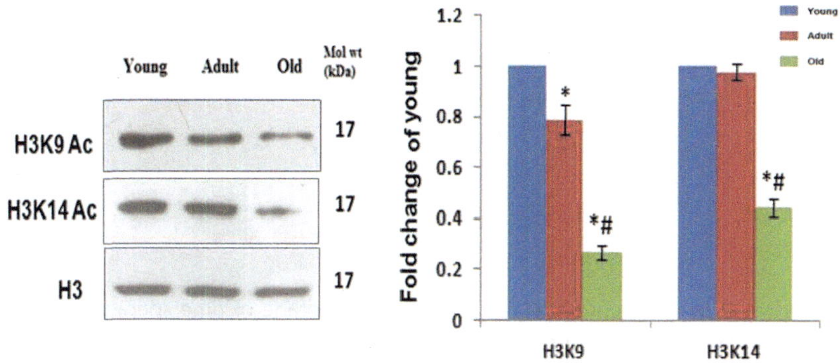

Fig. 3 Histone H3K9 and H3K14 acetylation level in the hippocampus of young, adult and old mice. *Image adapted from Singh and Thakur (2018).*

4. HDAC2 inhibition improved learning and memory in old mice

4.1 Animal

1. Animal Species – Swiss albino strain, male mice of 25–30 g of weight.
2. Age: Old (80 ± 2 weeks). Groups: Control, Sodium butyrate

4.2 Equipment

Plastic box (40 cm × 50 cm × 50 cm in size), camera, plastic objects, ANY-maze software.

4.3 Reagents

Ethanol, NaB, NaCl.

4.4 Procedure

1. Sodium butyrate was used to inhibit HDAC2 in old mice. Old mice were divided in to two groups i.e. Control and Sodium butyrate.
2. Control group was administered with vehicle (0.9 % saline) and Sodium butyrate group was administered with NaB (100 mg/kgBW) for 7 days.
3. The behavior of mice were analyzed by novel object recognition test.
4. During habituation, each mouse was allowed to explored the open box for 5 min with out any objects for two consecutive days.
5. During training, each mouse was allowed to interact with two similar objects for 5 min in the box for one day.
6. During test phase (48 hr after training), one familiar object was replaced with a novel object of similar dimension and different shape.

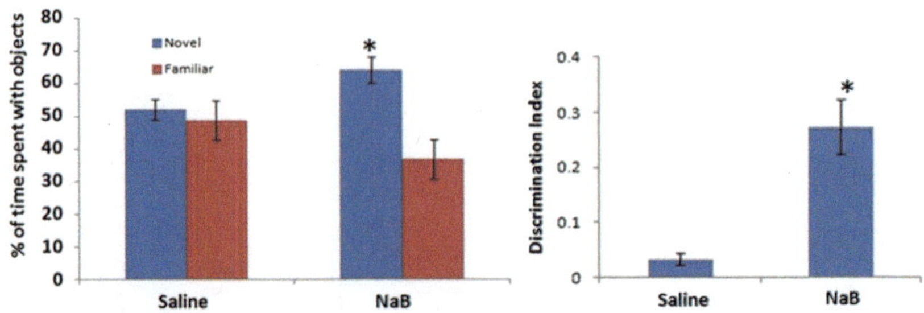

Fig. 4 Novel object recognition test showing percentage of time spent with objects and discrimination index in control and NaB treated old mice. *Image adapted from Singh and Thakur (2018).*

7. After that mice were placed in the box and allowed to interacts with both the objects (familiar and novel objects) for 5 min
8. To remove odor of mice, after each trial, the box and objects were cleaned with 70 % ethanol.
9. The data were analyzed by ANY-maze software. Percentage of time spent and the discrimination Index was calculated as mentioned earlier (Fig. 4).

5. HDAC2 inhibition increased histone acetylation level in the hippocampus of old mice

5.1 Animal

1. Animal Species – Swiss albino strain, male mice of 25–30 g of weight.
2. Age: Old (80 ± 2 weeks). Groups: Control, Sodium butyrate

5.2 Equipment

Surgical instruments, scissors, forceps, centrifuge tubes, tissue homogenizer, centrifuge, weighing balance, spectrophotometer.

5.3 Reagents

NaB, PBS, RIPA lysis buffer, PMSF, Bovine serum abumin, Bradford reagent, Tris.HCl, NaCl, SDS, Triton X-100, EDTA, PMSF, Laemmli buffer, acrylamide solution (acrylamide: bis-acrylamide, 29:1), APS, TEMED, glycine, PVDF membrane, Ponceau-S, fat free skimmed milk, primary antibodies (H3K9 Ac, H3K14 Ac and H3), HRP conjugated secondary antibodies, Tween 20, ECL reagents, X-ray film.

5.4 Procedure

Histone acetylation was analysed in the hippocampus of control and sodium butyrate treated old mice. After behavioral analysis, hippocampus was isolated and processed for histone acetylation analysis.

1. For histone acetylation analysis, protein was isolated from the hippocampus using RIPA lysis buffer with protease inhibitor.
2. Tissue homogenate was prepared using a Dounce homogenizer.
3. The tissue homogenate was centrifuged at 12,000 × *g* at 4 °C for 10 min
4. The pellet was discarded and supernatant was collected in separate centrifuge tubes.
5. Protein was estimated by Bradford methods (Bradford, 1976) using a spectrophotometer.
6. Proteins were separated using denaturing SDS-PAGE and transferred onto PVDF membrane.
7. The efficiency of protein transfer was checked by Ponceau-S and the membrane was blocked with 5 % (w/v) skimmed milk for 3 h at RT.
8. Thereafter, the membrane was incubated with primary antibody H3K9 Ac (1:1000 Millipore), H3K14 Ac (1:1000, Epigentek) at 4 °C overnight.
9. Next day, membrane was washed in PBST three times 5 min each and incubated with HRP conjugated secondary antibody (1:3000, Bangalore Genei) for 3 h at RT.
10. Membrane was washed in PBST three times 5 min each and signal was detected using ECL reagents on X-ray film.
11. Further, the membrane was stripped and incubated with primary H3 antibody (1:10,000, Sigma) followed HRP conjugated secondary antibody (1:3000, Bangalore Genei) for 3 h at room temperature.

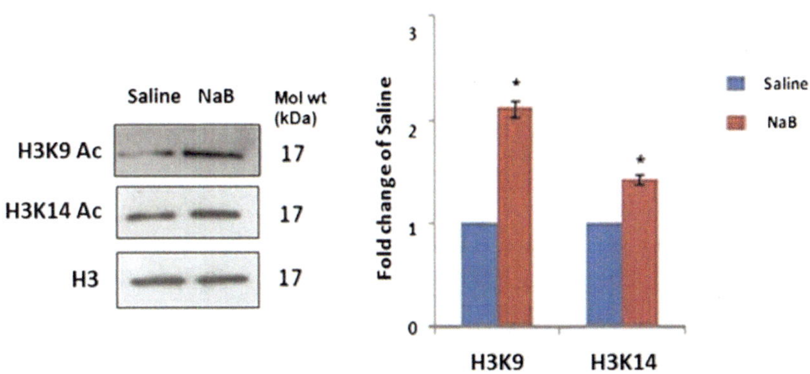

Fig. 5 Histone H3K9 and H3K14 acetylation level in the hippocampus of control and NaB treated old mice. *Image adapted from Singh and Thakur (2018).*

12. Membrane was washed in PBST three times 5 min each and signal was detected using ECL reagents on X-ray film.
13. Signal intensities were analyzed AlphaEaseFC software (Fig. 5).

6. HDAC2 inhibition increased histone acetylation level at the promoter of synaptic plasticity genes in the hippocampus of old mice

6.1 Animal

1. Animal Species – Swiss albino strain, male mice of 25−30 g of weight.
2. Age: Old (80 ± 2 weeks). Groups: Control, Sodium butyrate

6.2 Equipment

Surgical instruments, scissors, forceps, centrifuge tubes, tissue homogenizer, centrifuge, weighing balance, sonicator, spectrophotometer.

6.3 Reagents

NaB, Phosphate-buffered saline (PBS), formaldehyde, glycine, SDS, PIPES-KOH, KCl, NP-40, PMSF, Tris.HCl, EDTA, PMSF, antibodies (H3K9Ac, and H3K14Ac), protein A-Agarose bead, NaHCO3, NaCl, phenol–chloroform, promoter region specific primers.

6.4 Procedure

Histone acetylation at the promoter region of synaptic plasticity genes were analyzed by chromatin immunoprecipitation (ChIP) assay.

1. Hippocampus was isolated from Control and Sodium butyrate group after behavioral test.
2. Hippocampal tissue was chopped and cross-linked with 1 % formaldehyde for 15 min at RT. The reaction was stop by adding 0.125 M glycine the solution.
3. The tissue was homogenized in lysis buffer containing 5 mM PIPES-KOH (pH 8.0), 85 mM KCl, 0.5 % NP-40, 1 mM PMSF.
4. Tissue homogenate was centrifuged at 1000*g* for 5 min at 4 °C.
5. Supernatant was discarded and the pellet was resuspended in nuclear lysis buffer containing 50 mM Tris.Cl (pH 8.0), 1 % SDS, 10 mM EDTA, and 1 mM PMSF for 20 min at 4 °C.
6. For generating 500–1000 bp DNA fragments, chromatin was sonicated in a sonicator at 120 Hz for 30 s with 2 min interval at 4 °C.

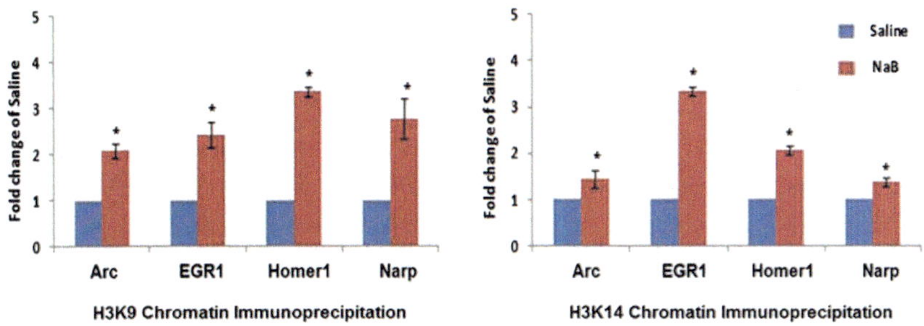

Fig. 6 Histone H3K9 and H3K14 acetylation level at the promoter of synaptic plasticity genes in the hippocampus of Control and NaB treated old mice. *Image adapted from Singh and Thakur (2018).*

7. The fragmented chromatin was incubated with 4 μg antibodies (H3K9Ac, or H3K14Ac) at 4 °C overnight.
8. Nextday, the antibody chromatin complex was incubated with protein A–Agarose bead slurry for 2 h at 4 °C.
9. Thereafter, the complex was centrifuged at 3500*g* at 4 °C for 10 min
10. The antibody chromatin complex was eluted using elution buffer containing 100 mM NaHCO3 and 1 % SDS at RT.
11. For reverse cross-linking, the protein–DNA complex was incubated with 200 mM NaCl at 65 °C for 5 h.
12. DNA was isolated from the protein–DNA complex by phenol–chloroform method.
13. The isolated DNA was used as template for the analysis of promoter region of synaptic plasticity genes.
14. Specific primers were used for the amplification of promoter region of Arc, EGR1, Homer1 and Narp in a real time PCR (Fig. 6).

7. Conclusion

HDACs are critical epigenetic regulators that control gene expression through the removal of acetyl groups from histone proteins, leading to chromatin condensation and transcriptional repression. In the central nervous system, HDAC-mediated regulation plays a fundamental role in neuronal development, synaptic plasticity, memory formation, and cellular stress responses. Increasing evidence indicates that dysregulation of HDAC activity contributes significantly to the pathogenesis of several neurodegenerative diseases, including AD, PD, HD, and ALS. These disorders are characterized by progressive neuronal loss, protein aggregation, and impaired cellular homeostasis, processes that are strongly influenced by epigenetic modifications. In many neurodegenerative conditions, an imbalance between histone acetylation and deacetylation alters transcriptional programs necessary for neuronal survival. Excessive HDAC activity can suppress genes involved in synaptic plasticity, neurotrophic signaling, and cellular repair mechanisms, ultimately contributing to neuronal dysfunction and degeneration. The presence of different HDAC isoforms and their distinct cellular localizations and biological functions, further increase their complex roles in the nervous system. Understanding their precise mechanisms is crucial for developing targeted therapies for a wide array of diseases, particularly those involving epigenetic dysregulation. In recent years, HDAC inhibitors have emerged as potential therapeutic agents for neurodegenerative diseases. These compounds promote histone hyperacetylation, neuronal survival and synaptic plasticity. Many currently available HDAC inhibitors lack specificity and can target multiple HDAC isoforms. Therefore, the development of isoform-specific inhibitors with improved brain penetration and reduced adverse effects represents an important direction for future research. While HDAC inhibitors offer promising therapeutic potential, further study is needed to better understand the specific roles of individual HDAC isoforms and to develop targeted strategies with improved safety and efficacy.

Acknowledgment

We acknowledge the Department of Biotechnology (DBT), Government of India for the DBT-Research Associate Fellowship to PS. We also acknowledge the National Academy of Sciences, India—Senior Scientist Fellowship (NASI-303/12/2021) to MKT.

References

Bradford, M. M. (1976). Rapid and sensitive method for the quantitation of microgram quantities of protein utilizing the principle of protein-dye binding. *Analytical Biochemistry, 72*, 248–254.

Camelo, S., Iglesias, A. H., Hwang, D., Due, B., Ryu, H., Smith, K., ... Dangond, F. (2005). Transcriptional therapy with the histone deacetylase inhibitor trichostatin A ameliorates experimental autoimmune encephalomyelitis. *Journal of Neuroimmunology, 164*(1-2), 10–21.

D'Mello, S. R. (2009). Histone deacetylases as targets for the treatment of human neurodegenerative diseases. *Drug News & Perspectives, 22*(9), 513–524.

Garcez, M. L., de Carvalho, C. A., Mina, F., Bellettini-Santos, T., Schiavo, G. L., da Silva, S., ... Budni, J. (2018). Sodium butyrate improves memory and modulates the activity of histone deacetylases in aged rats after the administration of d-galactose. *Experimental Gerontology, 113*, 209–217.

Govindarajan, N., Agis-Balboa, R. C., Walter, J., Sananbenesi, F., & Fischer, A. (2011). Sodium butyrate improves memory function in an Alzheimer's disease mouse model when administered at an advanced stage of disease progression. *Journal of Alzheimer's Disease: JAD, 26*(1), 187–197.

Gräff, J., Rei, D., Guan, J. S., Wang, W. Y., Seo, J., Hennig, K. M., ... Tsai, L. H. (2012). An epigenetic blockade of cognitive functions in the neurodegenerating brain. *Nature, 483*(7388), 222–226.

Guan, J. S., Haggarty, S. J., Giacometti, E., Dannenberg, J. H., Joseph, N., Gao, J., ... Tsai, L. H. (2009). HDAC2 negatively regulates memory formation and synaptic plasticity. *Nature, 459*(7243), 55–60.

Ishizuka, Y., Shimizu, H., Takagi, E., Kato, M., Yamagata, H., Mikuni, M., & Shirao, T. (2014). Histone deacetylase mediates the decrease in drebrin cluster density induced by amyloid beta oligomers. *Neurochemistry International, 76*, 114–121.

Ji, J., Xia, M., Xia, W., & Miao, G. (2025). Structural classification of HDAC inhibitors and cancer treatment prospects: A review. *Letters in Drug Design & Discovery, 22*(9), 100165.

Kandel, E. R. (2001). The molecular biology of memory storage: A dialogue between genes and synapses. *Science (New York, N. Y.), 294*(5544), 1030–1038.

Liu, D., Tang, H., Li, X. Y., Deng, M. F., Wei, N., Wang, X., ... Zhu, L. Q. (2017). Targeting the HDAC2/HNF-4A/miR-101b/AMPK pathway rescues tauopathy and dendritic abnormalities in Alzheimer's disease. *Molecular Therapy: The Journal of the American Society of Gene Therapy, 25*(3), 752–764.

Nakatsuka, D., Izumi, T., Tsukamoto, T., Oyama, M., Nishitomi, K., Deguchi, Y., ... Ogawa, K. (2021). Histone deacetylase 2 knockdown ameliorates morphological abnormalities of dendritic branches and spines to improve synaptic plasticity in an APP/PS1 transgenic mouse model. *Frontiers in Molecular Neuroscience, 14*, 782375.

Navakkode, S., & Kennedy, B. K. (2024). Neural ageing and synaptic plasticity: Prioritizing brain health in healthy longevity. *Frontiers in Aging Neuroscience, 16*, 1428244.

Pereira, B., Correia, F. P., Alves, I. A., Costa, M., Gameiro, M., Martins, A. P., & Saraiva, J. A. (2024). Epigenetic reprogramming as a key to reverse ageing and increase longevity. *Ageing Research Reviews, 95*, 102204.

Peserico, A., & Simone, C. (2011). Physical and functional HAT/HDAC interplay regulates protein acetylation balance. *Journal of Biomedicine & Biotechnology, 2011*, 371832.

Singh, P., & Thakur, M. K. (2014). Reduced recognition memory is correlated with decrease in DNA methyltransferase1 and increase in histone deacetylase2 protein expression in old male mice. *Biogerontology, 15*(4), 339–346.

Singh, P., & Thakur, M. K. (2018). Histone deacetylase 2 inhibition attenuates downregulation of hippocampal plasticity gene expression during aging. *Molecular Neurobiology, 55*(3), 2432–2442.

Robakis, N. K (2003). An Alzheimer's disease hypothesis based on transcriptional dysregulation. *Amyloid, 10*(2), 80–85.

Singh, P., Konar, A., Kumar, A., Srivas, S., & Thakur, M. K. (2015). Hippocampal chromatin-modifying enzymes are pivotal for scopolamine-induced synaptic plasticity gene expression changes and memory impairment. *Journal of Neurochemistry, 134*(4), 642–651.

Tan, Y., Delvaux, E., Nolz, J., Coleman, P. D., Chen, S., & Mastroeni, D. (2018). Upregulation of histone deacetylase 2 in laser capture nigral microglia in Parkinson's disease. *Neurobiology of Aging, 68*, 134–141.

Zhang, L. Y., Zhang, S. Y., Wen, R., Zhang, T. N., & Yang, N. (2024). Role of histone deacetylases and their inhibitors in neurological diseases. *Pharmacological Research, 208*, 107410.

CHAPTER TWELVE

Synergetic inhibition of deubiquitinases and acetyltransferases to regulate tau acetylation and aggregation in SH-SY5Y cell line model of Alzheimer's disease

Basavaraju K C[a,b], Harshitha D K[a,b], and Poornima Priyadarshini[a,b,*]

[a]Department of Molecular Nutrition, CSIR-Central Food Technological Research Institute (CSIR-CFTRI), Mysore, Karnataka, India

[b]Academy of Scientific and Innovative Research (AcSIR), Ghaziabad, India

*Corresponding author. e-mail address: ppriyadarshini@cftri.res.in; ppriyadarshini.cftri@csir.res.in; ppriyadarshini@gmail.com

Contents

1. Introduction 298
   1.1 Deubiquitinases 299
   1.2 Acetyltransferases and tau acetylation 302
2. Rationale 305
3. Methods 306
   3.1 Chemicals 306
   3.2 Procedure 307
   3.3 Notes 308
   3.4 Protocol for microscopy 308
   3.5 Gene expression studies 312
   3.6 cDNA synthesis: (iScript™ cDNA synthesis Kit-Biorad) 312
4. Notes 313
5. Conclusions 313
Acknowledgement 314
Funding 314
Data availability 314
Ethical approval and consent to participate 314
CRediT authorship contribution statement 314
Consent for publication 314
References 315

Methods in Enzymology, Volume 733
ISSN 0076-6879, https://doi.org/10.1016/bs.mie.2026.05.034

Abstract

Tau protein homeostasis depends on a balance between ubiquitination and acetylation, two key post-translational modifications (PTMs) which simultaneously compete for lysine residues that govern tau stability. Importantly, ubiquitination signals of proteins for degradation through the 26S proteasome, whereas acetylation prevents ubiquitin attachment, leading to tau stabilization and an increased risk of aggregation. In Alzheimer's disease (AD), this balance is disrupted by upregulation of acetyltransferase activity, such as p300/CBP, and increased deubiquitinase (DUBs) activity, which together hinder proteasomal clearance and accumulation of pathological tau. Since these pathways are interconnected, targeting only one may be insufficient to address tau-mediated neurotoxicity. In this book chapter, we highlight the importance of a dual approach aimed at inhibiting both aberrant acetylation and deubiquitinases activity, thereby restoring ubiquitination and promoting tau degradation. In addition, the development of standardized cell culture models in the SH-SY5Y neuronal system will be crucial to evaluating the therapeutic potential of this synergetic inhibition strategy.

1. Introduction

The ubiquitin-proteasome system (UPS) is one of the key cellular mechanisms that help to regulates protein homeostasis of the cell by maintaining the proper balance of intracellular proteins (Jia et al., 2023; Lee et al., 2013; Liu et al., 2021). This system involves attachment of a small regulatory protein called ubiquitin, to target substrate proteins (Cai et al., 2020). Ubiquitination primarily occurs on lysine residues, making their presence essential for proper UPS function (Wathan et al., 2025). The UPS consists of a coordinated series of ubiquitinating enzyme complexes, including the E1 activating enzyme, E2 conjugating enzyme, and E3 ligase enzyme (Ebadi et al., 2025; Liu et al., 2020; Wong & Kamarul, 2025). This system plays an important role in removing misfolded proteins via ubiquitination. However, the dysregulation of the UPS is also linked to the development of neurodegenerative disorders, such as Alzheimer's disease (AD) (Hol et al., 2006; Oddo, 2008; Weng & He, 2021). AD is characterised by the abnormal build-up of two important proteins: Amyloid beta (Aβ) and the microtubule-associated protein tau (Chen et al., 2012; Sakono & Zako, 2010; Xiao et al., 2015; Zhang et al., 2021). Under normal conditions, tau stabilizes microtubules, supports the transport of vesicles, mitochondria (Cheng & Bai, 2018), and regulates actin filament polarisation (Das et al., 2020; He et al., 2009; Hlushchenko et al., 2016). In AD brains, the dysregulated ubiquitination-driven degradation leads to an abnormal build-up of hyperphosphorylated and misfolded tau (Wang et al., 2021).

This accumulation triggers several deleterious processes, including oxidative stress (Pant et al., 2017; Varadarajan et al., 2000), microglial activation (Spanic et al., 2019), and synaptic dysregulation, ultimately resulting in widespread neuronal death (Colnaghi et al., 2020; Lasagna-Reeves et al., 2011; Pandya & Patani, 2021).

The disruption of these pathways is complex and is often caused by an imbalance of regulatory enzymes. It is characterised by increased activity of deubiquitinases (DUBs) (Bello et al., 2022; Lim et al., 2020; Wei et al., 2022) while certain acetyltransferases have also been linked to the disruption of proper tau ubiquitination (Balasubramanyam et al., 2003; Shahib et al., 2024), thereby protecting pathogenic tau from degradation. Therefore, understanding the interplay between these enzymes and their role in UPS dysfunction is crucial for elucidating the molecular mechanisms of neurodegeneration.

1.1 Deubiquitinases

Deubiquitinases play a key role in UPS dysregulation by reversing the ubiquitination of target proteins. In AD, an abnormal increase in certain DUBs, especially Ubiquitin-specific proteases (USP) such as USP7 (Volbracht & Fog, 2025), USP10 (Wei et al., 2022), USP11 (Jia & Fu, 2025), USP13 (Jia & Fu, 2025), OTU Deubiquitinase with Linear Linkage Specificity (OTULIN) (Tangavelou et al., 2025), and Ubiquitin carboxyl-terminal hydrolase L1 (UCH-L1) has been reported to accelerate tau pathology. Dysregulation of these DUBs pathways exposes lysine residues on proteins by removing ubiquitin. In the case of tau, once ubiquitin is removed, the free lysine residues on tau become susceptible to acetylation by acetyltransferases. This transition from ubiquitination to acetylation divert tau marked for degradation towards further stabilisation and aggregation, ultimately resulting in tau-related neurotoxicity. The functions of the other DUBs involved in AD pathogenesis include USP8, USP25, USP46, and UCH-L1, which influence the formation of neurofibrillary tangles (NFTs) by modulating APP ubiquitination. Meanwhile, the stabilization of phosphorylated tau (p-tau) is regulated by USP9, USP10, USP11, USP13, USP14, OTUB1 and UCH-L1/3/5. USP25 also plays a significant role in neuroinflammation associated with AD (Jia & Fu, 2025).

In addition, AD patients exhibited sex-specific expression patterns of USP9. Further, USP9X is believed to enhance tau phosphorylation through upregulation of MARK4 kinase activity, a known tau kinase, whereas USP9Y may play a more prominent role in regulating tau transcription.

Notably, USP9Y levels were markedly decreased in the brains of male AD patients. Knockdown of USP9 resulted in a significant reduction in tau levels in cellular and zebrafish models (Al-Hakim et al., 2008). In a study by Wei et al., USP10 levels were found to be elevated in post-mortem AD samples and in APP/PS1 mice, but not in P301S mice. Furthermore, in primary neuronal cultures, Aβ24 induced a dose-dependent increase in USP10 levels, resulting in an increase in the level of total and p-tau with a significant increase in the interaction between tau and USP10. This resulted in a significant reduction in tau ubiquitination and an increase in tau stability, eventually leading to tau aggregation and a delayed degradation process (Wei et al., 2022). Another study demonstrated that inhibition of USP7 and USP10 in cortical cultures of rTg4510 mice resulted in increased polyubiquitination of residual tau inclusions. These results indicated that Usp7 and Usp10 contribute to pathological tau accumulation by modulating ubiquitin-dependent degradation pathways (Volbracht & Fog, 2025).

Notably, research in this direction also indicates that this pathway may exhibit sex-specific effects, particularly involving the X-linked enzyme USP11. Increased levels of USP11 have been reported in AD, where it contributes to neurodegeneration by deubiquitinating tau at K281. This specific deubiquitination makes the site more susceptible to acetylation, promoting the formation of toxic tau aggregates. Furthermore, as USP11 partially escapes the X-inactivation, female brains often have higher levels of this enzyme. This may explain the differences in AD progression and cognitive decline between sexes (Yan et al., 2022). Yet another study on the knockdown of USP13 in animal models reported to reduce p-tau expression via proteasomal and autophagic pathways, resulting in reduced Aβ and increased tau ubiquitination. Thus, USP13 appears to regulate deubiquitination processes that influence the activity of proteasomes, ultimately reducing p-Tau and Aβ production (Liu et al., 2019). Furthermore, inhibition of USP25 has been reported to promote APP degradation during acute endoplasmic reticulum (ER) stress by activating ER-associated degradation (ERAD) and proteasome-dependent pathways (Zhu et al., 2017). However, the role of USP14 in AD remains unclear, with inconsistent results reported across cell-based and in vivo studies highlighting the need for further investigation (Jia & Fu, 2025).

OTULIN (OTU deubiquitinase with linear linkage specificity) removes linear polyubiquitin from proteins that have been modified by LUBAC (linear ubiquitin chain assembly complex). Notably, A study on the attenuation of OTULIN long non-coding RNA (OTULIN lncRNA) was found to be associated with elevated levels of OTULIN protein and p-tau at p-S202/p-T205 (AT8), p-T231 (AT180), and p-S396/p-S404 (PHF-1) in late sporadic AD (sAD2.1) inducible pluripotent stem cell (iPSC)-derived neurons. Pharmacological inhibition of OTULIN deubiquitinase activity using a small molecule UC495 resulted in reduced levels in both p-tau in iPSNs and SH-SY5Y cells. Conversely, knockout of OTULIN gene in sAD2.1 iPSNs by the CRISPR-Cas9 reduced the overall tau and p-tau levels (Tangavelou et al., 2025).

Yet another DUB, Ubiquitin carboxyl-terminal hydrolase L1 (UCH-L1), belonging to the UCH family, has been linked with intraneuronal aggregates and NFTs detected in AD. UCH-L1 constitutes up to ~2% of total brain protein and is widely distributed throughout the central nervous system. In neurons, it exists in two forms, a membrane-associated (UCH-L1M) form and a soluble cytoplasmic (UCH-L1C) form. In the hippocampus, UCH-L1 levels are reduced and, restoring UCH-L1 improve synaptic and cognitive functions in AD patients. UCH-L1 overexpression has been shown to upregulate ubiquitination and lysosomal degradation of APP (Gong et al., 2006). Further, it also inhibits memory impairments, reduces Aβ deposition, and prevents the formation of neuritic plaques, in an AD mouse model (Zhang et al., 2014) This indicates the positive effect of UCH-L1 in AD condition. UBB+1, a mutant form of ubiquitin containing a 19-amino-acid extension, has been found to coexist with neuritic plaques and NFTs in AD patients. (Fischer et al., 2003) In UBB + 1 transgenic mice, prolonged proteasome inhibition leads to AD-like behavioural abnormalities along with extracellular Aβ deposition and accumulation of insoluble p-tau aggregates (Maniv et al., 2023).

In addition to UCH-L1, UCH-L3 and UCH-L5 have also been identified as potential regulators of AD. UCH-L3 specifically cleaves the C-terminus of UBB + 1, generating a truncated form, which will be unable to form plaques and NFTs under normal conditions. (Maniv et al., 2023). In AD patients, UCH-L3 expression levels remain unchanged, comparable to those in healthy controls however, under oxidative conditions, commonly observed in neurological disorders, UCH-L3 loses its ability to efficiently process UBB + 1, thereby accelerating protein aggregation (Wagh et al., 2023). The UCH-L5 is an essential protein that plays a critical link between the histone acetyltransferase-associated module and the proteasome-associated module. A study in mice revealed that the absence of UCH-L5 led to perinatal lethality due to severe defects in central nervous system development. Thus, suggesting a reduced level of UCH-L5 may contribute to disruptions in both transcriptional regulation and protein degradation observed in AD (Al-Shami et al., 2010). These findings, together with other dysregulated deubiquitinases, are summarised in Table 1.

1.2 Acetyltransferases and tau acetylation

Post-translational modification of the tau, especially dysregulated acetylation, has emerged as a crucial contributor to tauopathy and neurodegeneration. This dysregulation results by an imbalance between acetyltransferases and deacetylases. The increased activity of acetyltransferases like p300 and CBP leads to excessive tau acetylation (Cohen et al., 2011; Dornan et al., 2003; Li et al., 2023). While, a decrease in the activity of deacetylases, including HDACs and SIRT1, further amplifies this pathological state (Chen et al., 2020; El-Maraghy et al., 2023; Yan et al., 2020). Notably, tau contains 23 lysine residues recognised as key acetylation sites, of which 13 are located within the microtubule-binding domain, a domain crucial for stabilising microtubules (Min et al., 2010b). Hyperacetylation of tau in particular significantly contributes to neurodegenerative processes by promoting pathological accumulation of tau into NFTs and impairing UPS

Table 1 Deubiquitinases regulating tau ubiquitination and aggregation through site-specific lysine deubiquitination of tau.

Deubiquitinases	Deubiquitination sites of tau	Mode of action	References
USP11	K281	USP11 enhances the abnormal aggregation of tau by deubiquitinating it at lysine-281, while allowing for enzymatic acetylation of tau at K281 and K274.	Yan et al. (2022)
Otub1	K48	Increasing tau aggregation and seeding generates pathological tau forms and impair its degradation	Wang et al. (2017)
USP7	K27, K48	Stabilizes tau by protecting it from CHIP-mediated ubiquitination	Kim et al. (2023)

Table 2 Impact of lysine acetylation in modulating aggregation and toxicity of amyloidogenic proteins (Aβ and Tau).

Protein	Lysine site (s)	Effect of acetylation	Aggregation behaviour	Toxicity level	Model/study	References
$A\beta_{42}$	K16	Disrupts salt-bridge formation, increases hydrophobicity	Forms amorphous, non-fibrillar aggregates	High (higher than WT or K28Ac)	Synthetic peptides, MD simulation, cytotoxicity assays	Adhikari et al. (2020), Kajava et al. (2010)
$A\beta_{42}$	K28	Neutralizes salt-bridge (normally stabilizes fibril β-sheet)	Slows fibrillization, fibril morphology similar to WT	Slightly increased or similar to WT	Synthetic peptides, fibrillation assays	Xiao et al. (2015)
$A\beta_{42}$	K16 + K28 (double)	Additive effects on hydrophobicity	Amorphous aggregates predominate	Very high cytotoxicity and ROS	Synthetic double-modified peptides, cell assays	Adhikari et al. (2020)
APP	K649 K650 K651	Accelerates the aggregation	Fibrillar aggregates	High	Mouse model	Cheng et al. (2024)

(continued)

Table 2 Impact of lysine acetylation in modulating aggregation and toxicity of amyloidogenic proteins (Aβ and Tau). (*cont'd*)

Protein	Lysine site (s)	Effect of acetylation	Aggregation behaviour	Toxicity level	Model/study	References
Tau	K280	Impairs microtubule binding; blocks ubiquitination sites	Enhances fibril and oligomer formation into PHFs	High, promotes pathological tau accumulation	Human AD brain, cell and mouse models	Kim et al. (2023)
Tau	K274, K281	Impairs microtubule binding, mislocalizes tau to dendrites	Enhanced oligomerization and mislocalization into the somatodendritic compartment	High, synaptic dysfunction, memory deficits	Human AD brain, tau-KQ mutants in neurons, transgenic mice	Rane et al. (2019)
Tau	K174	Slows turnover, reduces degradation (ubiquitin inability)	Accumulates in neuronal inclusions	Behavioural/ cognitive impairment in vivo	Mouse models, human AD brain samples	Chakraborty et al. (2023), Cohen et al. (2011), Min et al. (2010), Trzeciakiewicz et al. (2017)

(Ferreon et al., 2018; Gorsky et al., 2016; Rane et al., 2019). Thus, by disrupting the ubiquitination process, acetylation hinders the effective removal of misfolded tau, thereby creating a continuous cycle of tau accumulation in AD (Portillo et al., 2021).

Interestingly, the mammalian tau protein possesses intrinsic enzymatic activity that enables it to catalyse self-acetylation. A study evaluating soluble tau derived from frontal cortex homogenates of control and AD brains, found a comparable levels of *de novo* autoacetylation, as detected by anti-Ac-K280 (Cohen et al., 2013; Cohen et al., 2016). Notably, Phosphorylation is considered an early event in AD pathogenesis. Studies on Chronic Traumatic Encephalopathy (CTE) show that the acetylation of K280 may precede and facilitate subsequent phosphorylation (Lucke-Wold et al., 2017). Acetylation at lysine residues K164, K174, K180, and K280 has been associated with a defect in the degradation of p-tau, leading to its pathological accumulation (Gorsky et al., 2016; Min et al., 2010a). Conversely, pseudo-acetylation at residues K163, K280, K281, and K369, as well as modifications at the KIGS motif or K321, has been linked with decreased tau phosphorylation (Carlonnagno et al., 2017; Cook et al., 2014; Gorsky et al., 2017). Specifically, acetylation at KXGS motifs inhibits phosphorylation at S262 and S356 within these sequences (Cook et al., 2014). In addition to tau, the acetylation has also been linked to the aggregation of Aβ (Adhikari et al., 2020; Xiao et al., 2015), thereby exacerbating cellular degeneration, as summarized in Table 2.

2. Rationale

In Alzheimer's disease, increased DUB activity removes ubiquitin tags from tau, restoring lysine residues that would otherwise signal degradation. While this limits proteasomal clearance, it simultaneously increases the availability of lysine residues for ep300-mediated acetylation, resulting in enhanced tau stability and aggregation. Targeting a single component of this pathway is unlikely to be effective: inhibiting acetyltransferases alone may promote ubiquitination, but persistent DUB activity would counteract degradation, whereas blocking DUBs alone would not prevent acetylation-driven protection of tau from clearance. Therefore, a combinatorial approach targeting both ep300 and DUBs may more effectively maintain lysine residues in a ubiquitinated state, prevent aberrant acetylation, promote tau degradation, and ultimately reduce pathological aggregation in AD.

3. Methods

3.1 Chemicals

The following items were acquired from Sigma-Aldrich (USA): Dulbecco's Modified Eagle Medium (DMEM) powder, MEM (100x), Sodium bicarbonate, Phosphate-buffered saline (PBS), Okadaic acid, Penicillin-streptomycin, Retinoic acid (RA), RIPA buffer, Trypsin-EDTA, Trypan blue, DMSO, Fetal bovine serum (FBS) and RNA extraction diluent were supplied by Himedia (India). while Copper sulfate pentahydrate, Isopropanol, Chloroform, and Bovine serum albumin (BSA) was purchased from SRL (India). Protease inhibitor cocktail tablets were purchased from Roche Diagnostics (Germany).

3.1.1 Preparations

3.1.1.1 DMEM (10 % FBS)

Dulbecco's Modified Eagle Medium (DMEM) was prepared by dissolving DMEM powder in 880 mL of sterile autoclaved water and mixing thoroughly until completely dissolved. To this, 3.7 g sodium bicarbonate was added, followed by 10 mL penicillin–streptomycin solution and 10 mL MEM (100X). Subsequently, 100 mL of filter-sterilised fetal bovine serum (FBS) was added to obtain a final FBS concentration of 10 %. The complete medium was then filtered using a suction pump filtration unit under sterile conditions, aliquoted into 50 mL Falcon tubes, properly labelled, and stored at 4 °C until further use.

3.1.1.2 DMEM (0.5 % FBS)

DMEM powder was dissolved in 975 mL of sterile autoclaved water with thorough mixing. Sodium bicarbonate (3.7 g), 10 mL penicillin–streptomycin, and 10 mL MEM (100X) were added sequentially. Then, 5 mL of filter-sterilised FBS was added to achieve a final FBS concentration of 0.5 %. The medium was sterile-filtered using a suction pump filtration unit, aliquoted into 50 mL Falcon tubes, properly labelled, and stored at 4°C until further use.

3.1.1.3 1X PBS

Phosphate-buffered saline (PBS) was prepared by dissolving 8.58 g of PBS powder (Himedia) in 1 L of Milli-Q water. The solution was autoclaved for sterilisation, then filtered using a filtration unit. The sterile PBS was aliquoted and stored at 4°C until further use.

3.1.1.4 Retinoic acid (RA) stock and working solution

Retinoic acid (RA) stock solution was prepared by dissolving RA in DMSO to obtain a concentration of 3 mg/mL (10 mM). The solution was prepared and filter-sterilised under a laminar flow hood. For working concentration, 25 μL of RA stock (final 250 μM intermediate concentration) was added to 50 mL of 0.5 % FBS DMEM to achieve a final concentration of 5 μM per 1 mL of media. For experiments requiring 10 μM RA, 2 mL of this prepared medium was used.

3.1.1.5 Okadaic acid stock and working solution

Okadaic acid stock solution was prepared by dissolving in DMSO at a concentration of 2.41 mg/mL (3 mM) and filter-sterilised under sterile conditions. A secondary (intermediate) stock (200 μM) was prepared by adding 66.66 μL of the primary stock to 934 μL DMSO. From this secondary stock, 12.5 μL (equivalent to 2500 nM) was added to 50 mL of culture medium to obtain a final concentration of 50 nM per 1 mL. For experiments requiring 100 nM concentration, 2 mL of the prepared medium was used.

3.2 Procedure

1. Undifferentiated SH-SY5Y cells were maintained in DMEM supplemented with 10 % fetal bovine serum (FBS) under standard culture conditions.
2. For subculturing, the culture medium was aspirated, and cells were washed with 2 mL of 1X phosphate-buffered saline (PBS).
3. Cells were treated with 1 mL of 0.25 % trypsin and incubated until detachment was observed.
4. Trypsinisation was neutralised by adding 3 mL of complete medium, and the cell suspension was transferred to a 15 mL centrifuge tube.
5. The cells were centrifuged at 2500 rpm for 5 min, and the supernatant was discarded.
6. The cell pellet was resuspended in 1 mL of fresh medium.
7. For cell counting, 2 μL of the resuspended cells was mixed with 8 μL of trypan blue and loaded onto a hemocytometer.
8. Cell concentration was calculated using the formula: **Cells/mL = (number of cells in 16 squares) × dilution factor × 10 × 1000**.
9. Based on the cell count, an equal number of cells were seeded onto 6-well plates and allowed to grow until ~70 % confluency.
10. The experimental procedure was initiated on Day 1 with undifferentiated SH-SY5Y cells cultured in a 6-well plate with 10 % FBS medium.

11. From Day 2 to Day 9, cells were treated with 10 μM retinoic acid (RA) in DMEM supplemented with 0.5% FBS to induce neuronal differentiation. The RA-containing medium was replaced every 3 days.
12. On Day 10, differentiated cells were treated with 100 nM okadaic acid (OA) in 0.5 % FBS medium to induce tau hyperphosphorylation and mimic AD-like pathology. Fig. 1
13. On Day 11, cells were co-treated with selected inhibitors (drugs/phytochemicals) in 0.5 % FBS medium to evaluate their potential to activity against deubiquitinases and acetyltransferases. Some of the known inhibitors are listed in Tables 3 and 4.
14. On Day 12, cells were subjected to downstream biochemical and molecular analyses such as fluorescence microscopy and qPCR (as shown in Fig. 2) to assess efficacy of tested compounds on tau acetylation pathways.

3.3 Notes

1. Final DMSO concentration should not exceed 1 % (v/v),
2. Initiate RA treatment only when cells reach ~60–70 % confluency to ensure proper differentiation.
3. RA is light-sensitive, prepare and store in dark conditions, and avoid repeated freeze-thaw cycles.
4. OA is highly toxic, handle carefully and optimise exposure time to prevent excessive cell death.
5. Maintain low serum conditions (0.5 % FBS) during differentiation for optimal neuronal phenotype.

3.4 Protocol for microscopy

To evaluate the localisation and expression of proteins in the synergistic inhibition strategy.

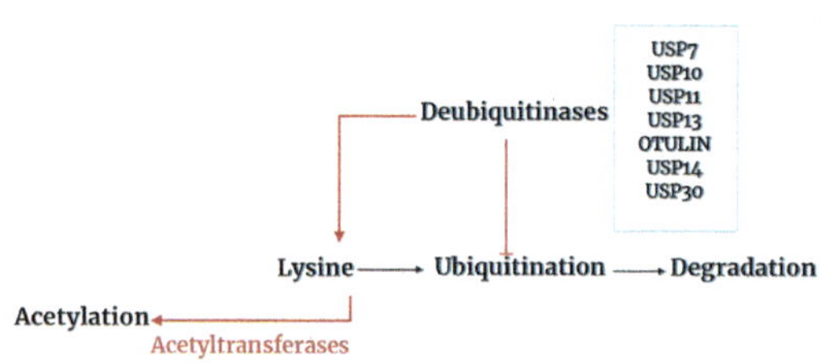

Fig. 1 Interplay between acetylation and ubiquitination at lysine residues regulated by acetyltransferases and deubiquitinases.

Table 4 List of Deubiquitinases inhibitors evaluated in SH-SY5Y cells.

Pharmacological drugs	Concentration	Duration (Hr)	Deubiquitinase enzyme	Mechanism of action	References
P22077	0–20 μM	24	USP7	Induces apoptosis in Neuroblastoma (NB) cells with an intact USP7-HDM2-p53 axis but not in NB cells with mutant p53 or without human homolog of MDM2 expression	Fan et al. (2013)
BK50118-A	IC50 = 0.11 nM	5	USP13	Decreases the alpha-synuclein level by increasing its ubiquitination	Liu et al. (2021)
BK50118-B,	IC50 = 2.13 nM				
BK50118-C	IC50 = 0.42 nM				
CL3-499	IC50 = 0.61 nM				
CL3-512	IC50 = 0.27 nM				
CL3-514	IC50 = 0.29 nM				
Spautin-1	IC50 = 600–700 nM				
$USP30_{inh}$ (Small-molecule benzosulfonamide-containing compound)	0–5 μM in protein lysate	1	USP30	Induces dramatic structural and conformational rearrangements, thereby preventing Ub binding and decrease DUB activity.	O'Brien et al. (2023)
UC495	2 μM	Overnight	OTULIN	Significantly decreases the levels of hyperphosphorylated and total tau, and OTULIN in SH-SY5Y	Tangavelou et al. (2025)

Table 3 List of ep300 inhibitors evaluated in SH-SY5Y cells.

Compounds	Concentration	Treatment time	Key finding	References
Pharmacological drugs				
A485	15uM	72 hr	*TSC1* loss increases tau acetylation and impairs clearance in SH-SY5Y cells	Alquezar et al. (2021)
C646	20 μM	60 h	Reduced H3K27 acetylation levels at target promoters in ChIP assays	Zhao et al. (2020)
Phytochemical				
Garcinol	0.5 μM, 1 μM	24 h	Reduced histone H3 acetylation through p300/PCAF inhibition in SH-SY5Y cells	Hegarty et al. (2016)

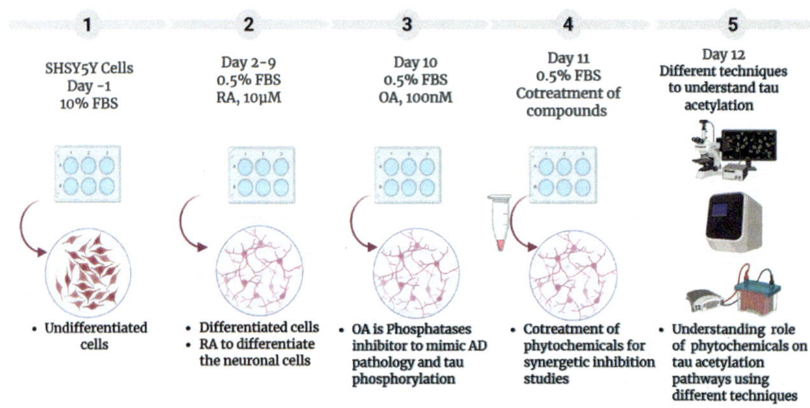

Fig. 2 Experimental workflow for modelling Alzheimer's disease pathology and phytochemical and pharmacological intervention in SH-SY5Y cells to understand the synergetic effects of tau acetylation and aggregation.

I. Preparations:

- Autoclaved cover slip
- 4 % paraformaldehyde (PFA)
- 0.1 % triton x 100
- Blocking solution (3 % BSA in 1X PBS)
- Primary antibody in 1: 200 in 3 % BSA. For each slide, take 20 µl.
- DAPI: 1: 100 in PBS. 10ul for each coverslip along with secondary antibody
- Secondary antibody: 1: 200 in 3 % BSA. For each slide, take 20 µl. add DAPI (10 µl), so total volume will be 30 µl.

I. Procedure:

1. Post differentiation and incubation with the selected inhibitors for 24 h remove the media from the 6-well plates and wash the cells with 1 mL of PBS
2. Fix the cells by adding 1 mL of 4 % PFA prepared in 1X PBS. Leave it for 20 min, then wash the cells three times with 1X PBS.
3. Permeabilize the cells by adding 1 mL of 0.1 % Triton X-100 in PBS for 5 min, followed by washing with 1X PBS.
4. Block the cells by incubating with 3 % BSA in 1X PBS for 30–45 min at room temperature. Wash the cells three times with 1X PBS.
5. Place a wet paper towel on the bench and cover it with parafilm.
6. Pipette 20 µL of the respective primary antibody (1:200) onto the parafilm and then place the coverslip onto it. Incubate for 4–6 h at 4 °C, then wash three times with PBS.
7. Incubate the cells with the secondary antibody (1:200) and DAPI (1:100) for 1 h at room temperature and wash three times with PBS.
8. After 3 final PBS washes, the coverslips were mounted on glass slides using a gold antifade mounting medium. The samples were then visualised with a fluorescence microscope to analyse protein expression and cellular morphology.

Notes:

1. Handle cells gently during PBS washes to prevent detachment and ensure cells do not dry at any stage.
2. Use freshly prepared or properly stored 4 % PFA and avoid over-fixation, as it can mask antigenic sites; perform fixation in a fume hood due to toxicity.
3. Protect secondary antibody and DAPI from light, as they are light-sensitive, perform incubation in the dark (cover with aluminum foil).
4. Mount coverslips carefully using antifade mounting medium, avoiding air bubbles and preserving fluorescence signals.

3.5 Gene expression studies

qPCR was performed using standardized protocols in accordance with the manufacturer's guidelines.

I. Preparations:

- 70 % ethanol
- cold isopropanol
- Primers: Stock- 100pmole/ul, working: Add 40 µl from the stock solution to 160 µl of autoclaved milli-Q water to obtain a 20 pmol/µl concentration.

I. Procedure:

1. Post differentiation and incubation with the selected inhibitors for 24 h mRNA after treating cells with the desired synergistic compounds, remove media from SH-SY5Y cells in a 6-well plate, wash with PBS (3 times)
2. Add 700 ul of Trizol reagent, and leave it for half an hour, homogenize by pipetting several times
3. Collect to 1.5 mL Eppendorf tube, and add 200ul of chloroform, vertex 1 min, incubate 15 min at room temperature, centrifuge at 13,000 g/10 min, and collect the supernatant,
4. To the supernatant add 600 µl of cold isopropanol, mix well by inversion, and incubate for 30 min at −20 °C
5. Centrifuge 13,000 g/20 min, remove the supernatant and check for pellet
6. Wash the pellet with 70 % ethanol, vertex briefly, spin at 8000 g for 5 min (repeat the wash 2 more times)
7. Air dry the pellet for 5−10 min. Dissolve in nuclease-free water (20 µl), and measure the RNA concentration using a nanodrop.

3.6 cDNA synthesis: (iScript™ cDNA synthesis Kit-Biorad)

Components	Volume
Nuclease-free water	Variable according to mRNA concentration
RNA	500 ng based on the nanodrop results
cDNA reaction mix	4 µl
Reverse transcriptase	1 µl
Total	20 µl

Take 500 ng of mRNA for the synthesis of cDNA

3.6.1 *Set up for qPCR reaction: Quant studio thermo qPCR system*

Components	Volume
Reverse primer	1ul
Forward primer	1ul
cDNA	4 μl
SYBR green mix	6ul
Total	12 μl

Reaction protocol:

Step	Time	Temp
Priming	5 min	25
Reverse transcription	20 min	46
RT inactivation	1	95
Optional step	Hold	4

4. Notes

1. Clean the working area thoroughly using RNase Zap to eliminate RNase contamination before handling RNA samples.
2. Mix the isolated RNA gently and thoroughly, measure concentration and purity using a Nanodrop spectrophotometer, take at least two readings, and use the average value for analysis.
3. If the A260/A230 ratio is lower than expected, improve RNA purity by increasing the 70% ethanol wash step from two to three washes during the isolation procedure.

5. Conclusions

Understanding the role of phytochemicals and drug inhibitors that enhance ubiquitination while lowering tau acetylation offers a promising approach for treating AD. In neurons, a strictly regulated balance between ubiquitination and acetylation is vital for maintaining steady tau homeostasis. Elevated acetylation of tau often disrupts ubiquitin-dependent breakdown, leading to stabilisation and accumulation of pathological tau. Additionally, upregulated DUBs activity further disrupts tau degradation. Therefore, compounds that can inhibit both DUBs and acetyltransferases may serve as an effective dual-target strategy to restore this balance and facilitate tau clearance. Such combinatorial strategies could promote ubiquitin-dependent degradation while preventing acetylation-mediated tau aggregation. The development and validation of suitable in vitro models, such as SH-SY5Y neurons, along with relevant in vivo animal models, will be crucial for assessing the effectiveness of these compounds and understanding their underlying molecular pathways. Gaining further understanding of the interplay between DUBs and acetyltransferases in regulating tau modifications will greatly benefit AD research and aid in designing targeted interventions to mitigate tau pathology and neurodegeneration.

Acknowledgement

The authors gratefully acknowledge all investigators, researchers, and funding agencies whose contributions have advanced the fields of neuroscience. We also sincerely appreciate the internal support by the Institute, CSIR-CFTRI and the Department of Molecular Nutrition, CSIR-CFTRI, Mysuru, Karnataka, India. Special thanks to Dr. Kunal Sharan, Department of Molecular Nutrition, CSIR-CFTRI, for the internal review of the manuscript. We also thank BioRender for providing the tools used to create the illustrations in this work.

Funding

The authors thank the financial assistance provided by the institute CSIR-CFTRI. Mr. Basavaraju K C acknowledge the fellowship support from the Department of Biotechnology (DBT), Regional Centre for Biotechnology, Biotech Science Cluster, 3rd Milestone, Faridabad-Gurugram Expressway, Faridabad – 121001, Haryana, India.

Data availability

All the data have been included in the manuscript.

Ethical approval and consent to participate

Not applicable.

CRediT authorship contribution statement

Basavaraju K C: Conceptualization, Formal analysis, Investigation, Writing – original draft, **Harshitha D, K:** Formal analysis, Investigation, Writing – original draft, **Poornima Priyadarshini C.G;** Conceptualization, Funding acquisition, Resources, Supervision, Visualization, Writing – review and editing.

Consent for publication

Not applicable.

References

Adhikari, R., Yang, M., Saikia, N., Dutta, C., Alharbi, W., Shan, Z., ... Tiwari, A. (2020). Acetylation of Aβ42 at lysine 16 disrupts amyloid formation. *ACS Chemical Neuroscience, 11*(8), 1178–1191. https://doi.org/10.1021/acschemneuro.0c00069.

Al-Hakim, A., Zagorska, A., Chapman, L., Deak, M., Peggie, M., & Alessi, D. (2008). Control of AMPK-related kinases by USP9X and atypical Lys29/Lys33-inked polyubiquitin chains. *Biochemical Journal, 411*, 249–260. https://doi.org/10.1042/BJ20080067.

Alquezar, C., Schoch, K., Geier, E., Ramos, E., Scrivo, A., Li, K., ... Kao, A. (2021). TSC1 loss increases risk for tauopathy by inducing tau acetylation and preventing tau clearance via chaperone-mediated autophagy. *Science Advances,* 7, eabg3897. https://doi.org/10.1126/sciadv.abg3897.

Al-Shami, A., Jhaver, K., Vogel, P., Wilkins, C., Humphries, J., Davis, J., ... Oravecz, T. (2010). Regulators of the proteasome pathway, Uch37 and Rpn13, play distinct roles in mouse development. *PLoS One, 5*(10), e13654. https://doi.org/10.1371/journal.pone.0013654.

Balasubramanyam, K., Swaminathan, V., Ranganathan, A., & Kundu, T. (2003). Small molecule modulators of histone acetyltransferase p300. *Journal of Biological Chemistry, 278*(21), 19134–19140. https://doi.org/10.1074/jbc.M301580200.

Bello, A., Goswami, R., Brown, S., Costanzo, K., Shores, T., Allan, S.,...Mohan, R. (2022). Deubiquitinases in Neurodegeneration. *CELLS, 11*(3), 556. https://doi.org/10.3390/cells11030556.

Cai, N., Chen, J., Bi, D., Gu, L., Yao, L., Li, X., ... Xu, X. (2020). Specific degradation of endogenous Tau protein and inhibition of Tau fibrillation by tanshinone IIA through the ubiquitin-pathway proteasome. *Journal of Agricultural and Food Chemistry, 68*(7), 2054–2062. https://doi.org/10.1021/acs.jafc.9b07022.

Carlonnagno, Y., Chung, D., Yue, M., Castanedes-Casey, M., Madden, B., Dunmore, J., ... Cook, C. (2017). An acetylation-phosphorylation switch that regulates tau aggregation propensity and function. *Journal of Biological Chemistry, 292*(37), 15277–15286. https://doi.org/10.1074/jbc.M117.794602.

Chakraborty, P., Riviere, G., Hebestreit, A., de Opakua, A. I., Vorberg, I. M., Andreas, L. B., & Zweckstetter, M. (2023). Acetylation discriminates disease-specific tau deposition. *Nature Communications, 14*(1), 5919. https://doi.org/10.1038/s41467-023-41672-1.

Chen, L., Yoo, S., Na, R., Liu, Y., & Ran, Q. (2012). Cognitive impairment and increased Aβ levels induced by paraquat exposure are attenuated by enhanced removal of mitochondrial H2O2. *Neurobiology of Aging, 33*(2), 432.e15. https://doi.org/10.1016/j.neurobiolaging.2011.01.008.

Chen, C., Zhou, M., Ge, Y., & Wang, X. (2020). SIRT1 and aging related signaling pathways. *Mechanisms of Ageing and Development, 187*, 111215. https://doi.org/10.1016/j.mad.2020.111215.

Cheng, R., Bai, N., Liu, S., Zhao, X., Jiang, N., Guo, W., ... Cao, L. (2024). The deacetylase SIRT6 reduces amyloid pathology and supports cognition in mice by reducing the stability of APP in neurons. *Science Signaling, 17*(866), eado1035. https://doi.org/10.1126/scisignal.ado1035.

Cohen, T., Constance, B., Hwang, A., James, M., & Yuan, C. (2016). Intrinsic Tau acetylation is coupled to auto-proteolytic Tau fragmentation. *PLoS One, 11*(7), e0158470. https://doi.org/10.1371/journal.pone.0158470.

Cohen, T., Friedmann, D., Hwang, A., Marmorstein, R., & Lee, V. (2013). The microtubule-associated tau protein has intrinsic acetyltransferase activity. *Nature Structural & Molecular Biology, 20*(6), 756. https://doi.org/10.1038/nsmb.2555.

Cohen, T., Guo, J., Hurtado, D., Kwong, L., Mills, I., Trojanowski, J., & Lee, V. (2011). The acetylation of tau inhibits its function and promotes pathological tau aggregation. *Nature Communications, 2*, 252. https://doi.org/10.1038/ncomms1255.

Cohen, T. J., Guo, J. L., Hurtado, D. E., Kwong, L. K., Mills, I. P., Trojanowski, J. Q., & Lee, V. M. Y. (2011). The acetylation of tau inhibits its function and promotes pathological tau aggregation. *Nature Communications, 2*, 252. https://doi.org/10.1038/ncomms1255.

Colnaghi, L., Rondelli, D., Muzi-Falconi, M., & Sertic, S. (2020). Tau and DNA damage in neurodegeneration. *Brain Sciences, 10*(12), 946. https://doi.org/10.3390/brainsci10120946.

Cook, C., Carlomagno, Y., Gendron, T., Dunmore, J., Scheffel, K., Stetler, C., ... Petrucelli, L. (2014). Acetylation of the KXGS motifs in tau is a critical determinant in modulation of tau aggregation and clearance. *Human Molecular Genetics, 23*(1), 104–116. https://doi.org/10.1093/hmg/ddt402.

Das, R., Balmik, A., & Chinnathambi, S. (2020). Phagocytosis of full-length Tau oligomers by actin-remodeling of activated microglia. *Journal of Neuroinflammation, 17*(1), 10. https://doi.org/10.1186/s12974-019-1694-y.

Dornan, D., Shimizu, H., Perkins, N., & Hupp, T. (2003). DNA-dependent acetylation of p53 by the transcription coactivator p300. *Journal of Biological Chemistry, 278*(15), 13431–13441. https://doi.org/10.1074/jbc.M211460200.

Ebadi, P., Stratton, C., & Olsen, S. (2025). E3 ubiquitin ligases in signaling, disease, and therapeutics. *Trends in Biochemical Sciences, 50*(11), 960–976. https://doi.org/10.1016/j.tibs.2025.07.009.

El-Maraghy, S., Reda, A., Essam, R., & Kortam, M. (2023). The citrus flavonoid "Nobiletin" impedes STZ-induced Alzheimer's disease in a mouse model through regulating autophagy mastered by SIRT1/FoxO3a mechanism. *Inflammopharmacology, 31*(5), 2701–2717. https://doi.org/10.1007/s10787-023-01292-z.

Fan, Y., Cheng, J., Vasudevan, S., Dou, J., Zhang, H., Patel, R., ... Yang, J. (2013). USP7 inhibitor P22077 inhibits neuroblastoma growth via inducing p53-mediated apoptosis. *Cell Death & Disease, 4*, e867. https://doi.org/10.1038/cddis.2013.400.

Ferreon, J., Jain, A., Choi, K., Tsoi, P., MacKenzie, K., Jung, S., & Ferreon, A. (2018). Acetylation disfavors Tau phase separation. *International Journal of Molecular Sciences, 19*(5), 1360. https://doi.org/10.3390/ijms19051360.

Fischer, D., De Vos, R., Van Dijk, R., De Vrij, F., Proper, E., Sonnemans, M., ... Van Leeuwen, F. (2003). Disease-specific accumulation of mutant ubiquitin as a marker for proteasomal dysfunction in the brain. *Faseb Journal, 17*(14), 2014–2024. https://doi.org/10.1096/fj.03-0205com.

Gong, B., Cao, Z., Zheng, P., Vitolo, O., Liu, S., Staniszewski, A., ... Arancio, O. (2006). Ubiquitin hydrolase Uch-L1 rescues β-amyloid-induced decreases in synaptic function and contextual memory. *Cell, 126*(4), 775–788. https://doi.org/10.1016/j.cell.2006.06.046.

Gorsky, M., Burnouf, S., Dols, J., Mandelkow, E., & Partridge, L. (2016). Acetylation mimic of lysine 280 exacerbates human Tau neurotoxicity in vivo. *Scientific Reports, 6*, 22685. https://doi.org/10.1038/srep22685.

Gorsky, M., Burnouf, S., Sofola-Adesakin, O., Dols, J., Augustin, H., Weigelt, C., ... Partridge, L. (2017). Pseudo-acetylation of multiple sites on human Tau proteins alters Tau phosphorylation and microtubule binding, and ameliorates amyloid beta toxicity. *Scientific Reports,* 7, 9984. https://doi.org/10.1038/s41598-017-10225-0.

He, H., Wang, X., Pan, R., Wang, D., Liu, M., & He, R. (2009). The proline-rich domain of tau plays a role in interactions with actin. *BMC Cell Biology, 10*, 81. https://doi.org/10.1186/1471-2121-10-81.

Hegarty, S. V., Sullivan, A. M., & O'Keeffe, G. W. (2016). Protocol for evaluation of neurotrophic strategies in Parkinson's disease-related dopaminergic and sympathetic neurons in vitro. *Journal of Biological Methods, 3*(3), e50. https://doi.org/10.14440/jbm.2016.124.

Hlushchenko, I., Koskinen, M., & Hotulainen, P. (2016). Dendritic spine actin dynamics in neuronal maturation and synaptic plasticity. *Cytoskeleton, 73*(9), 435–441. https://doi.org/10.1002/cm.21280.

Hol, E., Fischer, D., Ovaa, H., & Scheper, W. (2006). Ubiquitin proteasome system as a pharmacological target in neurodegeneration. *Expert Review of Neurotherapeutics, 6*(9), 1337–1347. https://doi.org/10.1586/14737175.6.9.1337.

Jia, F., & Fu, L. (2025). Roles of ubiquitin ligases and deubiquitylases in Alzheimer's disease. *Molecular Neurobiology, 62*(6), 7747–7761. https://doi.org/10.1007/s12035-025-04739-2.

Jia, W., Liu, G., Zhang, P., Li, H., Peng, Z., Wang, Y., ... Fu, D. (2023). The ubiquitin-26S proteasome pathway and its role in the ripening of fleshy fruits. *International Journal of Molecular Sciences, 24*(3), 2750. https://doi.org/10.3390/ijms24032750.

Kajava, A., Baxa, U., & Steven, A. (2010). β arcades: Recurring motifs in naturally occurring and disease-related amyloid fibrils. *FASEB Journal, 24*(5), 1311–1319. https://doi.org/10.1096/fj.09-145979.

Kim, J., de Haro, M., Al-Ramahi, I., Garaicoechea, L., Jeong, H., Sonn, J., ... Zoghbi, H. (2023). Evolutionarily conserved regulators of tau identify targets for new therapies. *Neuron, 111*(6), 824. https://doi.org/10.1016/j.neuron.2022.12.012.

Kim, M.S., Mun, Y.S., Lee, S.E., Cho, W.Y., Han, S.H., Kim, D.H., & Yoon, S.Y. (2023). Tau acetylation at K280 regulates tau phosphorylation. *International Journal of Neuroscience*. https://doi.org/10.1080/00207454.2022.2081165.

Lasagna-Reeves, C., Castillo-Carranza, D., Sengupta, U., Clos, A., Jackson, G., & Kayed, R. (2011). Tau oligomers impair memory and induce synaptic and mitochondrial dysfunction in wild-type mice. *Molecular Neurodegeneration, 6*, 39. https://doi.org/10.1186/1750-1326-6-39.

Lee, M., Lee, J., & Rubinsztein, D. (2013). Tau degradation: The ubiquitin-proteasome system versus the autophagy-lysosome system. *Progress in Neurobiology, 105*, 49–59. https://doi.org/10.1016/j.pneurobio.2013.03.001.

Li, L., Nguyen, B. A., Mullapudi, V., Li, Y., Saelices, L., & Joachimiak, L. A. (2023). Disease-associated patterns of acetylation stabilize tau fibril formation. *Structure (London, England: 1993), 31*(9), https://doi.org/10.1016/j.str.2023.05.020.

Lim, K., Joo, J., & Baek, K. (2020). The potential roles of deubiquitinating enzymes in brain diseases. *Ageing Research Reviews, 61*, 101088. https://doi.org/10.1016/j.arr.2020.101088.

Liu, X., Balaraman, K., Lynch, C., Hebron, M., Wolf, C., & Moussa, C. (2021). Novel ubiquitin specific protease-13 inhibitors alleviate neurodegenerative pathology. *Metabolites, 11*(9), 622. https://doi.org/10.3390/metabo11090622.

Liu, X., Hebron, M., Shi, W., Lonskaya, I., & Moussa, C. (2019). Ubiquitin specific protease-13 independently regulates parkin ubiquitination and alpha-synuclein clearance in alpha-synucleinopathies. *Human Molecular Genetics, 28*(4), 548–560. https://doi.org/10.1093/hmg/ddy365.

Liu, W., Tang, X., Qi, X., Fu, X., Ghimire, S., Ma, R., ... Si, H. (2020). The ubiquitin conjugating enzyme: An important ubiquitin transfer platform in ubiquitin-proteasome system. *International Journal of Molecular Sciences, 21*(8), 2894. https://doi.org/10.3390/ijms21082894.

Lucke-Wold, B., Seidel, K., Udo, R., Omalu, B., Ornstein, M., Nolan, R., ... Ross, J. (2017). Role of Tau acetylation in Alzheimer's disease and chronic traumatic encephalopathy: The way forward for successful treatment. *Journal of Neurology and Neurosurgery, 4*(2).

Maniv, I., Sarji, M., Bdarneh, A., Feldman, A., Ankawa, R., Koren, E., ... Glickman, M. (2023). Altered ubiquitin signaling induces Alzheimer's disease-like hallmarks in a three-dimensional human neural cell culture model. *Nature Communications, 14*(1), 5922. https://doi.org/10.1038/s41467-023-41545-7.

Min, S. W., Cho, S. H., Zhou, Y. G., Schroeder, S., Haroutunian, V., Seeley, W. W., ... Gan, L. (2010). Acetylation of Tau inhibits its degradation and contributes to tauopathy. *Neuron, 67*(6), 953–966. https://doi.org/10.1016/j.neuron.2010.08.044.

Min, S., Cho, S., Zhou, Y., Schroeder, S., Haroutunian, V., Seeley, W., ... Gan, L. (2010a). Acetylation of Tau inhibits its degradation and contributes to tauopathy. *Neuron, 67*(6), 953–966. https://doi.org/10.1016/j.neuron.2010.08.044.

Min, S., Cho, S., Zhou, Y., Schroeder, S., Haroutunian, V., Seeley, W., ... Gan, L. (2010b). Acetylation of Tau inhibits its degradation and contributes to tauopathy (vol 67, pg 953, 2010). *Neuron, 68*(4), 801. https://doi.org/10.1016/j.neuron.2010.11.013.

O'Brien, D., Jones, H., Guenther, F., Murphy, E., England, K., Vendrell, I., ... Pinto-Fernandez, A. (2023). Structural premise of selective deubiquitinase USP30 inhibition by small-molecule benzosulfonamides. *Molecular & Cellular Proteomics, 22*(8), 100609. https://doi.org/10.1016/j.mcpro.2023.100609.

Oddo, S. (2008). The ubiquitin-proteasome system in Alzheimer's disease. *Journal of Cellular and Molecular Medicine, 12*(2), 363–373. https://doi.org/10.1111/j.1582-4934.2008.00276.x.

Pandya, V., & Patani, R. (2021). Region-specific vulnerability in neurodegeneration: Lessons from normal ageing. *Ageing Research Reviews, 67*, 101311. https://doi.org/10.1016/j.arr.2021.101311.

Pant, R., Jangra, A., Kwatra, M., Singh, T., Kushwah, P., Bezbaruah, B., ... Phukan, S. (2017). Cognitive deficits induced by combined exposure of stress and alcohol mediated through oxidative stress-PARP pathway in the hippocampus. *Neuroscience Letters, 653*, 208–214. https://doi.org/10.1016/j.neulet.2017.05.058.

Portillo, M., Eremenko, E., Kaluski, S., Garcia-Venzor, A., Onn, L., Stein, D., ... Toiber, D. (2021). SIRT6-CBP-dependent nuclear Tau accumulation and its role in protein synthesis. *Cell Reports, 35*(4), 109035. https://doi.org/10.1016/j.celrep.2021.109035.

Rane, J., Kumari, A., & Panda, D. (2019). An acetylation mimicking mutation, K274Q, in tau imparts neurotoxicity by enhancing tau aggregation and inhibiting tubulin polymerization. *Biochemical Journal, 476*, 1401–1417. https://doi.org/10.1042/BCJ20190042.

Sakono, M., & Zako, T. (2010). Amyloid oligomers: Formation and toxicity of Aβ oligomers. *Febs Journal, 277*(6), 1348–1358. https://doi.org/10.1111/j.1742-4658.2010.07568.x.

Shahib, A., Rastegar, M., van Wijnen, A., & Davie, J. (2024). Neurodevelopmental functions and activities of the KAT3 class of lysine acetyltransferases. *Biochemistry and Cell Biology, 102*(6), 430–447. https://doi.org/10.1139/bcb-2024-0156.

Spanic, E., Horvat, L., Hof, P., & Simic, G. (2019). Role of Microglial Cells in Alzheimer's Disease Tau Propagation. *Frontiers in Aging Neuroscience, 11*, 271. https://doi.org/10.3389/fnagi.2019.00271.

Tangavelou, K., Bondu, V., Li, M., Li, W., Liao, F.-F., & Bhaskar, K. (2025). The deubiquitinase OTULIN regulates tau expression and RNA metabolism in neurons. *Genomic Psychiatry : Advancing Science from Genes to Society*. https://doi.org/10.1101/2025.04.09.648063.

Trzeciakiewicz, H., Tseng, J. H., Wander, C. M., Madden, V., Tripathy, A., Yuan, C. X., & Cohen, T. J. (2017). A dual pathogenic mechanism links Tau acetylation to sporadic tauopathy. *Scientific Reports*, 7, 44102. https://doi.org/10.1038/srep44102.

Varadarajan, S., Yatin, S., Aksenova, M., & Butterfield, D. A. (2000). Review: Alzheimer's amyloid beta-peptide-associated free radical oxidative stress and neurotoxicity. *Journal of Structural Biology, 130*(2–3), 184–208. https://doi.org/10.1006/jsbi.2000.4274.

Volbracht, C., & Fog, K. (2025). Deubiquitinating enzymes ubiquitin-specific proteases 7 and 10 regulate TAU aggregation. *International Journal of Molecular Sciences, 26*(22), 11062. https://doi.org/10.3390/ijms262211062.

Wagh, A., Sulakshane, P., & Glickman, M. (2023). Alzheimer's disease-associated mutant ubiquitin (UBB+1) is secreted through an autophagosome-like vesicle-mediated unconventional pathway. *BIochimica et Biophysica Acta-Gene Regulatory Mechanisms, 1866*(2), 194936. https://doi.org/10.1016/j.bbagrm.2023.194936.

Wang, P., Joberty, G., Buist, A., Vanoosthuyse, A., Stancu, I., Vasconcelos, B., ... Dewachter, I. (2017). Tau interactome mapping based identification of Otub1 as Tau deubiquitinase involved in accumulation of pathological Tau forms in vitro and in vivo. *Acta Neuropathologica, 133*(5), 731–749. https://doi.org/10.1007/s00401-016-1663-9.

Wang, X., Liu, Q., Li, X., Zhou, Q., Wu, D., Li, S., ... Wang, J. (2021). T217-Phosphorylation exacerbates Tau pathologies and Tau-induced cognitive impairment. *JOurnal of Alzheimers Disease, 81*, 1403–1418. https://doi.org/10.3233/JAD-210297.

Wathan, A., Deschene, N., Litz, J., & Sumner, I. (2025). The lysine deprotonation mechanism in a ubiquitin conjugating enzyme. *Journal of Physical Chemistry B, 129*(20), 4962–4968. https://doi.org/10.1021/acs.jpcb.5c01486.

Wei, Z., Zeng, K., Hu, J., Li, X., Huang, F., Zhang, B., ... Wang, X. (2022). USP10 deubiquitinates Tau, mediating its aggregation. *Cell Death & Disease, 13*(8), 726. https://doi.org/10.1038/s41419-022-05170-4.

Weng, F., & He, L. (2021). Disrupted ubiquitin proteasome system underlying tau accumulation in Alzheimer's disease. *Neurobiology of Aging, 99*, 79–85. https://doi.org/10.1016/j.neurobiolaging.2020.11.015.

Wong, P., & Kamarul, T. (2025). Targeting ubiquitin-proteasome system (UPS) in treating osteoarthritis. *European Journal of Pharmacology, 989*, 177237. https://doi.org/10.1016/j.ejphar.2024.177237.

Xiao, Y., Ma, B., McElheny, D., Parthasarathy, S., Long, F., Hoshi, M., ... Ishii, Y. (2015). Aβ(1-42) fibril structure illuminates self-recognition and replication of amyloid in Alzheimer's disease. *Nature Structural & Molecular Biology, 22*(6), 499–U497. https://doi.org/10.1038/nsmb.2991.

Yan, J., Luo, A., Sun, R., Tang, X., Zhao, Y., Zhang, J., ... Li, S. (2020). Resveratrol mitigates hippocampal Tau acetylation and cognitive deficit by activation SIRT1 in aged rats following anesthesia and surgery. *Oxidative Medicine and Cellular Longevity, 2020*, 4635163. https://doi.org/10.1155/2020/4635163.

Yan, Y., Wang, X., Chaput, D., Shin, M., Koh, Y., Gan, L., ... Kang, D. (2022). X-linked ubiquitin-specific peptidase 11 increases tauopathy vulnerability in women. *Cell, 185*(21), 3913. https://doi.org/10.1016/j.cell.2022.09.002.

Zhang, M., Cai, F., Zhang, S., Zhang, S., & Song, W. (2014). Overexpression of ubiquitin carboxyl-terminal hydrolase L1 (UCHL1) delays Alzheimer's progression in vivo. *Scientific Reports, 4*, 7298. https://doi.org/10.1038/srep07298.

Zhang, H., Wei, W., Zhao, M., Ma, L., Jiang, X., Pei, H., ... Li, H. (2021). Interaction between Aβ and Tau in the pathogenesis of Alzheimer's disease. *International Journal of Biological Sciences, 17*(9), 2181–2192. https://doi.org/10.7150/ijbs.57078.

Zhao, Y., Wang, Z., Mao, Y., Li, B., Zhu, Y., Zhang, S., ... Zhang, Y. (2020). NEAT1 regulates microtubule stabilization via FZD3/GSK3β/P-tau pathway in SH-SY5Y cells and APP/PS1 mice. *Aging-US, 12*(22), 23233–23250.

Zhu, B., Jiang, L., Huang, T., Zhao, Y., Liu, T., Zhong, Y., ... Xu, H. (2017). ER associated degradation regulates Alzheimer's amyloid pathology and memory function by modulating γ-secretase activity. *Nature Communications, 8*, 1472. https://doi.org/10.1038/s41467-017-01799-4.

CHAPTER THIRTEEN

Histone deacetylases in neurodegeneration and neuronal plasticity

Bhupesh Mehta*, Sinjitha Nambiar, Omkar Shirke, Satyarth Pandita, and Yogananda S. Markandeya
Department of Biophysics, National Institute of Mental Health and Neurosciences (NIMHANS), Bengaluru, India
*Corresponding author. e-mail address: bhupeshmehta@gmail.com

Contents

1. Introduction 322
 1.1 HDACs in the nervous system: An overview 322
 1.2 HDAC nucleocytoplasmic shuttling: How it works 323
 1.3 HDACs and neuronal plasticity 325
 1.4 Dysfunction of HDACs in neurodegenerative disease 326
 1.5 Scope and organization of this chapter 331
2. Core experimental methods for investigating HDAC function 332
 2.1 Enzymatic assays 332
 2.2 Cellular assays 334
 2.3 In vivo approaches 337
 2.4 Validation of HDAC subcellular localization 339
 2.5 CRISPR-based manipulation of HDAC localization 342
3. Data analysis and interpretation 352
 3.1 Quantification of HDAC localization 352
 3.2 Statistical considerations 353
 3.3 Integration of multi-modal data 354
4. Future methodological directions and outstanding challenges in HDAC research 355
 4.1 Emerging experimental technologies 355
 4.2 Translational and precision experimental strategies 356
 4.3 Unresolved mechanistic questions and experimental strategies 357
 4.4 Integrative perspective 358
5. Conclusion 358
References 359

Abstract

This chapter pulls together current research on how HDAC shuttling between the nucleus and cytoplasm affects neurodegenerative diseases like Alzheimer's, Parkinson's, Huntington's, and epilepsy. It takes a close look at why these shifts in HDAC localization

Methods in Enzymology, Volume 733
ISSN 0076-6879, https://doi.org/10.1016/bs.mie.2026.06.025

matter so much in brain disease and its implications for new treatments. Histone deacetylases (HDACs) are a big deal when it comes to gene regulation in the brain. They play key roles in both neurodegeneration and the brain's ability to adapt, working inside the nucleus and out in the cytoplasm. This chapter unpacks the molecular mechanisms behind HDAC trafficking—how they move around—highlights the different roles of HDAC isoforms, and compares localization-specific effects. It digs into how HDACs impact protein aggregation and synaptopathies. Some findings stand out: HDAC4 and HDAC1 are tightly controlled by phosphorylation signals, which change their cellular localization and influence neuronal mortality. For example, HDAC6 is majorly involved in cellular trafficking and clearing protein aggregates, whereas HDAC4 aggregation in the nucleus is responsible for driving neuronal toxicity. If HDAC1 undergoes nuclear export, it interacts with motor proteins to impact mitochondrial transport. Drugs that block HDAC6 look promising in preclinical models—they help restore neuronal transport systems and clear protein aggregation. Moving HDAC4 out of the nucleus seems to support better synaptic function and motor skills. As a general rule, HDAC accumulation in the nucleus shuts down genes that keep neurons alive, but keeping them in the cytoplasm helps preserve connections between neurons. You'll also find thorough, practical advice on how to study HDACs in brain research—covering everything from enzyme assays and cell experiments to live animal models, plasticity tracking, drug testing, and data analysis. A major innovation featured here is using CRISPR-based tricks to control exactly where HDACs go inside cells: forced targeting using dCas9 fusions, editing natural localization signals, and even using optogenetics for precise on-demand control. In short, the chapter is a hands-on guide for anyone trying to unravel HDAC mechanisms in diseases like Alzheimer's, Parkinson's, Huntington's, or in studies of brain plasticity. Some standout methods include tracking HDAC localization in the cells, measuring how phosphorylation affects their shuttling, and using HDAC2 inhibitors for cognitive boosts. It also covers isoform-specific approaches in Huntington's models, manipulating HDAC location with CRISPR for deeper insights, and combining live-cell imaging with biochemical and chromatin studies for robust validation. This chapter sheds light on the latest advances, with a strong focus on precision, quantitative results, and translating these findings into real-world applications.

1. Introduction

1.1 HDACs in the nervous system: An overview

Histone deacetylases, or HDACs, are a big family of enzymes. Their main task is to remove acetyl groups from lysine residues on both histone and non-histone proteins, which lets them control how tightly DNA is packed, which genes get switched on or off, and a heap of other crucial cell processes (Milazzo et al., 2020). Mammals have 18 different HDAC isoforms, grouped into four classes depending on their structure and what cofactors they need (Curcio et al., 2024). Classes I, II, and IV (like HDAC1, 2, 3, 4, etc.)

depend on zinc, while Class III (the sirtuins) need NAD+ . In the nervous system, expression of HDACs occurs everywhere. They are key players in brain development, learning, memory, and keeping neurons healthy (Graff, & Tsai, 2013a; Guan et al., 2009). However, each HDAC isoform is not expressed randomly—it has its own expression patterns, subcellular locations, and unique roles (Broide et al., 2007). For example, Class I HDACs are mostly found in the nucleus, where they regulate genes involved in neuron differentiation and survival (Montgomery et al., 2009). Class IIa HDACs, on the other hand, can move between the nucleus and cytoplasm, responding to neuron firing and calcium signals (Bolger & Yao, 2005; Chawla et al., 2003). Then there's HDAC6 from Class IIb, which is usually in the cytoplasm and keeps tabs on microtubules, assists in protein trafficking, and clears out protein aggregates through autophagy (Hubbert et al., 2002; Lee et al., 2010). When HDACs lose their rhythm—whether they are expressed in the wrong place at the wrong time or just not expressed in the right amount—it sets the stage for several neurodegenerative diseases. Alzheimer's, Parkinson's, Huntington's, Amyotrophic lateral sclerosis (ALS) (Chuang et al., 2009; Didonna, & Opal, 2015; Kazantsev & Thompson, 2008). Overactive or misplaced HDACs can switch off genes that protect neurons, mess with synaptic plasticity, and allow toxic protein clumps to pile up (Fischer et al., 2007; Jiang et al., 2011). So, investigating exactly how HDACs move around inside the cells and how their activity is regulated matters, especially to design drugs that target them precisely.

1.2 HDAC nucleocytoplasmic shuttling: How it works

A standout feature of Class IIa HDACs (HDAC4, 5, 7, 9) is their ability to shuttle between the nucleus and cytoplasm, tuning in to signals from inside and outside the cell (Chawla et al., 2003; McKinsey et al., 2000). This dynamic localization is regulated by phosphorylation of conserved serine residues in the N-terminal regulatory domain by calcium/calmodulin-dependent protein kinases (CaMKs) and other kinases (Backs et al., 2006; Zhao et al., 2001). Phosphorylation creates binding sites for 14-3-3 chaperone proteins, which mask the (NLS). CaMKII phosphorylation of S246/S467/S632 creates 14-3-3 binding sites, masking nuclear localization signal (NLS), exposing Nuclear Export Signals (NES) for CRM1-mediated nuclear export, and promote cytoplasmic retention (Grozinger & Schreiber, 2000; McKinsey et al., 2000). Dephosphorylation reverses shuttling, enabling nuclear import and MEF2-dependent transcriptional repression.

Dephosphorylation by protein phosphatases enables nuclear import and transcriptional repression of target genes by recruiting corepressor complexes (Paroni et al., 2004). In the dephosphorylated state, HDAC4/5 proteins enter the nucleus via importin-mediated recognition of the N-terminal nuclear localization signal (NLS), where they bind MEF2 transcription factors and repress target gene expression. Upon neuronal activity and calcium influx, CaMKII phosphorylates three conserved serine residues (S246, S467, and S632 in HDAC4), creating binding sites for 14-3-3 proteins (shown in green in Fig. 1). 14-3-3 binding masks the NLS and exposes the C-terminal NES, promoting CRM1 (exportin-1)-mediated nuclear export and cytoplasmic retention. This phosphorylation-dependent shuttling mechanism couples synaptic activity to transcriptional regulation and is critical for activity-dependent gene expression, dendritic growth, and synaptic plasticity (Backs et al., 2006; Chawla et al., 2003; Grozinger & Schreiber, 2000) (Fig. 1).

This Nucleocytoplasmic shuttling is how the cell couples electrical activity to new gene expression, which influences dendritic growth and synaptic changes (Backs et al., 2006; Chawla et al., 2003; Grozinger & Schreiber, 2000). Class I HDACs, especially HDAC1, can also exit the

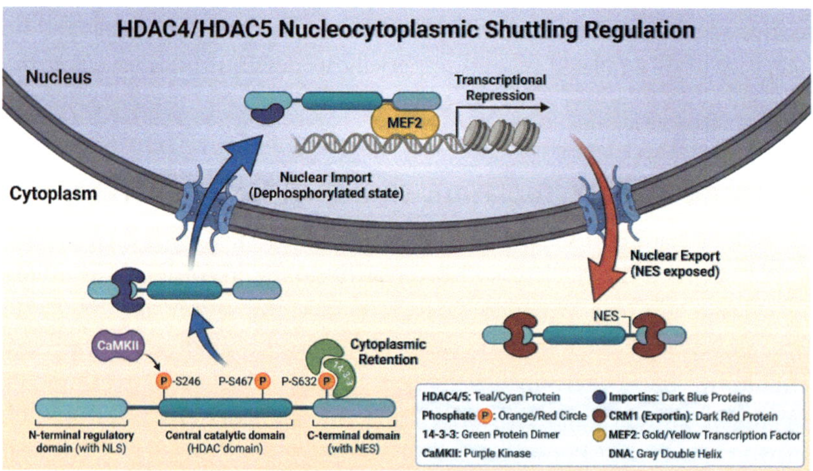

Fig. 1 HDAC4/HDAC5 nucleocytoplasmic shuttling regulation. Molecular mechanisms governing the dynamic subcellular localization of Class IIa HDACs. The figure illustrates the domain architecture of HDAC4/5, including the N-terminal regulatory domain (containing NLS), the central catalytic domain (HDAC domain), and the C-terminal domain (containing NES), along with key regulatory proteins (CaMKII, 14-3-3, importins, CRM1) and their interactions.

nucleus, but usually only in certain disorders or after injury (Kim et al., 2010). When this occurs, as observed in models of neurodegeneration, HDAC1 accumulates in the cytoplasm, damaging the mitochondria and leading to axonal degeneration (Bardai et al., 2012; Kim et al., 2010). The location of HDAC within a cell is not random. It determines both the target proteins it can regulate and the downstream effects. Inside the nucleus, HDACs act on histones and transcription factors, shaping gene activity (Kouzarides, 2007) In the cytoplasm, they target proteins involved in protein transport, the cytoskeleton, and quality control (Choudhary et al., 2009; Glozak et al., 2005).

Some HDACs work globally, but they can also act selectively at specific genes or chromatin locations (Haggarty & Tsai, 2011; Li et al., 2002). Researchers now design ways to direct HDACs to the regions of their interest, making it easier to distinguish local versus global effects (Kwon et al., 2017; Frank et al., 2016). Watching HDACs shuffle between nucleus and cytoplasm unlocks new ways to study how they work—using live imaging, fractionation, or chromatin assays, lets us link where HDACs are, how active they are, and what genes they affect (Reddy et al., 2025; Wang et al., 2014; Haggarty & Tsai, 2011). In addition, since HDAC movements can flip quickly, scientists use inducible systems and even optogenetics to track HDAC dynamics with more precision (Wang et al., 2014; Sando et al., 2012).

1.3 HDACs and neuronal plasticity

Synaptic plasticity—the way synapses strengthen or weaken over time—is at the heart of learning and memory. HDACs help steer these changes through both gene-level and other mechanisms (Graff, & Tsai, 2013a; Guan et al., 2009). Block HDACs, and you boost long-term potentiation (LTP), lock in memories better, and ramp up acetylation at genes needed for these processes (Kilgore et al., 2010; Levenson et al., 2004; Vecsey et al., 2007). Not all HDACs are created equal: HDAC2 puts the brakes on LTP and memory, whereas blocking HDAC3 actually makes memory stronger (Gräff et al., 2012; McQuown et al., 2011). HDACs also play a role in long-term depression (LTD), tuning down synaptic strength by repressing gene expression (Alchini et al., 2017). When it comes to the physical side—dendrite and axon growth, for example—HDAC2 limits spinogenesis, whereas HDAC6 tweaks microtubules to assist axonal transport (Gräff et al., 2012; Hubbert et al., 2002). The ability of Class IIa HDACs to move in response to activity tethers synaptic changes directly to transcription programs (Chawla et al., 2003; Sugo et al., 2010). Researchers can probe these rapid, activity-dependent shifts using live-cell imaging, stimulating neurons in different ways, or targeting HDACs at specific gene spots.

1.4 Dysfunction of HDACs in neurodegenerative disease

When HDACs accumulate in the wrong place or dysfunction, all kinds of trouble follow—especially in neurodegenerative diseases. Misplaced or overactive HDACs can smother gene activity needed for healthy synapses, throw off protein balance, and push neurons toward death. Here's what happens in specific diseases: Healthy neurons usually keep HDAC2 split between the nucleus and the cytoplasm. But in Alzheimer's disease, the balance tips—HDAC2 piles up in the nucleus (see Fig. 2), shutting down survival and synaptic genes. This links up with classic AD markers like amyloid plaques and tracks closely with synaptic loss and memory decline (Gräff et al., 2012; Graff & Tsai, 2013b). In Parkinson's disease (PD), healthy neurons spread HDAC4 and HDAC6 between the nucleus and cytoplasm. In PD, HDAC4 gets caught up in Lewy bodies (packed with α-synuclein), while HDAC6 colocalizes with α-synuclein aggregates. Both actions disrupt normal function—HDAC6 even deacetylates α-synuclein and tubulin, making it harder for the cell to clear these aggregates and keep axonal transport running smoothly (Du et al., 2014; Godena et al., 2014; Luo et al., 2016). In Huntington's disease, neurons trap HDAC4 inside the nucleus. This represses neuroprotective genes, and the mutant huntingtin protein (mHtt) sequesters HDAC6, wrecking microtubule-based transport. All of the above lead to cortical atrophy and neuronal malfunction (Bobrowska et al., 2012; Dompierre et al., 2007; Mielcarek et al., 2013). For ALS, HDAC4 does not just accumulate inside the nucleus—it is upregulated in both the nucleus and the cytoplasm, leading to the formation of cytoplasmic TDP-43 aggregates. These changes cause motor neuron degeneration and brain atrophy (Guo et al., 2017; Taes et al., 2013). In short, each disease has its own pattern of HDAC dysfunction, but the result always involves neurons losing their ability to adapt, survive, or maintain a healthy environment (Fig. 2).

1.4.1 Alzheimer's disease (AD)

AD brings progressive memory loss, synapse loss, and clusters of amyloid-β and tau tangles. One major culprit here is the accumulation of HDAC2 in the neuronal nuclei, especially in the hippocampus and the cortex (Gräff et al., 2012;

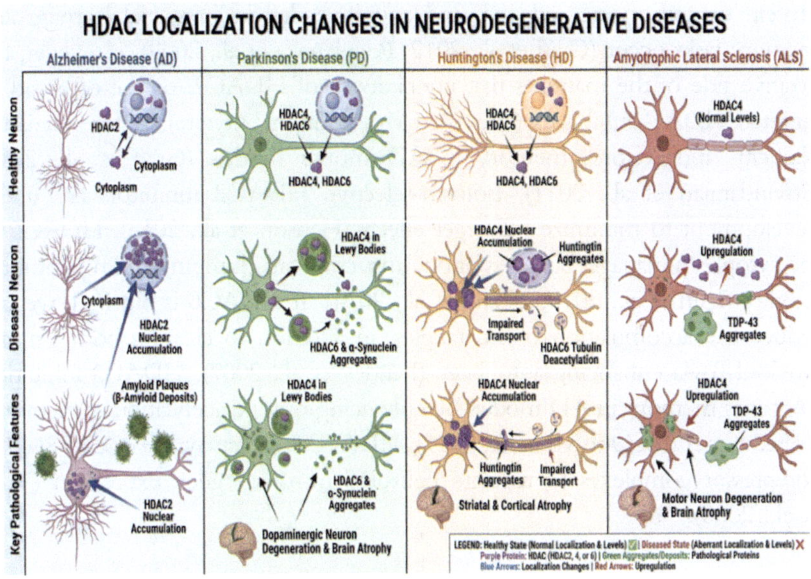

Fig. 2 HDAC localization changes in neurodegenerative diseases. Healthy neurons maintain HDAC levels balanced between the nucleus and cytoplasm. But in Alzheimer's, HDAC2 accumulates in the nucleus (blue arrow), switching off genes crucial for synaptic plasticity and survival. You end up with amyloid plaques (green), fewer synapses, and memory loss. With Parkinson's, HDAC4 gets trapped in the Lewy bodies—those aggregates of α-synuclein in the cytoplasm (green)—while HDAC6 is associated with α-synuclein aggregates (black arrows). HDAC6 strips acetyl groups from α-synuclein and tubulin, making it harder for cells to clear those junky aggregates and mess up axonal transport. This triggers degeneration of dopamine neurons and results in brain atrophy. In Huntington's disease, HDAC4 builds up in the nucleus (blue arrow), turning off crucial protective genes. Mutant huntingtin (mHtt) aggregates (purple) associate with HDAC6, so it cannot acetylate tubulin—this ruins microtubule transport and leads to atrophy in the striatum and cortex, resulting in neuronal degeneration. With ALS, HDAC4 expression increases (red arrow) and accumulates in both the nucleus and cytoplasm. TDP-43 aggregates (green) appear in the cytoplasm, interfering with gene regulation, killing motor neurons, and causing brain atrophy.

Graff & Tsai, 2013b). This spike lines up with less histone acetylation at genes that keep synapses healthy. Amyloid-β itself can increase HDAC2 expression, although we do not fully understand the mechanism (Narayan et al., 2015). Oxidative stress and inflammation also contribute to increased levels of HDAC2 (Saha & Pahan, 2006). Consequently, the genes that support synaptic plasticity (like Bdnf and Arc) show reduced expression, long-term potentiation (LTP) and long-term depression (LTD) are impaired, dendritic spine density and synaptic protein expression are reduced, and cognitive deficits in spatial learning and memory tasks occur (Gräff et al., 2012; Ricobaraza et al., 2009). However, the positive side of the matter is that knockdown of HDAC2, or use of broad or targeted HDAC inhibitors (like SAHA or sodium butyrate), restores synaptic plasticity and restores memory in AD mouse models (Gräff et al., 2012; Govindarajan et al., 2011). Isoform-selective HDAC2 inhibitors are under development to minimize off-target effects (Hanson et al., 2013). HDAC6 in the cytoplasm manages tau acetylation, impacting tau clumping and microtubule stability (Min et al., 2010; Zhang et al., 2014). If HDAC6 is blocked, we can reduce the accumulation of tau tangles, in addition to clearing both tau and amyloid-β via enhancing autophagy (Pandey et al., 2007). HDAC3 inhibition improves memory in AD models by enhancing histone acetylation at memory-related genes (McQuown et al., 2011). HDAC3 interacts with NCoR/SMRT corepressor complexes to regulate neuroinflammatory gene expression (Wen et al., 2000).

1.4.2 Parkinson's disease (PD)

The hallmark of Parkinson's disease is the loss of dopaminergic neurons in the substantia nigra, motor dysfunction, and accumulation of α-synuclein-containing Lewy bodies. Here, HDAC4 gets upregulated in surviving dopamine neurons and clusters inside Lewy bodies alongside α-synuclein (Luo et al., 2016; Outeiro et al., 2007). This causes increased neuronal death and weaker dopamine production (Luo et al., 2016; Rane et al., 2012). Oxidative stress and mitochondrial dysfunction shuttle HDAC4 into the nucleus, where it represses neuroprotective genes (e.g., *Bcl-2*, *Bdnf*) and survival genes by binding MEF2 (Flavell et al., 2008). The persistent recurrence of this process accelerates neuronal apoptosis (Luo et al., 2016). Interventions that keep HDAC4 in the cytoplasm (via 14-3-3 binding) or knock it down seem to protect dopamine neurons (Luo et al., 2016), while HDAC inhibitors show therapeutic potential. Meanwhile, HDAC6 is found alongside α-synuclein in Lewy bodies, where it deacetylates this protein and encourages its aggregation (Du et al., 2014). Furthermore, HDAC6 deacetylates tubulin, derailing axonal transport (Godena et al., 2014). When HDAC6 is inhibited, α-synuclein aggregation slows, axonal transport is restored, and movement improves in animal models (Du et al., 2014; Godena et al., 2014). Combining HDAC6 inhibition with L-DOPA therapy potentially boosts treatment efficacy, addressing both symptoms and disease progression better than L-DOPA alone (Monti et al., 2009).

1.4.3 Huntington's disease (HD)

Huntington's disease starts with a dysfunctional huntingtin gene, which produces mHtt protein that accumulates in neurons and degenerating cells in the striatum, followed by motor and cognitive deficits. Here, HDAC4 accumulates in the nucleus because mHtt interferes with the proteins that normally restrict HDAC4 in the cytoplasm (Mielcarek et al., 2013). This accumulation of HDAC4 in the nucleus reduces the activation of neuroprotective genes, ultimately leading to neuronal death. If HDAC4 is tethered back in the cytoplasm—by increasing the production of 14-3-3, for instance, striatal neurons are rescued (Mielcarek et al., 2013). Meanwhile, mHtt aggregates sequester HDAC6, which ruins its ability to regulate microtubules (Dompierre et al., 2007). However, blocking HDAC6 may promote the clearance of mHtt and can reduce aggregation—but results vary depending on the model and timing (Bobrowska et al., 2012; Kawaguchi et al., 2003; Jia et al., 2012).

1.4.4 Amyotrophic lateral sclerosis (ALS)

ALS is characterised by progressive loss of motor neurons, muscle atrophy, ultimately leading to paralysis. In ALS, neurons upregulate HDAC4 expression, accumulating in both the nucleus and cytoplasm (Guo et al., 2017; Taes et al., 2013). Meanwhile, the cytoplasm fills with toxic TDP-43 aggregates. This double insult causes transcriptional disturbance and helps drive neuronal death.

Researchers have found that HDAC4 protein accumulates in spinal cord motor neurons of people with ALS and in SOD1 mouse models (Guo et al., 2017). Higher levels of HDAC4 are associated with increased motor neuron death (Guo et al., 2017). HDAC4 is not found accumulated in one spot either—it is found in both the nucleus and the cytoplasm of ALS motor neurons (Taes et al., 2013).

Higher HDAC4 levels are driven by the synergistic effects of oxidative stress and excitotoxicity. (Guo et al., 2017). HDAC4 reduces the activity of genes necessary for motor neuron survival (Guo et al., 2017). It might also interact with TDP-43 and other proteins linked to ALS (Taes et al., 2013). Consequently, a higher number of motor neurons undergo apoptosis (Guo et al., 2017). As a result, neurotrophic factors and survival signals decline, leading to a gradual loss of motor function and eventual paralysis (Guo et al., 2017).

Therapeutic Implications: But if you knock down HDAC4, things take a positive turn—SOD1 ALS mice live longer and move better (Guo et al., 2017). Drugs that block HDACs are showing promise in preclinical studies of ALS (Rouaux et al., 2007).

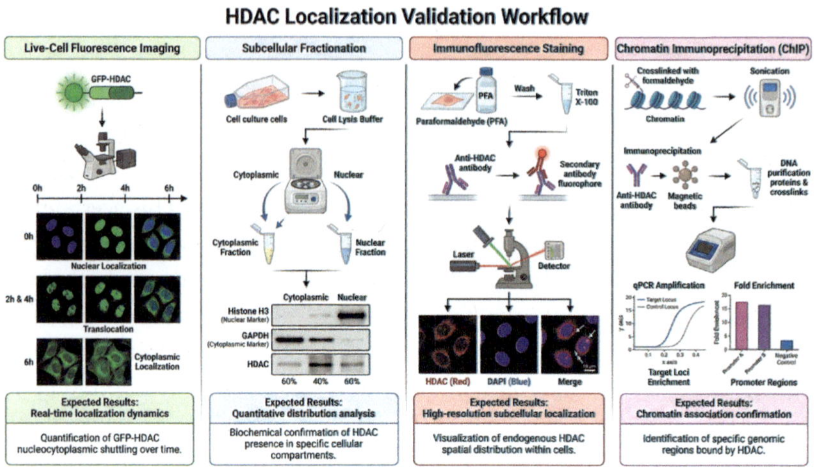

Fig. 3 HDAC localization validation workflow. Comprehensive experimental strategy for validating HDAC subcellular localization using four complementary approaches. Together, these provide temporal, quantitative, spatial, and functional validation of HDAC subcellular distribution.

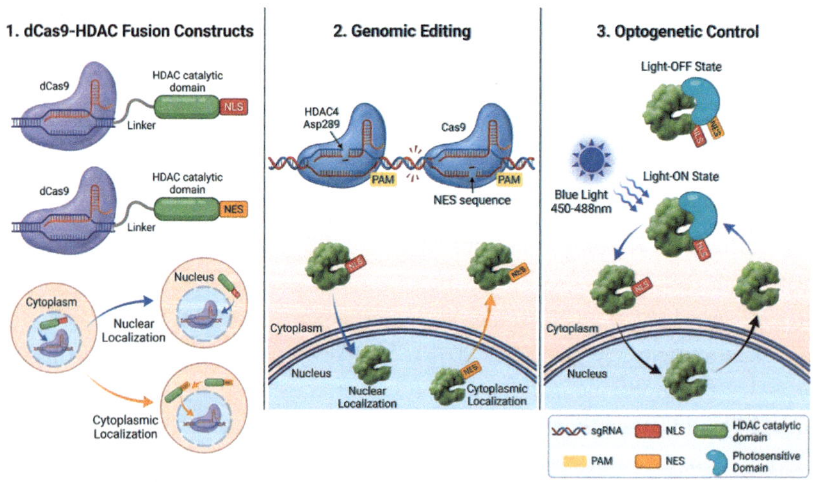

Fig. 4 CRISPR-based strategies for manipulating HDAC subcellular localization. Three complementary approaches: (1) dCas9-HDAC fusion constructs with NLS/NES tags, (2) genomic editing of endogenous HDAC localization signals via HDR, and (3) optogenetic control using light-sensitive protein domains for dynamic, reversible regulation of HDAC nuclear-cytoplasmic shuttling in neurons.

To summarize, we can understand that HDAC dysfunction—whether accumulated in the wrong place, increased expression, or both—shows up in every major neurodegenerative disease and aligns with specific types of neuronal deficits. Understanding and redirecting these enzymes isn't just academic—it's the next step in finding smarter, more effective therapies.

1.5 Scope and organization of this chapter

This chapter walks you through how to study HDACs in the nervous system, focusing on the nuts and bolts of deducing where these proteins are within the cells and how their location is controlled. Section 2 describes the main experimental methods: enzymatic assays, cellular and animal approaches, and a step-by-step guide to verify HDAC location in cells (see Fig. 3). The spotlight's also on new CRISPR-based tools for precisely moving HDACs around inside cells—think dCas9 fusion constructs (Fig. 4), gene

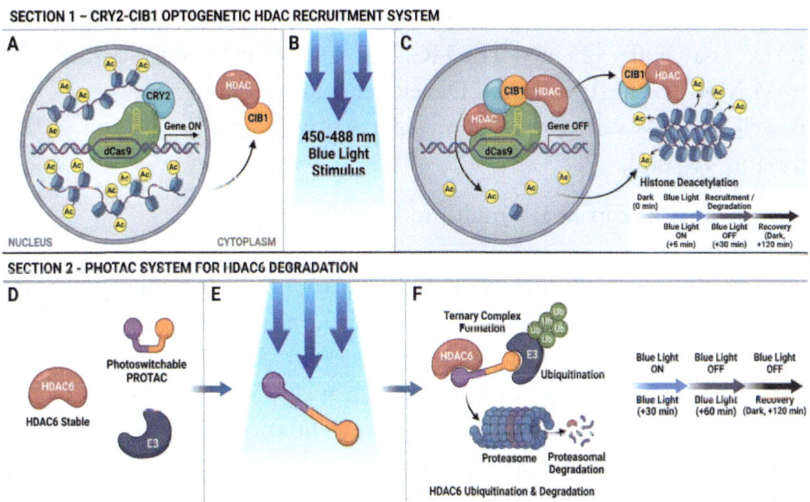

Fig. 5 Optogenetic systems for dynamic HDAC control. Two complementary light-controlled strategies for temporal regulation of HDAC activity- Two strategies: (Section 1) (A) In dark, the nuclus contains acetylated histones and dCas9-CRY2 is bound to target genomic loci. (B) Incident blue light of 450–488 nm. (C) In the presence of light, CRY2-CIB1 undergoes heterodimerization for reversible locus-specific HDAC recruitment, and (Section 2) PHOTAC system for light-activated HDAC6 degradation via ubiquitin-proteasome pathway, enabling temporal control of cytoplasmic HDAC6 levels. (D) In the absence of light, HDAC6, E3 protein, and Photoswitchable PROTAC are not bound to each other. (E) Incident light causes a conformational shift in the PROTAC molecule. (F) The conformational change in PROTAC brings HDAC6 and E3 protein in proximity causing HDAC6 ubiquitination and degradation.

editing, and optogenetic systems for real-time control (Fig. 5). Section 3 dives into analyzing and interpreting the data, with tips on quantification and statistics. Section 4 exposes gaps in the literature and discusses avenues for future investigation.

2. Core experimental methods for investigating HDAC function

2.1 Enzymatic assays

2.1.1 In vitro HDAC activity assays

Principle: HDAC enzymatic activity is measured using synthetic peptide substrates containing acetylated lysine residues. Deacetylation is detected by fluorescence or luminescence readouts.

Materials: Recombinant HDAC enzymes or nuclear/cytoplasmic extracts, Cell-permeable fluorogenic HDAC substrate (e.g., Boc-Lys(Ac)-AMC), HDAC assay buffer (25 mM Tris-HCl, pH 8.0, 137 mM NaCl, 2.7 mM KCl, 1 mM $MgCl_2$, 1 mg/mL BSA), Developer solution containing trypsin, Black 96-well or 384-well plates, Fluorescence plate reader (excitation 360 nm, emission 460 nm).

Equipment: Tecan Spark multimode reader.

Procedure

1. Prepare enzyme dilutions in assay buffer (optimize the concentration for a linear range; typically 0.1–10 μg/mL for recombinant enzymes or 10–50 μg protein for cell extracts).
2. Add 25 μL of enzyme solution to each well.
3. Add 25 μL substrate solution (final concentration 50–100 μM).
4. Incubate at 37 °C for 30–60 min.
5. Add 50 μL developer solution to stop the reaction and release the fluorophore.
6. Incubate at room temperature for 15 min.
7. Measure fluorescence intensity.
8. Calculate HDAC activity from standard curves using deacetylated peptide standards.

Controls:

- No-enzyme blank (substrate + buffer only)
- Positive control (known active HDAC enzyme)

- HDAC inhibitor controls (TSA 100 nM for Class I/II; nicotinamide 10 mM for sirtuins)

Isoform Selectivity: For isoform-specific assays, use selective inhibitors to subtract background activity or employ immunoprecipitation-coupled assays (Bantscheff et al., 2011; Bradner et al., 2010).

Notes:

- Enzyme concentration must be optimized to ensure linear kinetics
- Substrate concentration should be near Km for kinetic studies
- Temperature and pH significantly affect activity
- Some HDACs require cofactors (e.g., inositol phosphates for HDAC3) (Watson et al., 2012)

2.1.2 Substrate specificity profiling

Principle: Peptide arrays or mass spectrometry-based approaches identify preferred HDAC substrates and sites of deacetylation.

Equipment: Orbital shaker, Zeiss LSM 980 confocal microscope, Nano-LC system, High-Resolution Mass Spectrometer, Electrospray Ionization source, C18 spin columns.

Peptide Array Approach:

1. Synthesize a peptide library representing diverse acetylation sites (commercial or custom).
2. Incubate arrays with purified HDAC isoforms.
3. Detect deacetylation using anti-acetyl-lysine antibodies and fluorescence scanning.
4. Quantify signal reduction to determine substrate preferences (Gurard et al., 2010).

Mass Spectrometry Approach:

1. Incubate cell lysates or purified proteins with HDAC enzymes or inhibitors.
2. Perform tryptic digestion and acetyl-lysine enrichment (immunoprecipitation with anti-acetyl-lysine antibodies).
3. Analyze by LC-MS/MS to identify and quantify acetylation sites.
4. Compare acetylation levels between HDAC-treated and control samples (Choudhary et al., 2014; Weinert et al., 2013).

Applications:

- Identify novel HDAC substrates in neurons
- Determine isoform-specific substrate preferences
- Map acetylation dynamics in response to neuronal activity

Notes:

- Ensure peptides are projected away from the solid support using flexible spacers while also maintaining optimal spot density to prevent crowding.
- Perform a time-course and enzyme-titration pilot study to avoid reaction saturation.
- Use an N-terminal "loading control" fluorophore to normalize modification signals against the actual amount of peptide synthesized/spotted.

2.2 Cellular assays

2.2.1 Immunofluorescence microscopy

Principle: Visualize endogenous or tagged HDAC proteins in fixed cells using specific antibodies and fluorescent secondary detection.

Materials:

Primary neurons, neuronal cell lines, or brain tissue sections, HDAC-specific primary antibodies (validate specificity by knockdown/knockout controls), Fluorescent secondary antibodies (Alexa Fluor 488, 555, 647), Mounting medium with DAPI for nuclear counterstaining, Confocal or epifluorescence microscope.

Equipment: Zeiss LSM 980 confocal microscope.

Procedure

1. Fix cells or tissue sections with 4 % paraformaldehyde in PBS for 15 min at room temperature.
2. Permeabilize with 0.1–0.3 % Triton X-100 in PBS for 10 min.
3. Block with 5 % normal serum (from secondary antibody host species) +0.1 % Triton X-100 in PBS for 1 h at room temperature.
4. Incubate with primary antibody (diluted in blocking buffer) overnight at 4 °C.
5. Typical dilutions: 1:100–1:500 for HDAC antibodies
6. Include no-primary-antibody controls
7. Wash 3 × with PBS (5 min each).
8. Incubate with fluorescent secondary antibody (1:500–1:1000 in blocking buffer) for 1 h at room temperature in the dark.
9. Wash 3 × with PBS.
10. Mount with DAPI-containing mounting medium.
11. Image using confocal microscopy with appropriate laser lines and emission filters.

Quantification of Subcellular Localization:

1. Define nuclear and cytoplasmic regions of interest (ROIs) using DAPI or other nuclear markers
2. Measure mean fluorescence intensity in each compartment
3. Calculate nuclear/cytoplasmic (N/C) ratio for each cell
4. Analyze 50–100 cells per condition for statistical power (Nott et al., 2016)

Antibody Validation: Critical for accurate localization studies. Validate antibodies by:

- Western blot showing a single band at the expected molecular weight
- Signal loss in HDAC knockdown/knockout cells
- Comparison with tagged HDAC constructs (Uhlen et al., 2015)

Notes:

- Match the fixative (e.g., PFA vs. Methanol) to your specific target to preserve both cellular morphology and epitope accessibility.
- Include a "secondary-only" control and titrate antibodies to maximize the signal-to-noise ratio

2.2.2 Live-cell imaging of HDAC dynamics

Principle: Track HDAC subcellular localization in real time using fluorescent protein fusions.

Constructs:

HDAC-GFP or HDAC-mCherry fusion proteins, ensure functional validation (rescue of knockdown phenotypes, enzymatic activity), Include nuclear marker (e.g., H2B-mCherry) for reference.

Equipment: Zeiss LSM 980 confocal microscope.

Procedure

1. Transfect or transduce neurons with HDAC-fluorescent protein constructs.
2. Culture in imaging chambers with controlled temperature (37 °C) and CO_2 (5 %).
3. Image using confocal or spinning disk microscopy with appropriate time resolution (e.g., every 5–10 min for Class IIa HDAC shuttling).
4. Apply stimuli (e.g., KCl depolarization, glutamate, BDNF) to induce HDAC translocation.
5. Track nuclear and cytoplasmic fluorescence intensity over time.

Stimulation Protocols for Activity-Dependent Shuttling:

- KCl depolarization: 50 mM KCl for 5–10 min (Chawla et al., 1998)

- Glutamate stimulation: 50–100 μM glutamate for 5 min (Hardingham et al., 2001)
- BDNF treatment: 50–100 ng/mL for 30–60 min (Bardai & D'Mello, 2006)

Analysis:

1. Calculate N/C ratio at each time point
2. Determine kinetics of nuclear export/import (half-time, plateau levels)
3. Correlate with calcium imaging if performed simultaneously (Flavell et al., 2006)

Notes:

- Overexpression artifacts: use low expression levels or knock-in approaches
- Photobleaching: minimize laser power and exposure time
- Phototoxicity: use red-shifted fluorophores and reduce imaging frequency if needed

2.2.3 Biochemical fractionation

Principle: Separate nuclear and cytoplasmic fractions to quantify HDAC distribution by Western blotting.

Materials:

Cultured neurons or brain tissue, Hypotonic lysis buffer (10 mM HEPES pH 7.9, 10 mM KCl, 0.1 mM EDTA, 0.1 mM EGTA, 1 mM DTT, protease inhibitors, phosphatase inhibitors), Nuclear extraction buffer (20 mM HEPES pH 7.9, 400 mM NaCl, 1 mM EDTA, 1 mM EGTA, 1 mM DTT, protease inhibitors, phosphatase inhibitors), Dounce homogenizer or syringe with 27 G needle, Refrigerated centrifuge.

Equipment: Refrigerated centrifuge, Western blot apparatus, Chemi-doc.

Procedure

1. Harvest cells or tissue and wash with ice-cold PBS.
2. Resuspend pellet in 5 volumes of hypotonic lysis buffer.
3. Incubate on ice for 15 min to allow cell swelling.
4. Lyse cells by Dounce homogenization (10–15 strokes) or passage through a 27 G needle (10 times).
5. Centrifuge at 1000 g for 5 min at 4 °C to pellet nuclei.
6. Transfer supernatant (cytoplasmic fraction) to a new tube.
7. Wash nuclear pellet 2 × with hypotonic buffer.
8. Resuspend the nuclear pellet in the nuclear extraction buffer.
9. Incubate on ice for 30 min with occasional vortexing.
10. Centrifuge at 15,000 g for 10 min at 4 °C.
11. Transfer supernatant (nuclear fraction) to new tube.

12. Quantify protein concentration (BCA or Bradford assay).
13. Analyze equal amounts of protein (10–30 μg) by SDS-PAGE and Western blotting.

Validation of Fractionation Quality:
- Cytoplasmic marker: GAPDH, α-tubulin, or β-actin
- Nuclear marker: Histone H3, Lamin B1, or TATA-binding protein
- Minimal cross-contamination is essential for accurate quantification (Suzuki et al., 2010)

Quantification:
1. Densitometry of Western blot bands using ImageJ or similar software
2. Normalize HDAC signal to loading control in each fraction
3. Calculate percentage of total HDAC in nuclear vs. cytoplasmic fraction
4. Statistical analysis across biological replicates ($n \geq 3$)

Notes:
- Perform all homogenization and centrifugation steps at 4 °C with pre-chilled buffers to inhibit endogenous proteases and prevent thermal protein degradation.
- Supplement lysis buffers with a comprehensive protease and phosphatase inhibitor cocktail immediately before use to preserve the native state of the proteome.
- Always include "fraction-specific" markers in your Western Blot (e.g., GAPDH for cytosol, Histone H3 for nucleus, VDAC1 for mitochondria) to confirm clean separation.

2.3 In vivo approaches

2.3.1 Stereotactic viral delivery

Principle: Deliver HDAC constructs, shRNAs, or CRISPR components to specific brain regions using adeno-associated virus (AAV) or lentivirus.
Materials:

High-titer viral vectors ($\geq 10^{12}$ genome copies/mL for AAV), Stereotactic apparatus with digital coordinates, Hamilton syringe (10 μL) with 33 G needle, Anesthesia (isoflurane or ketamine/xylazine), Analgesics (buprenorphine).

Equipment: Stereotactic apparatus.
Procedure:
1. Anesthetize the animal and secure it in a stereotactic frame.
2. Make a midline scalp incision and expose the skull.

3. Drill a small burr hole at the target coordinates (e.g., hippocampus: AP −2.0 mm, ML ± 1.5 mm, DV −2.0 mm from bregma for mouse).
4. Lower the needle to the target depth.
5. Inject 0.5–2 μL virus at 0.1 μL/min.
6. Wait 5 min before slowly retracting the needle.
7. Close incision and monitor recovery.
8. Allow 2–4 weeks for transgene expression (AAV) or 1–2 weeks (lentivirus).

Viral Vector Selection:

- AAV serotypes: AAV9 for broad neuronal tropism, AAV-PHP.eB for enhanced CNS transduction, AAV1/2 for hippocampal neurons (Chan et al., 2017)
- Promoters: Synapsin for pan-neuronal expression, CaMKIIα for excitatory neurons, mDlx for interneurons (Dimidschstein et al., 2016)

Validation:

- Histological confirmation of injection site and spread
- Immunofluorescence or reporter fluorescence to assess transduction efficiency
- Functional validation (e.g., HDAC activity assays, behavioral phenotypes)

Notes:

- Strictly account for the bregma-lambda alignment and skull leveling to ensure that your X, Y, and Z coordinates accurately hit the target brain region.
- Carefully calculate the injection volume and flow rate (typically 50–100 nL/min) to prevent tissue displacement or "backflow" up the needle track.
- Monitor for cytotoxicity caused by high viral concentrations or local immune responses, which can alter the physiological state of the neurons you are trying to study.

2.3.2 In vivo imaging

Principle: Visualize HDAC localization and dynamics in living animals using two-photon microscopy.

Requirements:

Transgenic mice expressing fluorescent HDAC fusions or viral delivery of constructs, Cranial window surgery for chronic imaging, Two-photon microscope with appropriate lasers (e.g., 920 nm for GFP).

Equipment: Two-photon microscope.

Procedure

1. Perform cranial window surgery over target brain region (e.g., somatosensory cortex, hippocampus).
2. Allow 2–4 weeks for recovery and window stabilization.
3. Anesthetize animal (or use a head-fixed awake imaging setup).
4. Image through cranial window using two-photon microscopy.
5. Acquire z-stacks to capture neuronal populations at different depths.
6. Apply stimuli or behavioral paradigms and re-image to track HDAC localization changes.

Applications:

- Track HDAC shuttling during learning tasks (Ziv et al., 2013)
- Monitor HDAC localization in disease models over time (Holtmaat et al., 2009)
- Correlate HDAC dynamics with neuronal activity (using calcium indicators) (Chen et al., 2013)

Limitations:

- Requires surgical expertise and specialized equipment
- Limited imaging depth (typically $< 500\ \mu m$)
- Potential effects of anesthesia on HDAC dynamics

Notes:

- Utilize HDAC specific activity-based probes (ABPs) or FRET-based sensors with high two-photon absorption cross-sections to maximize signal-to-noise ratio.
- Precisely calibrate the femtosecond laser output to avoid photobleaching and photodamage
- Adjust the point spread function (PSF) and use long-wavelength excitation to penetrate the scattering layers of the brain while maintaining the cellular resolution necessary to distinguish nuclear vs. cytoplasmic HDAC localization.

2.4 Validation of HDAC subcellular localization

Accurate determination of HDAC subcellular localization requires multiple complementary approaches to avoid artifacts and ensure reproducibility. This section outlines a comprehensive validation workflow integrating live-cell imaging, biochemical fractionation, immunofluorescence, and chromatin immunoprecipitation (Fig. 3).

In Live-Cell Fluorescence Imaging (shown in the left panel of Fig. 3), GFP-tagged HDAC constructs are expressed in neurons and imaged over time (0 h, 2 h, 4 h, 6 h) to track real-time localization dynamics. Here, the representative images show nuclear localization at baseline (0 h), translocation at intermediate time-points (2 h, 4 h), and cytoplasmic localization after stimulation (6 h). This approach provides temporal resolution and avoids fixation artifacts.

In Subcellular Fractionation (shown in the second panel of Fig. 3), Cells are lysed under hypotonic conditions, and nuclei are separated from cytoplasm by differential centrifugation. Nuclear and cytoplasmic fractions are analyzed by Western blotting for HDAC distribution, with histone H3 (nuclear marker) and GAPDH (cytoplasmic marker) confirming fractionation quality. Quantification shows the percentage of HDAC in each compartment (e.g., 60 % cytoplasmic, 40 % nuclear). This biochemical approach provides quantitative validation of localization.

In Immunofluorescence staining (shown in the third panel of Fig. 3), fixed cells are stained with HDAC-specific antibodies (red) and DAPI (nuclear counterstain, (blue)) then imaged by confocal microscopy. High-resolution images reveal subcellular distribution of endogenous HDAC proteins without overexpression artifacts. Merged images confirm colocalization with nuclear or cytoplasmic compartments.

In Chromatin Immunoprecipitation (ChIP) (shown in the right panel of Fig. 3), Cells are crosslinked with formaldehyde to preserve protein-DNA interactions, chromatin is fragmented by sonication, and HDAC-bound DNA is immunoprecipitated with specific antibodies. After crosslink reversal and DNA purification, enrichment at target loci is quantified by qPCR (showing amplification curves for target vs. control loci) and fold enrichment over negative control regions. ChIP confirms functional nuclear localization and identifies specific genomic binding sites. Together, these four approaches provide robust, multi-modal validation of HDAC localization: live-cell imaging for dynamics, fractionation for quantitative distribution, immunofluorescence for spatial resolution, and ChIP for functional nuclear activity (Nott et al., 2016; Park, 2009; Schübeler, 2015; Suzuki et al., 2010).

2.4.1 Multi-modal validation strategy

Rationale: Each localization method has inherent limitations:

- Live-cell imaging may be affected by overexpression artifacts
- Biochemical fractionation can suffer from cross-contamination

- Immunofluorescence depends on antibody specificity and fixation conditions
- ChIP confirms functional nuclear localization but does not directly visualize the protein

Recommended Workflow:

1. **Initial screening:** Live-cell imaging of fluorescent HDAC fusions to establish baseline localization and stimulus-induced changes
2. **Biochemical validation:** Subcellular fractionation and Western blotting to quantify HDAC distribution
3. **Endogenous confirmation:** Immunofluorescence with validated antibodies to confirm the localization of endogenous HDAC
4. **Functional validation:** ChIP to confirm nuclear HDAC occupancy at target genomic loci
5. **Quality Control Checkpoints:**
   - Fractionation purity: $<5\%$ cross-contamination based on marker proteins
   - Antibody specificity: Single band on Western blot, signal loss in knockout/knockdown
   - Construct functionality: Rescue of knockdown phenotypes, preserved enzymatic activity
   - Reproducibility: Consistent results across ≥ 3 independent experiments

2.4.2 Quantitative image analysis

Software Tools:

- ImageJ/Fiji with custom macros for automated N/C ratio calculation (Schindelin et al., 2012)
- CellProfiler for high-throughput analysis of large image datasets (McQuin et al., 2018)
- Imaris for 3D reconstruction and colocalization analysis (Bolte & Cordelières, 2006)

Analysis Pipeline:

1. Segmentation: Define nuclear ROIs using DAPI or nuclear marker
2. Cytoplasmic ROI: Expand nuclear ROI by 2–5 μm or use cell membrane marker
3. Intensity measurement: Calculate mean fluorescence intensity in each compartment
4. Background subtraction: Subtract local background or use rolling ball algorithm

5. N/C ratio calculation: Divide nuclear by cytoplasmic intensity for each cell
6. Statistical analysis: Compare N/C ratios across conditions using appropriate tests

Normalization Strategies:

1. Express N/C ratio relative to control condition
2. Use internal reference (e.g., constitutively nuclear protein) to control for imaging variations
3. Correct for photobleaching in time-lapse experiments

2.5 CRISPR-based manipulation of HDAC localization

CRISPR-Cas9 technology has revolutionized the ability to manipulate HDAC function and localization with unprecedented precision. This section describes three complementary CRISPR-based strategies (Fig. 4):

- dCas9-HDAC fusion constructs for forced localization
- genomic editing of endogenous localization signals
- optogenetic control systems for dynamic temporal regulation.
- dCas9-HDAC fusion constructs for forced localization (Panel 1 of Fig. 4):

Precise spatial and locus-specific control of histone deacetylase (HDAC) activity can be acheived using engineered dCas9-based fusion constructs. In Panel 1 of Fig. 4, catalytically inactive Cas9 (dCas9) is shown fused to an HDAC catalytic domain via a flexible linker, together with engineered localization signals that determine its intracellular distribution. The upper construct contains a nuclear localization signal (NLS), directing the fusion protein to accumulate in the nucleus; lower construct incorporates a nuclear export signal (NES), promoting its retention in the cytoplasm. Single-guide RNAs (sgRNAs) guide these fusion proteins to specific genomic loci. As depicted, NLS-tagged constructs enable targeted histone deacetylation at defined nuclear sites, whereas NES-tagged constructs remain excluded from the nucleus, preventing chromatin interaction. Together, the figure highlights how these complementary designs permit controlled manipulation of HDAC localization and activity, providing a framework for dissecting compartment-specific epigenetic functions before exploring the detailed mechanisms described in the following sections.

- Genomic Editing of HDAC genes (Panel 2 of Fig. 4):

CRISPR-Cas9 is used to edit endogenous HDAC genes by inserting or deleting localization signals. For example, if the HDAC4 gene has to be edited at the Asp289 phosphorylation site, Cas9 (shown in blue in Panel 2 of Fig. 4) is guided by sgRNAs to create double-strand breaks at the target site (marked by PAM sequences). A donor DNA template encoding either an NLS or NES sequence is provided for homology-directed repair (HDR). The resulting edited HDAC4 proteins display altered localization: NLS insertion forces nuclear accumulation, while NES insertion promotes cytoplasmic retention. This strategy modifies endogenous HDAC expression and avoids overexpression artifacts (Doudna & Charpentier, 2014; Ran et al., 2013).

- Optogenetic Control for dynamic temporal regulation (Panel 3 of Fig. 3):

Light-sensitive protein domains enable reversible, spatiotemporal control of HDAC localization. In the light-OFF state (top of Panel 3 of Fig. 4), the HDAC protein (green) with attached photosensitive domain (cyan) and localization signals (NLS and NES tags) is in a default state. Upon illumination with blue light (450–488 nm, middle), the photosensitive domain undergoes a conformational change, exposing or masking localization signals to drive nuclear import or export. The schematic shows bidirectional shuttling between the cytoplasm and the nucleus in response to light. This approach enables dynamic, reversible control with high temporal precision for studying activity-dependent HDAC shuttling mechanisms (Nihongaki et al., 2015; Zetsche et al., 2015).

Strategies in detail.

2.5.1 dCas9-HDAC fusion constructs for forced localization

Principle: Catalytically inactive Cas9 (dCas9) is fused to HDAC catalytic domains and engineered localization signals (NLS or NES) to force nuclear or cytoplasmic targeting. sgRNAs direct the fusion protein to specific genomic loci, enabling locus-specific histone deacetylation (Hilton & Vockley, 2015; Kwon et al., 2017).

Three complementary approaches can be used to precisely control HDAC localization and activity, as shown in Fig. 4. Catalytically inactive Cas9 (dCas9, Panel 1, shown in blue) is fused to the HDAC catalytic domain via a flexible linker (green) and engineered localization signals to create dCas9-HDAC Fusion Constructs. The construct (seen at the top of panel 1 of Fig. 4) includes a nuclear localization signal (NLS, red tag) to force nuclear targeting, whereas the construct (seen at bottom of panel 1 of Fig. 4) contains a nuclear export signal (NES, orange tag) for cytoplasmic localization. sgRNAs guide the fusion protein to specific genomic loci. When these constructs are transfected into cells, NLS-tagged constructs accumulate in the nucleus and enable locus-specific histone deacetylation, while NES-tagged constructs are retained in the cytoplasm. This approach allows spatial control of HDAC activity and mechanistic dissection of compartment-specific functions (Hilton & Vockley, 2015; Kwon et al., 2017).

Construct Design:

Nuclear-Targeted dCas9-HDAC:

- [dCas9] – [Linker] – [HDAC catalytic domain] – [NLS]
- dCas9: D10A and H840A mutations abolish nuclease activity
- Linker: Flexible (e.g., $(GGSGGS)_3$ or rigid (e.g., $EAAAK)_4$ linker
- HDAC domain: Catalytic domain of HDAC1, 2, 3, or 8 (Class I) or HDAC4, 5, 7, 9 (Class IIa)
- NLS: Strong NLS (e.g., SV40 NLS: PKKKRKV) or multiple NLS copies for enhanced nuclear retention (Kalderon et al., 1984)

Cytoplasmic-Targeted dCas9-HDAC:

- [dCas9] – [Linker] – [HDAC catalytic domain] – [NES]
- NES: Strong NES (e.g., HIV Rev NES: LQLPPLERLTLD or PKI NES: LALKLAGLDI).

sgRNA Design:

1. Target promoter or enhancer regions of genes of interest
2. Design 3–5 sgRNAs per locus to ensure robust targeting
3. Avoid off-target sites using prediction tools (e.g., CRISPOR, Benchling)
4. Validate sgRNA efficiency by ChIP-qPCR or dCas9-GFP imaging (Concordet, & Haeussler, 2018)

Delivery Methods:

Plasmid Transfection:

- Suitable for cell lines and primary neurons (using Lipofectamine or Nucleofection)
- Co-transfect dCas9-HDAC plasmid + sgRNA expression plasmid
- Include a fluorescent reporter for transfection efficiency monitoring

Viral Transduction:

- AAV or lentivirus for in vivo delivery
- Use separate vectors for dCas9-HDAC and sgRNA, or all-in-one vectors (size permitting)

- AAV packaging limit: ~4.7 kb; split large constructs across two AAV vectors with trans-splicing (Truong et al., 2015)

Validation:

Localization confirmation:
- Immunofluorescence or live-cell imaging to confirm nuclear vs. cytoplasmic targeting
- Compare NLS vs. NES constructs

Locus-specific targeting:
- ChIP-qPCR to confirm dCas9-HDAC enrichment at target loci
- Use HA or FLAG tag on dCas9 for immunoprecipitation

Functional effects:
- ChIP-qPCR for histone acetylation (H3K9ac, H3K27ac) at target loci
- RT-qPCR for target gene expression changes
- RNA-seq for genome-wide transcriptional effects

Specificity controls:
- Non-targeting sgRNA control
- Catalytically inactive HDAC mutant (e.g., H976A for HDAC4) (Fischle et al., 2002)
- Compare on-target vs. off-target loci

Applications:
- Dissect nuclear vs. cytoplasmic HDAC functions by forcing localization
- Achieve locus-specific histone deacetylation at disease-relevant genes
- Model pathological HDAC mislocalization in disease contexts
- Identify genomic targets responsible for HDAC-mediated phenotypes

Advantages:
- Precise spatial control (nuclear vs. cytoplasmic)
- Locus-specific targeting via sgRNAs
- Reversible (remove doxycycline in inducible systems)

Limitations:
- Overexpression artifacts (use low-titer virus or inducible promoters)
- Off-target effects (validate sgRNA specificity)
- Limited temporal control (see optogenetic approaches below)

2.5.2 Genomic editing of localization signals

Principle: Use CRISPR-Cas9 to edit endogenous HDAC genes, inserting or deleting NLS/NES sequences or mutating phosphorylation sites that regulate shuttling. This approach modifies endogenous HDAC expression, avoiding overexpression artifacts (Doudna & Charpentier, 2014; Ran et al., 2013).

Target Sites for Editing:

Class IIa HDACs (HDAC4, 5, 7, 9):

Phosphorylation sites: Mutate conserved serines (e.g., S246A, S467A, S632A in HDAC4) to prevent 14-3-3 binding and cytoplasmic retention (Backs et al., 2011).

NES deletion: Remove or mutate NES sequence to prevent nuclear export (Miska et al., 1999).

NLS enhancement: Insert additional NLS sequences to promote nuclear accumulation.

Class I HDACs (HDAC1, 2, 3, 8):

NES insertion: Add NES to force cytoplasmic localization (model pathological export) (Kim et al., 2010).

NLS deletion: Remove NLS to reduce nuclear import.

CRISPR Editing Strategies:

Homology-Directed Repair (HDR):

1. Design sgRNA targeting insertion/mutation site
2. Prepare donor DNA template (ssDNA oligo or dsDNA plasmid) with desired edit flanked by 50–100 bp homology arms
3. Deliver Cas9 RNP (ribonucleoprotein) + sgRNA + donor template by electroporation
4. Screen clones by PCR and sequencing
5. Validate functional effects (localization, activity, phenotypes)

HDR Donor Template Design:

- Include silent mutations in PAM or seed sequence to prevent re-cutting
- Use asymmetric homology arms (longer on 3' side) for improved efficiency (Richardson et al., 2016)
- For knock-in of epitope tags or localization signals, maintain the reading frame

Base Editing:

- Use cytosine or adenine base editors to introduce point mutations without double-strand breaks (Komor et al., 2016)
- Suitable for phosphorylation site mutations (e.g., S→A conversions)
- Higher efficiency than HDR, especially in non-dividing neurons

Prime Editing:

- Insert, delete, or replace sequences up to ~100 bp without donor DNA or double-strand breaks (Anzalone et al., 2019)
- Useful for NLS/NES insertions or deletions
- Requires design of prime editing guide RNA (pegRNA) with desired edit encoded in extension

Delivery to Neurons:

In Vitro (Primary Neurons or iPSC-Derived Neurons):

- Nucleofection of Cas9 RNP + donor template (highest efficiency for HDR)
- Viral transduction (AAV or lentivirus) for base/prime editors
- Single-cell cloning and validation (for cell lines or iPSCs)

In Vivo:

- AAV delivery of Cas9 + sgRNA + donor template to neonatal or adult brain (Nishiyama et al., 2017)
- HDR efficiency is low in post-mitotic neurons; consider base/prime editing
- Use Cre-dependent systems for cell-type-specific editing (Platt et al., 2014)

Validation:

1. **Genotyping:**
   - PCR amplification of edited locus
   - Sanger sequencing to confirm precise edit
   - Quantify HDR efficiency (% edited alleles)
2. **Protein expression:**
   - Western blot to confirm edited HDAC protein expression
   - Immunofluorescence to assess localization
3. **Functional validation:**
   - Subcellular fractionation to quantify nuclear/cytoplasmic distribution
   - Phosphorylation-site mutants: test 14-3-3 binding by co-IP
   - Assess downstream phenotypes (gene expression, neuronal morphology, behavior)

Applications:

- Generate "localization-locked" HDAC mutants (constitutively nuclear or cytoplasmic)
- Model disease-associated HDAC mutations
- Create knock-in reporter lines (e.g., HDAC-GFP) for endogenous expression tracking

- Dissect phosphorylation-dependent shuttling mechanisms

Advantages:
- Edits endogenous locus (physiological expression levels)
- Avoids overexpression artifacts
- Permanent, heritable modifications (for germline editing)

Limitations:
- Low HDR efficiency in post-mitotic neurons (use base/prime editing)
- Potential off-target editing (validate with whole-genome sequencing)
- Time-consuming clone screening and validation

2.5.3 Optogenetic control of HDAC activity and localization

Principle: Light-sensitive protein domains enable reversible, spatio-temporal control of HDAC localization or activity. Two main strategies are: (1) light-induced recruitment of HDAC to specific subcellular compartments or genomic loci, and (2) light-induced degradation of HDAC proteins (Fig. 5) (Nihongaki et al., 2015; Zetsche et al., 2015).

For the temporal regulation of HDAC activity, two complementary light-controlled strategies are used. The first strategy involves CRY2-CIB1 Optogenetic HDAC Recruitment System that recruits HDAC reversibly to specific genomic loci or subcellular compartments using the Light-induced heterodimerization process (Section 1). In the dark (light-OFF state) (Panel A of Fig. 5), the dCas9-CRY2 fusion protein (shown in green) is bound to target genomic loci via sgRNA, and the nucleus contains high levels of acetylated histones (Ac, yellow circles) (gene expression switched ON). HDAC protein (shown in pink) fused to CIB1 (shown in orange) remains in the cytoplasm, separated from nuclear substrates. Blue light illumination (450–488 nm) triggers CRY2-CIB1 heterodimerization (as shown in Panel B of Fig. 5). In the light-ON state, CIB1-HDAC is recruited to dCas9-CRY2 at the target loci, resulting in local histone deacetylation and gene repression (gene expression switched OFF) (as seen in Panel C of Fig. 5).

Thus, under dark conditions (0 min), the system remains inactive; blue-light stimulation triggers fast recruitment of CIB1-HDAC and histone deacetylation within ~5 min. Switching off the light (~30 min) leads to dissociation and gradual recovery, with full restoration after prolonged darkness (~120 min). Overall, this system enables reversible, locus-specific HDAC regulation with a temporal resolution of approximately 5–30 min (Kennedy et al., 2010; Taslimi et al., 2014).

Panel C of Fig. 5, Section 1, represents a timeline of kinetics from 0 min to 120 min: dark baseline (0 min), blue light ON (+5 min) induces rapid recruitment and deacetylation, blue light OFF (+30 min) allows dissociation and recovery, with full recovery after an extended dark period (+120 min).

Another strategy for temporal control of HDAC protein levels involves light-activated protein degradation, known as the PHOTAC System (Section 2 of Fig. 5). In the stable state, HDAC6 protein (shown in pink) is expressed at normal levels. A photoswitchable PROTAC molecule (shown as purple and orange domains) is inactive. E3 ubiquitin ligase (dark blue) is present but not engaged (as shown in Panel D of Fig. 5). Blue-light illumination induces conformational change in the photoswitchable PROTAC by activating its binding domains (Panel E of Fig. 5). The activated PROTAC forms a ternary complex, bringing HDAC6 into proximity with the E3 ligase. This triggers ubiquitination (Ub, green circles) of HDAC6, targeting it for proteasomal degradation (Panel F of Fig. 5).

Panel F of Fig. 5 represents a timeline of kinetics from 0 min to 120 min. The timeline shows that blue light ON (+30 min) initiates PHOTAC activation and ubiquitination, blue light OFF (+60 min) leads to proteasomal degradation and HDAC6 depletion, and recovery in the dark (+120 min) allows new HDAC6 synthesis. This approach enables reversible, light-controlled depletion of cytoplasmic HDAC6 to study its roles in axonal transport and aggregate clearance (Naro et al., 2020; Wurnig et al., 2024)

Strategy 1: Light-Induced HDAC Recruitment (CRY2-CIB1 System).

Components:

CRY2 (Cryptochrome 2): Blue-light sensitive domain from *Arabidopsis thaliana*, CIB1 (Cryptochrome-Interacting Basic-helix-loop-helix 1): CRY2-binding partner, Blue light (450–488 nm) induces rapid, reversible CRY2-CIB1 heterodimerization (Kennedy et al., 2010).

Construct Design:

For Locus-Specific HDAC Recruitment:

- [dCas9] - [CRY2] + [sgRNA] [CIB1] - [HDAC catalytic domain]
- Blue light recruits CIB1-HDAC to dCas9-CRY2 at target genomic loci
- Enables temporal control of locus-specific histone deacetylation

For Subcellular Compartment Recruitment:

- [Mitochondrial/ER/Membrane targeting sequence] – [CRY2] [CIB1] – [HDAC]

- Blue light recruits HDAC to specific organelles or membrane compartments
- Study compartment-specific HDAC substrates and functions

Procedure

1. **Construct preparation:**
   - **a.** Clone CRY2 and CIB1 fusion constructs into mammalian expression vectors
   - **b.** Include fluorescent tags (e.g., mCherry-CRY2, GFP-CIB1) for visualization
   - **c.** Validate expression and light-responsiveness in HEK293 cells before neuronal experiments
2. **Neuronal delivery:**
   - **a.** Co-transfect or co-transduce neurons with both constructs
   - **b.** Use bicistronic vectors or dual AAV for in vivo delivery
   - **c.** Allow 3–7 days for expression
3. **Light stimulation:**
   - **a.** Use blue LED (470 nm) or laser (488 nm) for stimulation
   - **b.** Optimize light intensity (typically 0.1–1 mW/mm^2) and duration (5–30 min)
   - **c.** Pulsed illumination (e.g., 5 s ON / 10 s OFF) reduces phototoxicity
4. **Validation:**
   - **a.** Live-cell imaging to confirm light-induced recruitment
   - **b.** ChIP-qPCR to measure histone acetylation changes at target loci
   - **c.** RT-qPCR to assess gene expression changes
   - **d.** Reversibility: turn off light and monitor recovery (typically 30–120 min)

Applications:

- Temporal control of HDAC activity during specific behavioral epochs (e.g., learning, consolidation)
- Dissect acute vs. chronic effects of HDAC localization changes
- Correlate HDAC recruitment with neuronal activity or calcium dynamics
- Study activity-dependent HDAC shuttling mechanisms with precise temporal resolution

Advantages:

- Reversible (unlike genetic manipulations)
- High temporal resolution (minutes)

- Spatial control (illuminate specific brain regions or subcellular compartments)
- Orthogonal to endogenous signaling pathways

Limitations:

- Requires light delivery (fiber optics for in vivo, limited tissue penetration)
- Potential phototoxicity (minimize light intensity and duration)
- Incomplete recruitment (optimize construct design and expression levels)
- CRY2 oligomerization at high light intensities (use monomeric CRY2 variants) (Taslimi et al., 2014)

Strategy 2: Light-Induced HDAC Degradation (PHOTAC System).

Principle: Photoswitchable PROTACs (proteolysis-targeting chimeras) enable light-controlled ubiquitination and proteasomal degradation of HDAC proteins (Naro et al., 2020; Wurnig et al., 2024).

PHOTAC Design:

- HDAC-binding moiety: HDAC inhibitor (e.g., tubastatin A for HDAC6)
- Photoswitchable linker: Azobenzene or other photoisomerizable group
- E3 ligase-binding moiety: Ligand for E3 ubiquitin ligase (e.g., VHL, CRBN)
- Blue light induces conformational change, bringing HDAC into proximity with E3 ligase for ubiquitination

Procedure-

PHOTAC synthesis:

- Chemical synthesis of PHOTAC molecule (requires organic chemistry expertise)
- Validate photoswitching properties (UV-Vis spectroscopy)
- Test HDAC binding affinity in light-ON vs. light-OFF states

1. **Cell treatment:**
   - **a.** Add PHOTAC to culture medium (typically 1–10 μM)
   - **b.** Incubate in dark for 1–2 h to allow cell uptake
   - **c.** Illuminate with blue light (450–488 nm) to activate degradation
2. **Validation:**
   - **a.** Western blot to confirm HDAC protein depletion (time course: 0, 1, 2, 4, 8 h post-illumination)
   - **b.** Immunofluorescence to visualize HDAC loss
   - **c.** Functional assays (e.g., acetylation of HDAC6 substrates like α-tubulin)
   - **d.** Reversibility: remove PHOTAC and monitor protein recovery (typically 12–24 h)

Applications:
- Temporal control of HDAC protein levels (complementary to inhibitors, which block activity but not scaffolding functions)
- Study HDAC6 roles in axonal transport and aggregate clearance with temporal precision
- Dissect acute effects of HDAC depletion vs. chronic knockout phenotypes

Advantages:
- Degrades entire protein (eliminates both catalytic and scaffolding functions)
- Reversible (protein resynthesis after PHOTAC removal)
- Cell-permeable small molecule (easier delivery than genetic tools)

Limitations:
- Requires custom chemical synthesis
- Potential off-target effects (validate specificity with rescue experiments)
- Incomplete degradation (optimize PHOTAC concentration and light dose)
- Currently limited to HDAC6 (expand to other isoforms requires new PHOTAC designs)

Comparison of Optogenetic Strategies:

Feature	CRY2-CIB1 recruitment	PHOTAC degradation
Temporal resolution	Minutes (5–30 min)	Hours (2–8 h)
Reversibility	Fast (30–120 min)	Slow (12–24 h)
Delivery	Genetic (plasmid/virus)	Chemical (small molecule)
Specificity	High (sgRNA-directed)	Moderate (depends on inhibitor selectivity)
In vivo applicability	Yes (requires fiber optics)	Yes (systemic or local injection)
Current isoform coverage	All HDACs (customizable)	HDAC6 (expandable)

3. Data analysis and interpretation

3.1 Quantification of HDAC localization

3.1.1 Nuclear/cytoplasmic ratio calculation

Image Analysis Workflow:

1. Segmentation: Use DAPI or nuclear marker to define nuclear ROIs

2. Cytoplasmic ROI: Expand nuclear ROI by 2–5 μm or use cell membrane marker
3. Intensity measurement: Calculate mean fluorescence intensity in each compartment
4. Background subtraction: Subtract local background or use rolling ball algorithm
5. N/C ratio: Divide nuclear by cytoplasmic intensity for each cell

Normalization:

- Express N/C ratio relative to control condition
- Use internal reference (e.g., constitutively nuclear protein) to control for imaging variations

Statistical Analysis:

1. Compare N/C ratios across conditions using *t*-test or ANOVA
2. Analyze 50–100 cells per condition for statistical power
3. Report mean ± SEM or median with interquartile range

3.1.2 Fractionation quantification

Western Blot Densitometry:

1. Acquire high-quality Western blot images (avoid saturation)
2. Use ImageJ or similar software for densitometry
3. Normalize HDAC signal to loading control in each fraction
4. Calculate percentage of total HDAC in nuclear vs. cytoplasmic fraction
5. Statistical analysis across biological replicates ($n \geq 3$)

Quality Control:

- Validate fractionation purity with marker proteins
- Ensure linear range of detection (dilution series)
- Include positive and negative controls

3.2 Statistical considerations

3.2.1 Experimental design

Sample Size:

- Power analysis to determine required sample size (typically n = 3–6 biological replicates for cell culture, n = 6–12 animals for in vivo studies)
- Consider effect size, variability, and desired statistical power (typically 0.8)

Randomization and Blinding:

1. Randomize animals to treatment groups

2. Blind experimenters to treatment conditions during data collection and analysis
3. Unblind only after data analysis is complete

Controls:

- Include appropriate positive and negative controls
- Use vehicle-treated or sham-operated controls for in vivo studies
- Include technical replicates to assess assay variability

3.2.2 Statistical tests

Choosing Appropriate Tests:

- Two groups: *t*-test (parametric) or Mann-Whitney *U* test (non-parametric)
- Multiple groups: ANOVA (parametric) or Kruskal-Wallis test (non-parametric)
- Repeated measures: Repeated-measures ANOVA or mixed-effects model
- Time course data: Two-way ANOVA with time and treatment as factors

Post-Hoc Tests:

- Use appropriate post-hoc tests after ANOVA (e.g., Tukey, Bonferroni, Dunnett)
- Correct for multiple comparisons to control false discovery rate

Reporting:

- Report exact p-values (not just "$p < 0.05$")
- Report effect sizes (e.g., Cohen's d) in addition to p-values
- Include measures of variability (SEM or SD) in figures and text

3.3 Integration of multi-modal data

3.3.1 Correlating localization with function

Approach:

1. Measure HDAC localization (imaging, fractionation) and functional outcomes (gene expression, behavior) in the same samples
2. Perform correlation analysis to assess the relationship between localization and function
3. Use multivariate analysis (e.g., PCA, clustering) to identify patterns across multiple variables

Example:

- Correlate HDAC4 N/C ratio with expression of MEF2 target genes
- Assess whether changes in HDAC localization predict behavioral outcomes

3.3.2 Integrating omics data

Transcriptomics:

- RNA-seq to identify genome-wide transcriptional changes following HDAC manipulation
- Integrate with ChIP-seq data to identify direct HDAC target genes
- Use pathway analysis to identify enriched biological processes

Proteomics:

- Mass spectrometry to identify HDAC substrates and interacting proteins
- Acetylome profiling to map acetylation changes following HDAC inhibition
- Integrate with transcriptomics to distinguish direct vs. indirect effects

Epigenomics:

- ChIP-seq for histone modifications (e.g., H3K9ac, H3K27ac) to map chromatin changes
- ATAC-seq to assess chromatin accessibility
- Integrate with transcriptomics to link chromatin changes to gene expression

4. Future methodological directions and outstanding challenges in HDAC research

Even though scientists have learned a lot about histone deacetylases (HDACs) in neurodegeneration and brain plasticity, there's still plenty we do not understand—both in terms of the actual science and the approach towards it. In the future, it will be imperative to integrate genetic as well as biochemical technological advances with an exploration of critical unanswered questions. Below, we discuss promising experimental strategies and highlight key challenges that next-generation methods aim to tackle.

4.1 Emerging experimental technologies

4.1.1 Single-cell and high-resolution approaches

Techniques operating at the single-cell level provide new ways to decipher the role of HDACs in the brain. Single-cell RNA sequencing (scRNA-seq), for example, stands out at selecting individual cell types that express certain HDACs and the modifications in the HDAC distribution patterns when we manipulate them. We can finally see exactly where these enzymes are, right down to the specific neurons or glial cells (Zeisel et al., 2018). Additionally, we can track their responses at the gene level in real time (Saunders et al., 2018). On the imaging front, single-cell approaches let researchers track individual HDAC molecules moving within the neurons. You can accurately see HDACs travel from the nucleus to the cytoplasm and back, and see how this changes depending on what the neuron's up to (Packer et al., 2015; Chen et al., 2013). Compared to bulk methods, it is like switching from a blurry group photo to a crystal-clear candid shot.

4.1.2 Advanced CRISPR-based technologies

When it comes to gene editing, new CRISPR variants allow us to be more precise than before. Tools like base and prime editing allow researchers to swap DNA nucleotides without breaking both strands, making it easier—and safer—to model specific disease-linked HDAC mutations (Komor et al., 2016; Anzalone et al., 2019; Gaudelli et al., 2017). Then there's genome-wide CRISPR screening, which can help us discover genes that control HDAC translocation within a cell and the magnitude of their activity (Shao et al., 2018). Furthermore, researchers can also investigate gene partners that make HDAC inhibitors more (or less) effective (King et al., 2021). Essentially, this signifies bigger, faster, and more targeted explorations of HDAC biology.

4.1.3 Improved optogenetic and chemogenetic tools

Scientists are refining optogenetics and chemogenetics to tweak HDAC functionality with greater specificity. Red-shifted light-sensing proteins help researchers reach deeper brain layers, and when paired with blue-light tools, they can flip specific HDACs on or off in complex combinations (Chuong et al., 2014). Chemogenetic variants such as DREADDs can keep HDAC pathways switched on or off for longer, and can be layered with optogenetics for greater precision (Roth, 2016). In short, scientists are getting better than ever at controlling exactly when and where HDACs function.

4.2 Translational and precision experimental strategies

4.2.1 Biomarker development and patient stratification

Future therapies depend on translating bench science directly to patient care. This requires creating biomarkers to measure activity of HDAC in humans. For example, measuring histone acetylation in spinal fluid or blood can be a window into HDAC activity (Sharma & Taliyan, 2015). PET tracers take this further, allowing imaging of HDACs in a living brain. Further, genetic screening can spot disease-linked HDAC variants (Bester et al., 2018), and patient-derived iPSCs let scientists test new drugs in cells that match a specific patient's genetics (Roth, 2016). These strategies make research more tailored—and more likely to benefit actual patients.

4.2.2 Personalized and combination therapeutic approaches

Isoform-selective HDAC inhibitors are another big step. Rather than targeting all HDACs broadly, new drugs are being designed to target the specific isoforms (Eckschlager et al., 2017). The idea is to match the right drug to the right patient based on their HDAC profile (Boyden et al., 2005). There is growing interest in combination therapies—combining HDAC inhibitors with other treatments to boost their overall effect (Xu et al., 2011). Systems biology tools can help map out these combinations, aiming for synergy rather than adding pharmacological load.

4.3 Unresolved mechanistic questions and experimental strategies

4.3.1 Mechanisms of HDAC shuttling

A lot of questions remain unanswered in this field. For instance, What are the complete sets of kinases and phosphatases regulating Class IIa HDAC shuttling? How do different patterns of nerve cell firing (like bursts or steady tonic signals) affect where HDACs are in the cell? And what drives HDAC1 out of the nucleus in neurodegenerative disease? To tackle these, scientists lean on phospho-proteomics (to track every phosphorylation tweak), live-cell imaging that keeps tabs on calcium levels, and CRISPR screens to track upstream players.

4.3.2 Non-histone substrates

We know HDACs act on more than just histones, but the full list of other targets is still unknown. How does deacetylation affect all the non-histone proteins—and which matter most for neurodegenerative disease? Answering these questions means deep-diving into acetylome profiling with mass spectrometry and using substrate-trapping methods to identify transient HDAC-substrate interactions. After selecting likely targets, researchers make acetylation mimetics and acetylation-deficient mutants allowing for direct assessment of phenotypic shifts.

4.3.3 Therapeutic window

As timing is everything for treatment or prevention, many questions on this aspect arise, like- What is the best window for using HDAC inhibitors? Can we prevent neurodegenerative diseases, or just slow them down? And what happens to brain function if patients keep taking these drugs for years?

To figure it out, researchers run longitudinal studies in animals with treatment initiation times, biomarker-driven trials, and keep a close eye on safety over time in animals.

4.4 Integrative perspective

Moving the field forward means combining these technologies—single-cell sequencing, CRISPR editing, optogenetics, and big-picture systems biology—while rigorously validating the results. The big hurdle? Linking HDAC dynamics to neuronal function and disease. That's going to take experimental setups that cover time, space, quantity, and function all at once. It's a big ask, but it's where the breakthroughs will come from.

5. Conclusion

Histone deacetylases (HDACs) shape neurodegeneration and neuronal plasticity in many ways, both inside the nucleus and out in the cytoplasm. They don't just sit at one place, either—the way HDACs, especially Class IIa isoforms, move between the nucleus and cytoplasm ties neuronal activity directly to changes in gene expression. When this movement or activity goes off track, all sorts of problems can follow. Issues with HDACs are now linked to Alzheimer's, Parkinson's, Huntington's, and ALS, affecting everything from how genes switch on and off to how proteins stay stable and travel through neurons. In this chapter, we have provided detailed information on how to study HDACs in the nervous system, with a real focus on tracking exactly where these proteins go inside cells and how to control that movement. When you combine traditional biochemistry and imaging with CRISPR-based tools, you can tweak HDAC locations with incredible accuracy and decipher what each isoform really does. The workflow covered here—mixing live-cell imaging, fractionation, immunofluorescence, and chromatin immunoprecipitation—gives researchers a strong blueprint for digging deep into localization. Therapeutically, the emergence of isoform-specific HDAC inhibitors is a big step forward. Drugs that just target HDAC6 look especially promising for improving axonal transport and clearing protein clumps in models of Parkinson's and Huntington's, while compounds that target HDAC2 could boost memory in Alzheimer's. Still, there are hurdles: many drugs cannot cross the blood-brain barrier easily, biomarkers are still lacking, and researchers haven't nailed down a treatment regimen for the best results.

Looking ahead, single-cell methods will help sort out how HDACs work in different types of neurons, and CRISPR tools will drive sharper editing and faster screening. Optogenetics and chemogenetics offer new ways to control HDACs in real time. Precision medicine is also on the rise, helping tailor treatments to each patient. But a lot of questions remain—like what exactly controls HDAC shuttling, what non-histone targets HDACs act on in neurons, and how wide (or narrow) the treatment window is for these drugs in different diseases. By bringing together all these new techniques with careful validation, the field sits on the verge of turning detailed molecular insight into real treatments for neurodegenerative disorders.

References

Alchini, R., Sato, H., Matsumoto, N., et al. (2017). Nucleocytoplasmic shuttling of histone deacetylase 9 controls activity-dependent thalamocortical axon branching. *Scientific Reports, 7*. https://doi.org/10.1038/s41598-017-06243-7.

Anzalone, A. V., Randolph, P. B., Davis, J. R., et al. (2019). Search-and-replace genome editing without double-strand breaks or donor DNA. *Nature, 576*, 149–157. https://doi.org/10.1038/s41586-019-1711-4.

Backs, J., Song, K., Bezprozvannaya, S., & Olson, E. N. (2006). CaM kinase II selectively signals to histone deacetylase 4 during cardiomyocyte hypertrophy. *Journal of Clinical Investigation, 116*, 1853–1864. https://doi.org/10.1172/jci27438.

Backs, J., Worst, B. C., Lehmann, L. H., et al. (2011). Selective repression of MEF2 activity by PKA-dependent proteolysis of HDAC4. *Journal of Cell Biology, 195*, 403–415. https://doi.org/10.1083/jcb.201105063.

Bantscheff, M., Hopf, C., Savitski, M. M., et al. (2011). Chemoproteomics profiling of HDAC inhibitors reveals selective targeting of HDAC complexes. *Nature Biotechnology, 29*, 255–265. https://doi.org/10.1038/nbt.1759.

Bardai, F. H., & D'Mello, S. R. (2006). Intracellular trafficking of histone deacetylase 4 regulates neuronal cell death. *Journal of Neuroscience, 26*(3), 907–916.

Bardai, F. H., Price, V., Zaayman, M., & D'Mello, S. R. (2012). Histone deacetylase-1 (HDAC1) is a molecular switch between neuronal survival and death. *Journal of Biological Chemistry, 287*, 35444–35453. https://doi.org/10.1074/jbc.m112.394544

Bester, A. C., Lee, J. D., Chavez, A., Lee, Y. R., Nachmani, D., Vora, S., ... Pandolfi, P. P. (2018). An integrated genome-wide CRISPRa approach to functionalize lncRNAs in drug resistance. *Cell, 173*(3), 649–664.e620. https://doi.org/10.1016/j.cell.2018.03.052.

Bobrowska, A., Donmez, G., Weiss, A., & Bates, B. (2012). SIRT2 ablation has no effect on tubulin acetylation in brain, cholesterol biosynthesis or the progression of Huntington's disease phenotypes in vivo. *PLoS One, 7*. https://doi.org/10.1371/journal.pone.0034805.

Bolger, T. A., & Yao, T. P. (2005). Intracellular trafficking of histone deacetylase 4 regulates neuronal cell death. *Journal of Neuroscience, 25*, 9544–9553. https://doi.org/10.1523/jneurosci.1826-05.2005.

Bolte, S., & Cordelières, F. P. (2006). A guided tour into subcellular colocalization analysis in light microscopy. *Journal of Microscopy, 224*, 213–232. https://doi.org/10.1111/j.1365-2818.2006.01706.x.

Boyden, E. S., Zhang, F., Bamberg, E., Nagel, G., & Deisseroth, K. (2005). Millisecond-timescale, genetically targeted optical control of neural activity. *Nature Neuroscience, 8*(9), 1263–1268. https://doi.org/10.1038/nn1525.

Bradner, J. E., West, N., Grachan, M. L., et al. (2010). Chemical phylogenetics of histone deacetylases. *Nature Chemical Biology, 6*, 238–243. https://doi.org/10.1038/nchembio.313.

Broide, R. S., Redwine, J. M., Aftahi, N., & Winrow, C. J. (2007). Distribution of histone deacetylases 1-11 in the rat brain. *Journal of Molecular Neuroscience, 31*, 47–58. https://doi.org/10.1007/bf02686117.

Chan, K. Y., Jang, M. J., Yoo, B. B., et al. (2017). Engineered AAVs for efficient non-invasive gene delivery to the central and peripheral nervous systems. *Nature Neuroscience, 20*, 1172–1179. https://doi.org/10.1038/nn.4593.

Chawla, S., Hardingham, G. E., Quinn, D. R., & Bading, H. (1998). CBP: A signal-regulated transcriptional coactivator controlled by nuclear calcium and CaM kinase IV. *Science (New York, N. Y.), 281*, 1505–1509. https://doi.org/10.1126/science.281.5382.1505.

Chawla, S., Vanhoutte, P., Arnold, F. J., & Bading, H. (2003). Neuronal activity-dependent nucleocytoplasmic shuttling of HDAC4 and HDAC5. *Journal of Neurochemistry, 85*, 151–159. https://doi.org/10.1046/j.1471-4159.2003.01648.x.

Chen, T. W., Wardill, T. J., Sun, Y., et al. (2013). Ultrasensitive fluorescent proteins for imaging neuronal activity. *Nature, 499*, 295–300. https://doi.org/10.1038/nature12354.

Choudhary, C., Kumar, C., Gnad, F., et al. (2009). Lysine acetylation targets protein complexes and co-regulates major cellular functions. *Science (New York, N. Y.), 325*, 834–840. https://doi.org/10.3410/f.1163482.625816.

Choudhary, C., Weinert, B. T., Nishida, Y., & Mann, M. (2014). The growing landscape of lysine acetylation links metabolism and cell signalling. *Nature Reviews. Molecular Cell Biology, 15*, 536–550. https://doi.org/10.1038/nrm3841.

Chuang, D. M., Leng, Y., Marinova, Z., & Chiu, C. T. (2009). Multiple roles of HDAC inhibition in neurodegenerative conditions. *Trends in Neurosciences, 32*, 591–601. https://doi.org/10.1016/j.tins.2009.06.002.

Chuong, A. S., Miri, M. L., Busskamp, V., Matthews, G. A. C., Acker, L. C., Sørensen, A. T., et al. (2014). Noninvasive optical inhibition with a red-shifted microbial rhodopsin. *Nature Neuroscience, 17*(8), 1123–1129.

Concordet, J. P., & Haeussler, M. (2018). CRISPOR: Intuitive guide selection for CRISPR/Cas9 genome editing experiments and screens. *Nucleic Acids Research, 46*. https://doi.org/10.1093/nar/gky354.

Curcio, A., Rocca, R., Alcaro, S., & Artese, A. (2024). The histone deacetylase family: Structural features and application of combined computational methods. *Pharmaceuticals (Basel), 17*(5), https://doi.org/10.3390/ph17050620.

Didonna, A., & Opal, P. (2015). The promise and perils of HDAC inhibitors in neurodegeneration. *Annals of Clinical and Translational Neurology, 2*, 79–101. https://doi.org/10.1002/acn3.147.

Dimidschstein, J., Chen, Q., Tremblay, R., et al. (2016). A viral strategy for targeting and manipulating interneurons across vertebrate species. *Nature Neuroscience, 19*, 1743–1749. https://doi.org/10.1038/nn.4430.

Dompierre, J. P., Godin, J. D., Charrin, B. C., et al. (2007). Histone deacetylase 6 inhibition compensates for the transport deficit in Huntington's disease by increasing tubulin acetylation. *Journal of Neuroscience, 27*, 3571–3583. https://doi.org/10.1523/jneurosci.0037-07.2007.

Doudna, J. A., & Charpentier, E. (2014). Genome editing. The new frontier of genome engineering with CRISPR-Cas9. *Science (New York, N. Y.), 346*. https://doi.org/10.1126/science.1258096.

Du, Y., Wang, F., Zou, J., et al. (2014). Histone deacetylase 6 regulates cytotoxic α-synuclein accumulation through induction of the heat shock response. *Neurobiology of Aging, 35*, 2316–2328. https://doi.org/10.1016/j.neurobiolaging.2014.04.029.

Eckschlager, T., Plch, J., Stiborova, M., & Hrabeta, J. (2017). Histone deacetylase inhibitors as anticancer drugs. *International Journal of Molecular Sciences, 18*. https://doi.org/10.3390/ijms18071414.

Fischer, A., Sananbenesi, F., Wang, X., & Tsai, L. H. (2007). Recovery of learning and memory is associated with chromatin remodelling. *Nature, 447*, 178–182. https://doi.org/10.1038/nature05772.

Fischle, W., Dequiedt, F., Hendzel, M. J., et al. (2002). Enzymatic activity associated with class II HDACs is dependent on a multiprotein complex containing HDAC3 and SMRT/N-CoR. *Molecular Cell, 9*, 45–57. https://doi.org/10.1016/s1097-2765(01)00429-4.

Flavell, S. W., Cowan, C. W., Kim, T. K., et al. (2006). Activity-dependent regulation of MEF2 transcription factors suppresses excitatory synapse number. *Science (New York, N. Y.), 311*, 1008–1012. https://doi.org/10.1126/science.1122511.

Flavell, S. W., Kim, T. K., Gray, J. M., et al. (2008). Genome-wide analysis of MEF2 transcriptional program reveals synaptic target genes and neuronal activity-dependent polyadenylation site selection. *Neuron, 60*, 1022–1038. https://doi.org/10.1016/j.neuron.2008.11.029.

Frank, C. L., Manandhar, D., Gordân, R., et al. (2016). HDAC inhibitors cause site-specific chromatin remodeling at PU.1-bound enhancers in K562 cells. *Epigenetics & Chromatin, 9*, 15. https://doi.org/10.1186/s13072-016-0065-5.

Gaudelli, N. M., Komor, A. C., Rees, H. A., Packer, M. S., Badran, A. H., Bryson, D. I., & Liu, D. R. (2017). Programmable base editing of A*T to G*C in genomic DNA without DNA cleavage. *Nature, 551*(7681), 464–471. https://doi.org/10.1038/nature24644.

Glozak, M. A., Sengupta, N., Zhang, X., & Seto, E. (2005). Acetylation and deacetylation of non-histone proteins. *Gene, 363*, 15–23. https://doi.org/10.1016/j.gene.2005.09.010.

Godena, V. K., Brookes, H. N., Moller, A., et al. (2014). Increasing microtubule acetylation rescues axonal transport and locomotor deficits caused by LRRK2 Roc-COR domain mutations. *Nature Communications, 5*. https://doi.org/10.1038/ncomms6245.

Govindarajan, N., Agis, B. R. C., Walter, J., Sananbenesi, F., & Fischer, A. (2011). Sodium butyrate improves memory function in an Alzheimer's disease mouse model when administered at an advanced stage of disease progression. *Journal of Alzheimer's Disease, 26*, 187–197. https://doi.org/10.3233/jad-2011-110080.

Graff, J., & Tsai, L. H. (2013a). Histone acetylation: Molecular mnemonics on the chromatin. *Nature Reviews. Neuroscience, 14*, 97–111. https://doi.org/10.1038/nrn3427.

Graff, J., & Tsai, L. H. (2013b). The potential of HDAC inhibitors as cognitive enhancers. *Annual Review of Pharmacology and Toxicology, 53*, 311–330. https://doi.org/10.1146/annurev-pharmtox-011112-140216.

Gräff, J., Rei, D., Guan, J. S., et al. (2012). An epigenetic blockade of cognitive functions in the neurodegenerating brain. *Nature, 483*, 222–226. https://doi.org/10.1038/nature10849.

Grozinger, C. M., & Schreiber, S. L. (2000). Regulation of histone deacetylase 4 and 5 and transcriptional activity by 14-3-3-dependent cellular localization. *Proceedings of the National Academy of Sciences USA, 97*, 7835–7840. https://doi.org/10.1073/pnas.140199597.

Guan, J. S., Haggarty, S. J., Giacometti, E., et al. (2009). HDAC2 negatively regulates memory formation and synaptic plasticity. *Nature, 459*, 55–60. https://doi.org/10.3410/f.1160338.620669.

Guo, W., Naujock, M., Fumagalli, L., et al. (2017). HDAC6 inhibition reverses axonal transport defects in motor neurons derived from FUS-ALS patients. *Nature Communications, 8*. https://doi.org/10.1038/s41467-017-00911-y.

Gurard, L. Z. A., Kilian, K. A., Kim, J. B., & Mrksich, M. (2010). Peptide arrays identify isoform-selective substrates for profiling endogenous lysine deacetylase activity. *ACS Chemical Biology, 5*, 863–873. https://doi.org/10.1021/cb100088g.

Haggarty, S. J., & Tsai, L. H. (2011). Probing the role of HDACs and mechanisms of chromatin-mediated neuroplasticity. *Neurobiology of Learning and Memory, 96*(1), 41–52. https://doi.org/10.1016/j.nlm.2011.04.009.

Hanson, J. E., La, H., Plise, E., et al. (2013). SAHA enhances synaptic function and plasticity in vitro but has limited brain availability in vivo and does not impact cognition. *PLoS One, 8*. https://doi.org/10.1371/journal.pone.0069964.

Hardingham, G. E., Arnold, F. J., & Bading, H. (2001). Nuclear calcium signaling controls CREB-mediated gene expression triggered by synaptic activity. *Nature Neuroscience, 4*, 261–267. https://doi.org/10.1038/85109.

Hilton, I. B., & Vockley, C. M. (2015). Epigenome editing by a CRISPR-Cas9-based acetyltransferase activates genes from promoters and enhancers. *Nature Biotechnology, 33*, 510–517. https://doi.org/10.1038/nbt.3199.

Holtmaat, A., Bonhoeffer, T., Chow, D. K., et al. (2009). Long-term, high-resolution imaging in the mouse neocortex through a chronic cranial window. *Nature Protocols, 4*, 1128–1144. https://doi.org/10.1038/nprot.2009.89.

Hubbert, C., Guardiola, A., Shao, R., et al. (2002). HDAC6 is a microtubule-associated deacetylase. *Nature, 417*, 455–458. https://doi.org/10.1038/417455a.

Jia, H., Pallos, J., Jacques, V., et al. (2012). Histone deacetylase (HDAC) inhibitors targeting HDAC3 and HDAC1 ameliorate polyglutamine-elicited phenotypes in model systems of Huntington's disease. *Neurobiology of Disease, 46*, 351–361. https://doi.org/10.1016/j.nbd.2012.01.016.

Jiang, M., Wang, J., Fu, J., et al. (2011). Neuroprotective role of Sirt1 in mammalian models of Huntington's disease through activation of multiple Sirt1 targets. *Nature Medicine, 18*, 153–158. https://doi.org/10.3410/f.13498956.14878056.

Kalderon, D., Roberts, B. L., Richardson, W. D., & Smith, A. E. (1984). A short amino acid sequence able to specify nuclear location. *Cell, 39*, 499–509. https://doi.org/10.1016/0092-8674(84)90457-4.

Kawaguchi, Y., Kovacs, J. J., McLaurin, A., & Lto, A. (2003). The deacetylase HDAC6 regulates aggresome formation and cell viability in response to misfolded protein stress. *Cell, 115*, 727–738. https://doi.org/10.1016/s0092-8674(03)00939-5.

Kazantsev, A. G., & Thompson, L. M. (2008). Therapeutic application of histone deacetylase inhibitors for central nervous system disorders. *Nature Reviews. Drug Discovery, 7*, 854–868. https://doi.org/10.1038/nrd2681.

Kennedy, M. J., Hughes, R. M., Peteya, L. A., et al. (2010). Rapid blue-light-mediated induction of protein interactions in living cells. *Nature Methods, 7*, 973–975. https://doi.org/10.1038/nmeth.1524.

Kilgore, M., Miller, C. A., Fass, D. M., et al. (2010). Inhibitors of class 1 histone deacetylases reverse contextual memory deficits in a mouse model of Alzheimer's disease. *Neuropsychopharmacology: Official Publication of the American College of Neuropsychopharmacology, 35*, 870–880. https://doi.org/10.1038/npp.2009.197.

Kim, J. Y., Shen, S., Dietz, K., et al. (2010). HDAC1 nuclear export induced by pathological conditions is essential for the onset of axonal damage. *Nature Neuroscience, 13*, 180–189. https://doi.org/10.3410/f.1593956.1085054.

King, J., Patel, M., & Chandrasekaran, S. (2021). Metabolism, HDACs, and HDAC inhibitors: A systems biology perspective. *Metabolites, 11*(11), https://doi.org/10.3390/metabo11110792.

Komor, A. C., Kim, Y. B., Packer, M. S., & Liu, D. R. (2016). Programmable editing of a target base in genomic DNA without double-stranded DNA cleavage. *Nature, 533*, 420–424. https://doi.org/10.1038/nature17946.

Kouzarides, T. (2007). Chromatin modifications and their function. *Cell, 128*, 693–705. https://doi.org/10.1016/j.cell.2007.02.005.

Kwon, D. Y., Zhao, Y. T., Lamonica, J. M., & Zhou, Z. (2017). Locus-specific histone deacetylation using a synthetic CRISPR-Cas9-based HDAC. *Nature Communications, 8*. https://doi.org/10.1038/ncomms15315.

Lee, J. Y., Koga, H., Kawaguchi, Y., et al. (2010). HDAC6 controls autophagosome maturation essential for ubiquitin-selective quality-control autophagy. *EMBO Journal, 29*, 969–980. https://doi.org/10.1038/emboj.2009.405.

Levenson, J. M., O'Riordan, K. J., Brown, K. D., Trinh, M. A., Molfese, D. L., & Sweatt, J. D. (2004). Regulation of histone acetylation during memory formation in the hippocampus. *Journal of Biological Chemistry, 279*, 40545–40559. https://doi.org/10.1074/jbc.m402229200.

Li, J., Lin, Q., Wang, W., Wade, P., & Wong, J. (2002). Specific targeting and constitutive association of histone deacetylase complexes during transcriptional repression. *Genes and Development, 16*(6), 687–692. https://doi.org/10.1101/gad.962502.

Luo, Y., He, Y., & Zhao, Q. (2016). The neuroprotective effects of minocycline against Parkinson's disease through inhibition of HDAC4. *Neuroscience Letters, 635*, 12–18. https://doi.org/10.1097/nrl.0000000000000160.

McKinsey, T. A., Zhang, C. L., Lu, J., & Olson, E. N. (2000). Signal-dependent nuclear export of a histone deacetylase regulates muscle differentiation. *Nature, 408*, 106–111. https://doi.org/10.1038/35040593.

McQuin, C., Goodman, A., Chernyshev, V., et al. (2018). CellProfiler 3.0: Next-generation image processing for biology. *PLoS Biology, 16*. https://doi.org/10.1371/journal.pbio.2005970.

McQuown, S. C., Barrett, R. M., Matheos, D. P., et al. (2011). HDAC3 is a critical negative regulator of long-term memory formation. *Journal of Neuroscience, 31*, 764–774. https://doi.org/10.1523/jneurosci.5052-10.2011.

Mielcarek, M., Landles, C., Weiss, A., et al. (2013). HDAC4 reduction: A novel therapeutic strategy to target cytoplasmic huntingtin and ameliorate neurodegeneration. *PLoS Biology, 11*. https://doi.org/10.1371/journal.pbio.1001717.

Milazzo, G., Mercatelli, D., Di Muzio, G., Triboli, L., De Rosa, P., Perini, G., & Giorgi, F. M. (2020). Histone deacetylases (HDACs): Evolution, specificity, role in transcriptional complexes, and pharmacological actionability. *Genes (Basel), 11*(5), https://doi.org/10.3390/genes11050556.

Min, S. W., Cho, S. H., Zhou, Y., et al. (2010). Acetylation of tau inhibits its degradation and contributes to tauopathy. *Neuron, 67*, 953–966. https://doi.org/10.1016/j.neuron.2010.11.013.

Miska, E. A., Karlsson, C., Langley, E., & Pines, J. (1999). HDAC4 deacetylase associates with and represses the MEF2 transcription factor. *EMBO Journal, 18*, 5099–5107. https://doi.org/10.1093/emboj/18.18.5099.

Montgomery, R. L., Hsieh, J., Barbosa, A. C., & Olson, E. N. (2009). Histone deacetylases 1 and 2 control the progression of neural precursors to neurons during brain development. *Proceedings of the National Academy of Sciences USA, 106*, 7876–7881. https://doi.org/10.1073/pnas.0902750106.

Monti, B., Polazzi, E., & Contestabile, A. (2009). Biochemical, molecular and epigenetic mechanisms of valproic acid neuroprotection. *Current Molecular Pharmacology, 2*, 95–109. https://doi.org/10.2174/1874467210902010095.

Narayan, P. J., Lill, C., Faull, R., & Dragunow, M. (2015). Increased acetyl and total histone levels in post-mortem Alzheimer's disease brain. *Neurobiology of Disease, 74*, 281–294. https://doi.org/10.1016/j.nbd.2014.11.023.

Naro, Y., Darrah, K., & Deiters, A. (2020). Optical control of small molecule-induced protein degradation. *Journal of the American Chemical Society, 142*, 2193–2197. https://doi.org/10.26434/chemrxiv.8216714.v1.

Nihongaki, Y., Kawano, F., Nakajima, T., & Sato, M. (2015). Photoactivatable CRISPR-Cas9 for optogenetic genome editing. *Nature Biotechnology, 33*, 755–760. https://doi.org/10.1038/nbt.3245.

Nishiyama, J., Mikuni, T., & Yasuda, R. (2017). Virus-mediated genome editing via homology-directed repair in mitotic and postmitotic cells in mammalian brain. *Neuron, 96*, 755–768. https://doi.org/10.1016/j.neuron.2017.10.004.

Nott, A., Cheng, J., Gao, F., et al. (2016). Histone deacetylase 3 associates with MeCP2 to regulate FOXO and social behavior. *Nature Neuroscience, 19*, 1497–1505. https://doi.org/10.1038/nn.4347.

Outeiro, T. F., Kontopoulos, E., Altmann, S. M., et al. (2007). Sirtuin 2 inhibitors rescue alpha-synuclein-mediated toxicity in models of Parkinson's disease. *Science (New York, N. Y.), 317*, 516–519. https://doi.org/10.1126/science.1143780.

Packer, A. M., Russell, L. E., & Dalgleish, H. W. H. M. (2015). Simultaneous all-optical manipulation and recording of neural circuit activity with cellular resolution in vivo. *Nature Methods, 12*, 140–146. https://doi.org/10.1038/nmeth.3217.

Pandey, U. B., Nie, Z., Batlevi, Y., et al. (2007). HDAC6 rescues neurodegeneration and provides an essential link between autophagy and the UPS. *Nature, 447*, 859–863. https://doi.org/10.1038/nature05853.

Park, P. J. (2009). ChIP-seq: Advantages and challenges of a maturing technology. *Nature Reviews. Genetics, 10*, 669–680. https://doi.org/10.1038/nrg2641.

Paroni, G., Mizzau, M., Henderson, C., & Brancolini, C. (2004). Caspase-dependent regulation of histone deacetylase 4 nuclear-cytoplasmic shuttling promotes apoptosis. *Molecular Biology of the Cell, 15*, 2804–2818. https://doi.org/10.1091/mbc.e03-08-0624.

Platt, R. J., Chen, S., Zhou, Y., et al. (2014). CRISPR-Cas9 knockin mice for genome editing and cancer modeling. *Cell, 159*, 440–455. https://doi.org/10.1016/j.cell.2014.09.014.

Ran, F. A., Hsu, P. D., Wright, J., & Scott, D. A. (2013). Genome engineering using the CRISPR-Cas9 system. *Nature Protocols, 8*, 2281–2308. https://doi.org/10.1007/978-1-4939-1862-1_10.

Rane, P., Shields, J., Heffernan, M., et al. (2012). The histone deacetylase inhibitor, sodium butyrate, alleviates cognitive deficits in pre-motor stage PD. *Neuropharmacology, 62*, 2409–2412. https://doi.org/10.1016/j.neuropharm.2012.01.026.

Reddy, D. S., Boyd, H., Karnati, V., Kim, J., Ramakrishnan, S., & Wu, X. (2025). Epigenetic histone deacetylases activity assay in the brain and peripheral organ tissues. *Current Protocols, 5*(6), e70143. https://doi.org/10.1002/cpz1.70143.

Richardson, C. D., Ray, G. J., DeWitt, M. A., & Jcorn, E. (2016). Enhancing homology-directed genome editing by catalytically active and inactive CRISPR-Cas9 using asymmetric donor DNA. *Nature Biotechnology, 34*, 339–344. https://doi.org/10.1038/nbt.3481.

Ricobaraza, A., Cuadrado, T., Frechilla, M. P.-M. A., & García, O. (2009). Phenylbutyrate ameliorates cognitive deficit and reduces tau pathology in an Alzheimer's disease mouse model. *Neuropsychopharmacology: Official Publication of the American College of Neuropsychopharmacology, 34*, 1721–1732. https://doi.org/10.1038/npp.2008.229.

Roth, B. L. (2016). DREADDs for neuroscientists. *Neuron, 89*(4), 683–694. https://doi.org/10.1016/j.neuron.2016.01.040.

Rouaux, C., Panteleeva, I., René, F., et al. (2007). Sodium valproate exerts neuroprotective effects in vivo through CREB-binding protein-dependent mechanisms but does not improve survival in an amyotrophic lateral sclerosis mouse model. *Journal of Neuroscience, 27*, 5535–5545. https://doi.org/10.1523/jneurosci.1139-07.2007.

Saha, R. N., & Pahan, K. (2006). HATs and HDACs in neurodegeneration: A tale of disconcerted acetylation homeostasis. *Cell Death and Differentiation, 13*, 539–550. https://doi.org/10.1038/sj.cdd.4401769.

Sando, R., Gounko, N., Pieraut, S., & Yates, J. (2012). HDAC4 governs a transcriptional program essential for synaptic plasticity and memory. *Cell, 151*, 821–834. https://doi.org/10.1016/j.cell.2012.09.037.

Saunders, A., Macosko, E. Z., Wysoker, A., et al. (2018). Molecular diversity and specializations among the cells of the adult mouse brain. *Cell, 174*, 1015–1030. https://doi.org/10.1016/j.cell.2018.07.028.

Schindelin, J., Arganda, C. I., Frise, E., et al. (2012). Fiji: An open-source platform for biological-image analysis. *Nature Methods, 9*, 676–682. https://doi.org/10.1038/nmeth.2019.

Schübeler, D. (2015). Function and information content of DNA methylation. *Nature, 517*, 321–326. https://doi.org/10.1038/nature14192.

Shao, J., Wang, M., Yu, G., Zhu, S., Yu, Y., Heng, B. C., & Ye, H. (2018). Synthetic far-red light-mediated CRISPR-dCas9 device for inducing functional neuronal differentiation. *Proc Natl Acad Sci U S A, 115*(29), E6722–E6730. https://doi.org/10.1073/pnas.1802448115.

Sharma, S., & Taliyan, R. (2015). Targeting histone deacetylases: A novel approach in Parkinson's disease. *Parkinson's disease, 2015*. https://doi.org/10.1155/2015/303294.

Sugo, N., Oshiro, H., Takemura, M., et al. (2010). Nucleocytoplasmic translocation of HDAC9 regulates gene expression and dendritic growth in developing cortical neurons. *European Journal of Neuroscience, 31*, 1521–1532. https://doi.org/10.1111/j.1460-9568.2010.07218.x.

Suzuki, K., Bose, P., Leong, Q. R. Y., & Riabowol, K. (2010). REAP: A two minute cell fractionation method. *BMC Research Notes*, 3. https://doi.org/10.1186/1756-0500-3-294.

Taes, I., Timmers, M., Hersmus, N., et al. (2013). Hdac6 deletion delays disease progression in the SOD1G93A mouse model of ALS. *Human Molecular Genetics, 22*, 1783–1790. https://doi.org/10.1093/hmg/ddt028.

Taslimi, A., Vrana, J. D., Chen, D., et al. (2014). An optimized optogenetic clustering tool for probing protein interaction and function. *Nature Communications, 5*. https://doi.org/10.1038/ncomms5925.

Truong, D. J. J., et al. (2015). Development of an intein-mediated split-Cas9 system for gene therapy. *Nucleic Acids Research, 43*, 6450–6458. https://doi.org/10.1093/nar/gkv601.

Uhlen, M., Fagerberg, L., Hallström, B. M., et al. (2015). Proteomics. Tissue-based map of the human proteome. *Science*, 347. https://doi.org/10.3410/f.725324774.793520862.

Vecsey, C. G., Hawk, J. D., Lattal, K. M., et al. (2007). Histone deacetylase inhibitors enhance memory and synaptic plasticity via CREB:CBP-dependent transcriptional activation. *Journal of Neuroscience, 27*, 6128–6140. https://doi.org/10.1523/jneurosci.0296-07.2007.

Wang, C., Schroeder, F. A., & Hooker, J. M. (2014). Visualizing epigenetics: Current advances and advantages in HDAC PET imaging techniques. *Neuroscience, 264*, 186–197. https://doi.org/10.1016/j.neuroscience.2013.09.018.

Watson, P. J., Fairall, L., Santos, G. M., & Schwabe, J. W. (2012). Structure of HDAC3 bound to co-repressor and inositol tetraphosphate. *Nature, 481*, 335–340. https://doi.org/10.1038/nature10728.

Weinert, B. T., Schölz, C., Wagner, S. A., et al. (2013). Lysine succinylation is a frequently occurring modification in prokaryotes and eukaryotes and extensively overlaps with acetylation. *Cell Reports, 4*, 842–851. https://doi.org/10.1016/j.celrep.2013.07.024.

Wen, Y. D., Perissi, V., Staszewski, L. M., et al. (2000). The histone deacetylase-3 complex contains nuclear receptor corepressors. *Proceedings of the National Academy of Sciences USA, 97*, 7202–7207. https://doi.org/10.1073/pnas.97.13.7202.

Wurnig, M. C., R., U. H., Grünwald, A., et al. (2024). Light-activatable photochemi cally targeting chimeras (PHOTACs) enable the optical control of targeted protein degradation of HDAC6. *ChemRxiv*. https://doi.org/10.26434/chemrxiv-2024-cs18b.

Xu, K., Dai, X. L., Huang, H. C., & Jiang, Z. F. (2011). Targeting HDACs: A promising therapy for Alzheimer's disease. *Oxidative Medicine and Cellular Longevity, 2011*. https://doi.org/10.1155/2011/143269.

Zeisel, A., Hochgerner, H., Lönnerberg, P., et al. (2018). Molecular architecture of the mouse nervous system. *Cell, 174*, 999–1014. https://doi.org/10.1101/294918.

Zetsche, B., Volz, S. E., & Zhang, F. (2015). A split-Cas9 architecture for inducible genome editing and transcription modulation. *Nature Biotechnology, 33*, 139–142. https://doi.org/10.1038/nbt.3149.

Zhang, L., Liu, C., Wu, J., et al. (2014). Tubastatin A/ACY-1215 improves cognition in Alzheimer's disease transgenic mice. *Journal of Alzheimer's Disease, 41*, 1193–1205. https://doi.org/10.3233/jad-140066.

Zhao, X., Ito, A., Kane, C. D., et al. (2001). The modular nature of histone deacetylase HDAC4 confers phosphorylation-dependent intracellular trafficking. *Journal of Biological Chemistry, 276*, 35042–35048. https://doi.org/10.1074/jbc.m105086200.

Ziv, Y., Burns, L. D., Cocker, E. D., et al. (2013). Long-term dynamics of CA1 hippocampal place codes. *Nature Neuroscience, 16*, 264–266. https://doi.org/10.1038/nn.3329.

CHAPTER FOURTEEN

ChIPmentation-based mapping of histone modifications in aging: Experimental design and analytical considerations

Amalanandan Johna[a], Sreedarsanam Sreya[a], Sengupta Tiyas[a,1], Pavithra Elumalai[a,1], Avik Sengupta[b], Rahul Kumar[b], and Mayilaadumveettil Nishana[a,*]

[a]School of Biology, Indian Institute of Science Education and Research Thiruvananthapuram, Thiruvananthapuram, India

[b]Department of Biotechnology, Indian Institute of Technology Hyderabad, Kandi, Telangana, India

*Corresponding author. e-mail address: nishana@iisertvm.ac.in

Contents

1. Introduction 368
2. Equipments 372
3. Consumables 373
4. Procedure 374
   4.1 Crosslinking of cells 374
   4.2 Cell lysis and chromatin preparation 375
   4.3 Chromatin sonication 375
   4.4 Immunoprecipitation 376
   4.5 Washing of immunocomplexes 376
   4.6 ChIPmentation 376
   4.7 Elution, reverse crosslinking, and protein digestion 376
   4.8 DNA purification 377
   4.9 Library amplification 377
   4.10 Library size selection 378
   4.11 Quality control and quantification 378
   4.12 Library pooling and sequencing 378
   4.13 Computational analysis 378
5. Conclusion and future perspectives 379
Acknowledgments 380
Ethical declaration 380
Data availability 380
References 380

[1] These authors contributed equally

Methods in Enzymology, Volume 733
ISSN 0076-6879, https://doi.org/10.1016/bs.mie.2026.05.032

Abstract

Aging is accompanied by progressive alterations in chromatin organization that contribute to transcriptional dysregulation, genomic instability, and loss of cellular identity. Among the epigenetic mechanisms implicated, histone modifications play a central role in regulating chromatin accessibility and gene expression during aging. Changes in key histone marks, including H3K9me3, H3K27me3, H4K20me3, and histone acetylation signatures, are associated with heterochromatin loss, increased transcriptional noise, and altered cellular function across diverse tissues and organisms. These modifications are dynamically regulated by histone-modifying enzymes and are responsive to metabolic and stress signaling pathways, linking environmental cues to chromatin state and longevity.

Genome-wide profiling of histone modifications is essential to understand these age-associated chromatin changes. ChIPmentation, a streamlined approach that integrates chromatin immunoprecipitation with transposase-mediated tagmentation, enables rapid and sensitive mapping of histone marks with reduced input requirements and simplified library preparation. This method is particularly advantageous for aging studies, where sample availability is often limiting.

In this chapter, we outline the principles and experimental workflow of ChIPmentation, including crosslinking, chromatin fragmentation, immunoprecipitation, tagmentation, library preparation, sequencing, and data analysis. Additionally, we discuss computational strategies for processing and analyzing ChIPmentation data, including alignment, normalization, peak calling, and reproducibility assessment.

Together, this chapter provides a comprehensive methodological framework for studying histone modification landscapes in aging systems, enabling robust investigation of epigenetic mechanisms underlying age-associated phenotypes.

1. Introduction

Aging is characterized by the progressive decline of biological function at the cellular, tissue, and organismal level, leading to an increased vulnerability to environmental stress and disease after the peak of reproductive and physiological capacity is achieved. Since aging directly impacts both lifespan and health span, understanding its molecular basis is essential for promoting healthy aging (Wang et al., 2018). Among the central mechanisms implicated in this process are age-associated alterations in chromatin organization and histone modifications.

Cellular senescence is accompanied by a progressive reduction in heterochromatin, forming the basis of the 'heterochromatin loss theory of aging' (Villeponteau, 1997). This model proposes that aging involves global heterochromatin loss and redistribution, resulting in genomic instability and dysregulation of gene expression (Tsurumi & Li, 2012). Genome-wide profiling further demonstrates extensive chromatin remodeling and transcriptional dysregulation associated with age, leading to increased epigenetic variability, stochastic gene expression, and loss of cellular identity (Cheung et al., 2018; Martinez-Jimenez et al., 2017). These findings establish epigenetic dysregulation as a hallmark of aging.

At the molecular level, chromatin consists of DNA wrapped around histone octamers composed of H2A, H2B, H3, and H4 (Luger et al., 1997). The N-terminal histone tails undergo reversible post-translational modifications, including acetylation, methylation, and phosphorylation, which dynamically modulate chromatin compaction and transcriptional accessibility (Cheung et al., 2000; Hebbes et al., 1988; Strahl & Allis, 2000). Table 1 depicts an overview of these histone modifications.

Histone methylation plays a critical regulatory role in aging across various species (Huidobro et al., 2013). Depending on the histone residue and the methylation state, it functions either as transcriptional activator or repressor and is dynamically regulated my Histone Methyl Transferases (HMTs) and Histone Demethylases (HDMs) (Bannister & Kouzarides, 2011; Nottke et al., 2009). Perturbations of H3K4me3 regulatory pathways significantly extend lifespan and stress resistance in *Caenorhabditis elegans*, supporting a causal role of histone methylation in longevity control (Greer et al., 2010; Li et al., 2010). Likewise, perturbation of metabolic and stress signalling pathways influences survival through remodelling of histone modification landscapes (McColl et al., 2008). Notably, age-associated depletion of heterochromatic marks such as H3K9me3, H4K20me3, and H3K27me3 compromises genome integrity and increases transcriptional noise, as observed in human stem cells and Hutchinson-Gilford progeria syndrome (HGPS) (McCord et al., 2013; Scaffidi & Misteli, 2008; Shumaker et al., 2006; Zhang et al., 2015).

Histone acetylation similarly modulates chromatin accessibility by weakening DNA-histone interactions and promoting transcription. Aging-associated changes in acetylation marks, including H4K12ac, H4K16ac, and H3K56ac are linked to lifespan regulation and cognitive decline (Dang et al., 2009; Peleg et al., 2016). In the aging brain, coordinated alterations in active and repressive methylation and acetylation marks reorganize chromatin domains and alter neuronal gene expression associated with cognitive decline and neurodegeneration (Klein et al., 2019; Persico et al., 2022). Similar epigenetic remodelling is observed in other tissues including muscle satellite cells and adipose

Table 1 Histone modifications and their functional role on genome regulation.

Histone modification	Functional role	References
Acetylation (H3K9ac, H4K12ac)	Gene transcription	Wang et al. (2008)
Acetylation (H4K16ac)	Decondenses chromatin structure, DNA repair	Rea et al. (2007)
Methylation (H3K27me, H3K9me, H4K20me, H3K79me, H2BK5me)	Gene activation	Barski et al. (2007)
Methylation (H3K27me3, H3K9me3, H3K79me3)	Gene repression	Barski et al. (2007), Sidler et al. (2017)
Phosphorylation (H3S10ph)	Stimulates the elongation of transcription	Zippo et al. (2009)
Phosphorylation (γH2AXS-139ph)	Response for double strand break repair	Fagagna et al. (2003), Kinner et al. (2008)
Ubiquitination (H2BK123ub)	Gene activation	Zhou et al. (2008)
Ubiquitination (H2AK119ub)	Polycomb silencing	Zhou et al. (2008)
Sumoylation (H4K12su)	Indirectly inhibits transcription of genes	Leonen et al. (2021)
Citrullination (H3R26cit)	Helps chromatin de-condensation	Zhang et al. (2012)
PARylation (ADP-ribosylation (ADPr))	DNA damage response	Leidecker et al. (2016)
Crotonylation (H3K9Cr)	Gene transcription repression, DNA damage repair	Xie et al. (2024)
Propionylation	Recruits of proteins and enzymes involved in gene regulation	Chen et al. (2007)
Succinnylation	Affects DNA and histone interaction	Xie et al. (2012)

Butyrylation	Recruits of regulatory proteins and enzymes	Chen et al. (2007)
Malonylation	Chromatin structural changes, cell lethality	Xie et al. (2012)
Lactylation	Chromatin confirmation is altered	Peng and Du (2025)
Formylation	Deregulation of chromatin function	Wiśniewski et al. (2008)
Hydroxylation	Enhances the demethylation of lysine	Liu et al. (2022)
Biotinylation	Regulates gene expression, DNA damage response	Chew et al. (2006, 2008)

derived stem cells, indicating that chromatin alterations are systemic features of aging (Afarideh et al., 2021; McCord et al., 2013). Table 2 summarizes the histone modifications and their dynamics during aging. Importantly, histone modifications are reversible, rendering the epigenome as a promising target for interventions aimed at extending health span and mitigating age-related diseases.

To study genome-wide profiles of histone modifications, several profiling techniques have been developed, among which ChIPmentation has emerged as a rapid and efficient approach. ChIPmentation integrates chromatin immunoprecipitation (ChIP) with simultaneous DNA fragmentation and sequencing by means of adapter insertion using Tn5 transposase-mediated tagmentation. While maintaining high resolution mapping of histone modifications, this method significantly reduces library preparation time, sample handling, and input material requirements compared with conventional ChIPseq approaches (Schmidl et al., 2015). Here, we summarize the principles and key steps of ChIPmentation, and discuss its applications in studying histone medication landscapes. The method involves a series of coordinated steps designed to preserve chromatin integrity, selectively enrich DNA associated with specific histone marks, and prepare sequencing-ready libraries, the major steps of which are summarised below.

Table 2 Age associated histone modifications and their functional consequences.

Histone modifications	Age associated change	Functional consequences	References
H4K16ac	Increases (yeast)	Gene expression is altered	Dang et al. (2009)
H3K56ac	Decreases (yeast)	Genome instability	Feser et al. (2010)
H3K9ac	Decreases (Rat)	Decrease in transcription of genes	Kawakami et al. (2009)
H3K27ac	Global loss in genome (mouse neurons)	Gene expression is altered	Signal et al. (2024)
H3K9me3	Decreases	Loss of heterochromatin regions, genome instability	Scaffidi and Misteli (2006)
H3K27me3	Decreases	Loss of Polycomb repression	Bracken et al. (2007)
H4K20me3	Decreases	Change in heterochromatin regions	O'Sullivan et al. (2010)
H3S10ph	Increases (rat)	Increased transcription of associated genes	Kawakami et al. (2009)

2. Equipments

1. Next Generation Sequencer
2. TapeStation system (Agilent)
3. Real-time PCR (qPCR) machine
4. Thermomixer (temperature-controlled with shaking)
5. Sonicator (e.g., Diagenode, Bioruptor)
6. PCR thermal cycler
7. Qubit fluorometer
8. Spectrophotometer (e.g., NanoDrop)
9. Refrigerated centrifuge (4 °C)
10. Benchtop centrifuge
11. Tube rotator for gentle mixing at room temperature

12. Vortex mixer
13. Biological safety cabinet
14. Liquid nitrogen storage system
15. −20 °C and −80 °C freezers
16. Magnetic rack for microcentrifuge tubes
17. Ice buckets
18. Pipettes and filter tips
19. High-performance computing system or workstation
20. Data storage system

3. Consumables

1. Cell samples
2. Phosphate-buffered saline (PBS), 1X
3. Trypsin
4. BSA (molecular biology grade)
5. Formaldehyde
6. EGS ethylene glycol bis(succinimidyl succinate)
7. Glycine
8. DMSO (Dimethyl Sulfoxide)
9. ChIP Lysis Buffer 3 (LB3) + protease inhibitors: The buffer contains Tris-HCl (10 mM, pH 8.0), NaCl (100 mM), EDTA (1 pH 8.0), EGTA (0.5 Mm pH 8.0), sodium deoxycholate (0.1 %), and N-lauroylsarcosine (0.5 %). Protease inhibitors should be added fresh immediately before use to a final concentration of 1X.
10. TF-WBI (Low-salt wash buffer): Contains Tris-HCl (20 mM, pH 7.5), NaCl (150 mM), EDTA (2 mM), SDS (0.1 %), and Triton X-100 (1 %).
11. TF-WBIII (High-stringency wash buffer): Contains Tris-HCl (10 mM, pH 8.0), LiCl (250 mM), and EDTA (1 mM, pH 8.0)
12. Triton X-100
13. Sodium deoxycholate
14. Dynabeads Protein A or G
15. Antibody of interest
16. Antibody conjugated beads
17. Optional: Spike-in chromatin (*Drosophila*) and spike-in antibody (H2Av)
18. Tn5 transposase

19. Tagmentation buffer: Containing Tris-HCl (10 mM, pH 8.0), $MgCl_2$ (5 mM), and 10 % (v/v) dimethylformamide
20. TET buffer: Contains Tris-HCl (10 mM, pH 8.0), EDTA (1 mM), and Tween-20 (0.2 %).
21. Elution buffer: Contains SDS (0.5 %), NaCl (300 mM), EDTA (5 mM), and Tris-HCl (10 mM, pH 8.0).
22. Tris-HCl buffer (10 mM, pH 8.0)
23. RNase A
24. Proteinase K
25. DNA purification kit (e.g., MinElute, Qiagen)
26. PCR reagents
27. AMPure XP beads
28. Ethanol (80 %)
29. Nuclease-free water
30. Qubit dsDNA HS Assay Kit
31. TapeStation High Sensitivity D1000 reagents and ScreenTape
32. Library quantification kit
33. 1.5 ml microcentrifuge tubes (DNA- and protein- low-binding tubes)
34. 15 ml or 50 ml conical tubes
35. Low-retention pipette tips
36. Universal Primer Ad1
37. Indexed primers (each sample must have a unique index)

4. Procedure

4.1 Crosslinking of cells

1. Cells should be collected, counted and pelleted by centrifugation under conditions appropriate for the specific cell type.
2. The cell pellet should be washed once with 1X PBS (at a density of approximately 1 ml per 1 million cells) at room temperature and centrifuged again.
3. For standard crosslinking, cells should be resuspended in PBS and formaldehyde should be added to a final concentration of 1 %, followed by incubation at room temperature for 10 min with gentle rotation.
4. Crosslinking should be quenched by adding glycine to a final concentration of 125 mM and incubating for 5 min at room temperature.
5. Cells should then be pelleted, washed twice with cold PBS containing 0.5 % BSA, and kept on ice.

6. The final pellet should be resuspended, transferred to low-binding tubes, centrifuged, snap-frozen in liquid nitrogen, and stored at −80 °C.

Note: In the classical one-step crosslinking, formaldehyde forms bridges between amino acids and DNA bases, thereby capturing DNA-protein interaction within close proximity (Hoffman et al., 2015; Kennedy-Darling & Smith, 2014; Solomon & Varshavsky, 1985) (Fig. 1). For double crosslinking, cells may first be incubated with EGS (final concentration ~1.5 mM) for 30 min at room temperature, followed by formaldehyde crosslinking as described above.

4.2 Cell lysis and chromatin preparation

1. Cell pellets should be resuspended in ChIP Lysis Buffer 3 (LB3) supplemented with fresh protease inhibitors.
2. Samples should be incubated at 4 °C for 10 min with gentle rotation to ensure efficient lysis.
3. A small aliquot may be reserved as a non-sonicated control if required.

4.3 Chromatin sonication

1. Chromatin should be sonicated using a Bioruptor or equivalent instrument with alternating cycles (e.g., 30 s ON/OFF) to obtain DNA fragments in the range of ~100–500 bp.
2. After partial sonication, Triton X-100 may be added to a final concentration of 1 %, and sonication continued if necessary.
3. Samples should be centrifuged at high speed at 4 °C to remove debris, and the supernatant containing fragmented chromatin should be transferred to fresh protein low binding tubes.

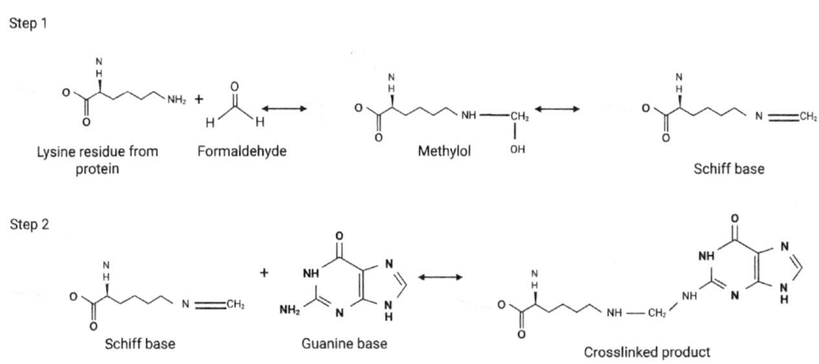

Fig. 1 Chemical reaction behind crosslinking.

4. A fraction of the sample may be reserved as input control if required.
5. Chromatin concentration should be estimated, and samples kept at 4 °C or stored at −80 °C until further use.
6. Optional spike-in chromatin (e.g., *Drosophila*) may be added at ~2 % of total chromatin for normalization purposes (Egan et al., 2016).

4.4 Immunoprecipitation

1. Magnetic beads (Protein A or G) should be washed and blocked with PBS containing 0.5 % BSA.
2. The antibody of interest, along with optional spike-in antibody, should be incubated with the beads for at least 1 h at room temperature with rotation.
3. Antibody-conjugated beads should then be added to the chromatin samples.
4. Samples should be incubated overnight at 4 °C with gentle rotation to allow immunoprecipitation.

4.5 Washing of immunocomplexes

1. Beads should be collected using a magnetic rack and washed sequentially with low-salt buffer (TF-WBI) and high-salt buffer (TF-WBIII) to remove non-specific interactions.
2. A final wash with Tris buffer should be performed to remove residual salts and detergents.

4.6 ChIPmentation

1. Beads should be resuspended in tagmentation reaction buffer containing $MgCl_2$ and dimethylformamide.
2. Tn5 transposase should be added, and the reaction incubated at 37 °C for approximately 10 min to simultaneously fragment DNA and insert sequencing adapters.
3. Following tagmentation, beads should be washed to remove residual enzyme and reaction components (Schmidl et al., 2015) (Fig. 2).

4.7 Elution, reverse crosslinking, and protein digestion

1. DNA should be eluted from beads using elution buffer containing SDS, NaCl, EDTA, and Tris-HCl.
2. Proteinase K should be added, and samples incubated at 55 °C followed by prolonged incubation at 65 °C (typically overnight) to reverse crosslinks and digest proteins.

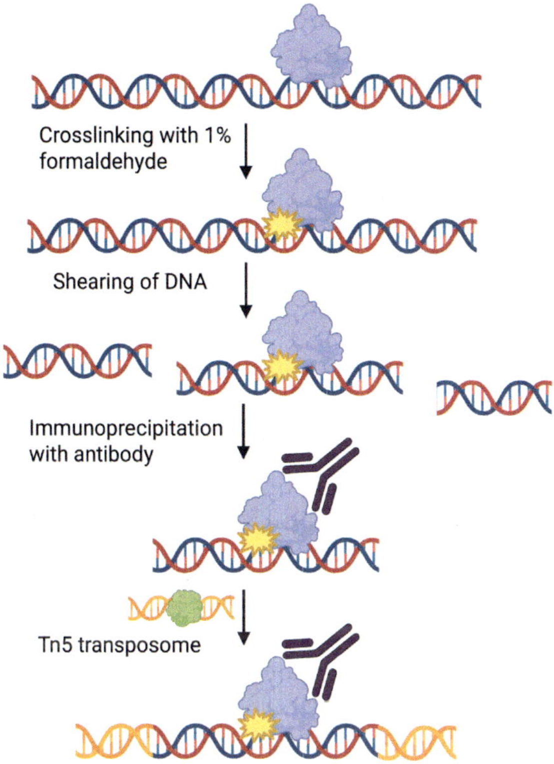

Fig. 2 Schematic representation of steps involving crosslinking, DNA shearing, immunoprecipitation and tagmentation.

4.8 DNA purification

1. The supernatant containing DNA should be separated from beads using a magnetic rack.
2. DNA should be purified using a column-based cleanup kit according to the manufacturer's instructions.
3. The purified DNA should be eluted in a small volume (e.g., ~20 μl) of elution buffer.

4.9 Library amplification

1. Purified DNA should be amplified using a high-fidelity PCR mix with indexed primers to generate sequencing-ready libraries.
2. PCR conditions should be optimized (typically ~12–14 cycles) to avoid over-amplification.

4.10 Library size selection

1. Amplified libraries should be size-selected using AMPure XP beads to retain fragments typically between ~100−700 bp.
2. The final libraries should be eluted in nuclease-free water and stored at −20 °C.

4.11 Quality control and quantification

1. Library concentration should be measured using a fluorometric assay (e.g., Qubit).
2. Fragment size distribution should be assessed using a microfluidics-based system (e.g., TapeStation), ensuring a peak typically between ~150–750 bp.
3. Library molarity should be accurately determined using qPCR-based quantification.

4.12 Library pooling and sequencing

1. Libraries should be normalized and pooled in equimolar amounts based on concentration and fragment size.
2. The pooled library should be quantified again, with particular emphasis on qPCR-based methods to ensure accurate molarity (typically ~2 nM).
3. Sequencing should be performed on a high-throughput platform such as the Illumina NovaSeq 6000, generating paired-end reads (e.g., 2 × 150 bp) with sufficient depth depending on experimental requirements.

4.13 Computational analysis

1. Raw sequencing reads should be adapter-trimmed using Skewer (Jiang et al., 2014).
2. Trimmed reads should be aligned to the reference genome (hg19 or hg38) using Bowtie2 with sensitive alignment parameters (Langmead et al., 2009, 2019; Langmead & Salzberg, 2012).
3. PCR duplicates should be removed using Picard (https://broadinstitute.github.io/picard/)
4. A Tn5-specific shift (+4 bp/−5 bp) should be applied to correct insertion positions (Schmidl et al., 2015).
5. Genome-wide coverage tracks should be generated using BEDTools and converted to BigWig format using UCSC bedGraph To BigWig (Quinlan & Hall, 2010).
6. Peaks should be identified using MACS2, with parameters adjusted for narrow or broad marks as required (Zhang et al., 2008).

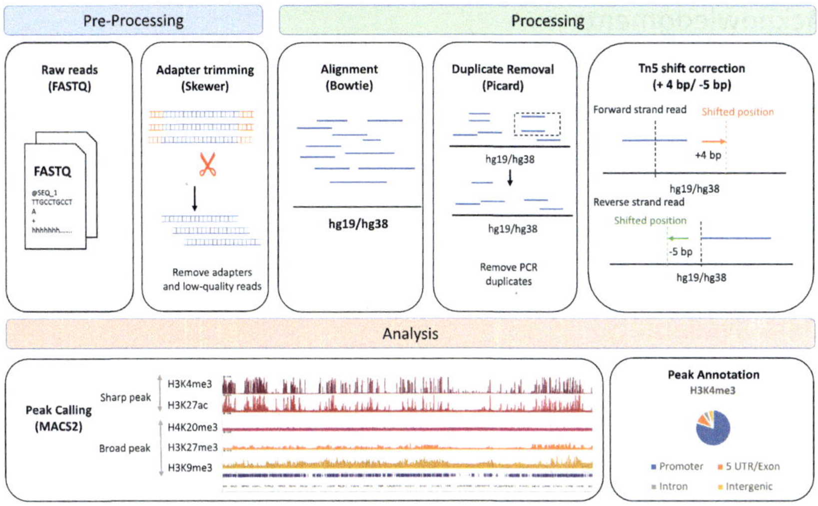

Fig. 3 Schematic representation of computational workflow for analysing ChIPmentation data.

7. Reproducibility should be assessed using peak overlap and correlation analyses, excluding problematic regions such as ENCODE blacklist (Amemiya et al., 2019).
8. Data quality should be evaluated using metrics such as FRiP (Fraction of Reads in Peaks) and inter-sample correlation (Fig. 3).

5. Conclusion and future perspectives

In summary, age-associated alterations in histone modifications represent a fundamental layer of epigenetic regulation underlying chromatin reorganization and functional decline during aging. Advances in genome-wide profiling approaches, particularly ChIPmentation, have enabled high-resolution mapping of these modifications with reduced input requirements and improved efficiency. The integration of optimized experimental workflows with robust computational analyses provides a powerful framework to dissect chromatin dynamics across diverse aging systems. As methodologies continue to evolve, combining ChIPmentation with complementary epigenomic and transcriptomic approaches will further enhance our understanding of the causal relationships between chromatin state and aging phenotypes. Ultimately, such insights may facilitate the development of targeted epigenetic interventions to promote healthy aging and mitigate age-associated diseases.

Acknowledgments

We acknowledge the financial support provided to NM by the Indian Council of Medical Research (ICMR) (IIRPSG-2024–01-02098) and IISER Thiruvananthapuram. JA, TS and SS are supported by the University Grants Commission-Junior Research Fellowships (UGC-JRF: 241610064377, 221610130748 and 241610141449, respectively). We also acknowledge Biorender for providing a platform for creating the figures.

Ethical declaration

The authors declare no competing interests.

Data availability

No datasets were generated during the current study.

References

Afarideh, M., Thaler, R., Khani, F., Tang, H., Jordan, K. L., Conley, S. M., ... Lerman, L. O. (2021). Global epigenetic alterations of mesenchymal stem cells in obesity: The role of vitamin C reprogramming. *Epigenetics: Official Journal of the DNA Methylation Society, 16*(7), 705–717. https://doi.org/10.1080/15592294.2020.1819663.

Amemiya, H. M., Kundaje, A., & Boyle, A. P. (2019). The ENCODE blacklist: Identification of problematic regions of the genome. *Scientific Reports, 9*(1), 9354. https://doi.org/10.1038/s41598-019-45839-z.

Bannister, A. J., & Kouzarides, T. (2011). Regulation of chromatin by histone modifications. *Cell Research, 21*(3), 381–395. https://doi.org/10.1038/cr.2011.22.

Barski, A., Cuddapah, S., Cui, K., Roh, T.-Y., Schones, D. E., Wang, Z., ... Zhao, K. (2007). High-resolution profiling of histone methylations in the human genome. *Cell, 129*(4), 823–837. https://doi.org/10.1016/j.cell.2007.05.009.

Bracken, A. P., Kleine-Kohlbrecher, D., Dietrich, N., Pasini, D., Gargiulo, G., Beekman, C., ... Helin, K. (2007). The Polycomb group proteins bind throughout the *INK4A-ARF* locus and are disassociated in senescent cells. *Genes & Development, 21*(5), 525–530. https://doi.org/10.1101/gad.415507.

Chen, Y., Sprung, R., Tang, Y., Ball, H., Sangras, B., Kim, S. C., ... Zhao, Y. (2007). Lysine propionylation and butyrylation are novel post-translational modifications in histones. *Molecular & Cellular Proteomics, 6*(5), 812–819. https://doi.org/10.1074/mcp.M700021-MCP200.

Cheung, P., Allis, C. D., & Sassone-Corsi, P. (2000). Signaling to chromatin through histone modifications. *Cell, 103*(2), 263–271. https://doi.org/10.1016/S0092-8674(00)00118-5.

Cheung, P., Vallania, F., Warsinske, H. C., Donato, M., Schaffert, S., Chang, S. E., ... Kuo, A. J. (2018). Single-cell chromatin modification profiling reveals increased epigenetic variations with aging. *Cell, 173*(6), 1385–1397.e14. https://doi.org/10.1016/j.cell.2018.03.079.

Chew, Y. C., Camporeale, G., Kothapalli, N., Sarath, G., & Zempleni, J. (2006). Lysine residues in N-terminal and C-terminal regions of human histone H2A are targets for biotinylation by biotinidase. *The Journal of Nutritional Biochemistry, 17*(4), 225–233. https://doi.org/10.1016/j.jnutbio.2005.05.003.

Chew, Y. C., West, J. T., Kratzer, S. J., Ilvarsonn, A. M., Eissenberg, J. C., Dave, B. J., ... Zempleni, J. (2008). Biotinylation of histones represses transposable elements in human and mouse cells and cell lines and in Drosophila melanogaster3. *The Journal of Nutrition, 138*(12), 2316–2322. https://doi.org/10.3945/jn.108.098673.

Dang, W., Steffen, K. K., Perry, R., Dorsey, J. A., Johnson, F. B., Shilatifard, A., ... Berger, S. L. (2009). Histone H4 lysine 16 acetylation regulates cellular lifespan. *Nature, 459*(7248), 802–807. https://doi.org/10.1038/nature08085.

Egan, B., Yuan, C.-C., Craske, M. L., Labhart, P., Guler, G. D., Arnott, D., ... Trojer, P. (2016). An alternative approach to ChIP-Seq normalization enables detection of genome-wide changes in histone H3 lysine 27 trimethylation upon EZH2 inhibition. *PLoS One, 11*(11), e0166438. https://doi.org/10.1371/journal.pone.0166438.

Fagagna, F. d'A. D., Reaper, P. M., Clay-Farrace, L., Fiegler, H., Carr, P., Von Zglinicki, T., ... Jackson, S. P. (2003). A DNA damage checkpoint response in telomere-initiated senescence. *Nature, 426*(6963), 194–198. https://doi.org/10.1038/nature02118.

Feser, J., Truong, D., Das, C., Carson, J. J., Kieft, J., Harkness, T., & Tyler, J. K. (2010). Elevated histone expression promotes life span extension. *Molecular Cell, 39*(5), 724–735. https://doi.org/10.1016/j.molcel.2010.08.015.

Greer, E. L., Maures, T. J., Hauswirth, A. G., Green, E. M., Leeman, D. S., Maro, G. S., ... Brunet, A. (2010). Members of the H3K4 trimethylation complex regulate lifespan in a germline-dependent manner in C. elegans. *Nature, 466*(7304), 383–387. https://doi.org/10.1038/nature09195.

Hebbes, T. R., Thorne, A. W., & Crane-Robinson, C. (1988). A direct link between core histone acetylation and transcriptionally active chromatin. *The EMBO Journal,* 7(5), 1395–1402. https://doi.org/10.1002/j.1460-2075.1988.tb02956.x.

Hoffman, E. A., Frey, B. L., Smith, L. M., & Auble, D. T. (2015). Formaldehyde cross-linking: A tool for the study of chromatin complexes. *The Journal of Biological Chemistry, 290*(44), 26404–26411. https://doi.org/10.1074/jbc.R115.651679.

Huidobro, C., Fernandez, A. F., & Fraga, M. F. (2013). Aging epigenetics: Causes and consequences. *Molecular Aspects of Medicine, 34*(4), 765–781. https://doi.org/10.1016/j.mam.2012.06.006.

Jiang, H., Lei, R., Ding, S.-W., & Zhu, S. (2014). Skewer: A fast and accurate adapter trimmer for next-generation sequencing paired-end reads. *BMC Bioinformatics, 15*(1), 182. https://doi.org/10.1186/1471-2105-15-182.

Kawakami, K., Nakamura, A., Ishigami, A., Goto, S., & Takahashi, R. (2009). Age-related difference of site-specific histone modifications in rat liver. *Biogerontology, 10*(4), 415–421. https://doi.org/10.1007/s10522-008-9176-0.

Kennedy-Darling, J., & Smith, L. M. (2014). Measuring the formaldehyde protein–DNA cross-link reversal rate. *Analytical Chemistry, 86*(12), 5678–5681. https://doi.org/10.1021/ac501354y.

Kinner, A., Wu, W., Staudt, C., & Iliakis, G. (2008). -H2AX in recognition and signaling of DNA double-strand breaks in the context of chromatin. *Nucleic Acids Research, 36*(17), 5678–5694. https://doi.org/10.1093/nar/gkn550.

Klein, H.-U., McCabe, C., Gjoneska, E., Sullivan, S. E., Kaskow, B. J., Tang, A., ... De Jager, P. L. (2019). Epigenome-wide study uncovers large-scale changes in histone acetylation driven by tau pathology in aging and Alzheimer's human brains. *Nature Neuroscience, 22*(1), 37–46. https://doi.org/10.1038/s41593-018-0291-1.

Langmead, B., & Salzberg, S. L. (2012). Fast gapped-read alignment with Bowtie 2. *Nature Methods, 9*(4), 357–359. https://doi.org/10.1038/nmeth.1923.

Langmead, B., Trapnell, C., Pop, M., & Salzberg, S. L. (2009). Ultrafast and memory-efficient alignment of short DNA sequences to the human genome. *Genome Biology, 10*(3), R25. https://doi.org/10.1186/gb-2009-10-3-r25.

Langmead, B., Wilks, C., Antonescu, V., & Charles, R. (2019). Scaling read aligners to hundreds of threads on general-purpose processors. *Bioinformatics (Oxford, England), 35*(3), 421–432. https://doi.org/10.1093/bioinformatics/bty648.

Leidecker, O., Bonfiglio, J. J., Colby, T., Zhang, Q., Atanassov, I., Zaja, R., ... Matic, I. (2016). Serine is a new target residue for endogenous ADP-ribosylation on histones. *Nature Chemical Biology, 12*(12), 998–1000. https://doi.org/10.1038/nchembio.2180.

Leonen, C. J. A., Shimada, M., Weller, C. E., Nakadai, T., Hsu, P. L., Tyson, E. L., ... Chatterjee, C. (2021). Sumoylation of the human histone H4 tail inhibits p300-mediated transcription by RNA polymerase II in cellular extracts. *eLife, 10*, e67952. https://doi.org/10.7554/eLife.67952.

Li, L., Greer, C., Eisenman, R. N., & Secombe, J. (2010). Essential functions of the histone demethylase lid. *PLoS Genetics, 6*(11), e1001221. https://doi.org/10.1371/journal.pgen.1001221.

Liu, X., Wang, J., Boyer, J. A., Gong, W., Zhao, S., Xie, L., ... Zhang, Q. (2022). Histone H3 proline 16 hydroxylation regulates mammalian gene expression. *Nature Genetics, 54*(11), 1721–1735. https://doi.org/10.1038/s41588-022-01212-x.

Luger, K., Mäder, A. W., Richmond, R. K., Sargent, D. F., & Richmond, T. J. (1997). Crystal structure of the nucleosome core particle at 2.8 Å resolution. *Nature, 389*(6648), 251–260. https://doi.org/10.1038/38444.

Martinez-Jimenez, C. P., Eling, N., Chen, H.-C., Vallejos, C. A., Kolodziejczyk, A. A., Connor, F., ... Odom, D. T. (2017). Aging increases cell-to-cell transcriptional variability upon immune stimulation. *Science (New York, N. Y.), 355*(6332), 1433–1436. https://doi.org/10.1126/science.aah4115.

McColl, G., Killilea, D. W., Hubbard, A. E., Vantipalli, M. C., Melov, S., & Lithgow, G. J. (2008). Pharmacogenetic analysis of lithium-induced delayed aging in Caenorhabditis elegans. *Journal of Biological Chemistry, 283*(1), 350–357. https://doi.org/10.1074/jbc.M705028200.

McCord, R. P., Nazario-Toole, A., Zhang, H., Chines, P. S., Zhan, Y., Erdos, M. R., ... Cao, K. (2013). Correlated alterations in genome organization, histone methylation, and DNA–lamin A/C interactions in Hutchinson-Gilford progeria syndrome. *Genome Research, 23*(2), 260–269. https://doi.org/10.1101/gr.138032.112.

Nottke, A., Colaiácovo, M. P., & Shi, Y. (2009). Developmental roles of the histone lysine demethylases. *Development (Cambridge, England), 136*(6), 879–889. https://doi.org/10.1242/dev.020966.

O'Sullivan, R. J., Kubicek, S., Schreiber, S. L., & Karlseder, J. (2010). Reduced histone biosynthesis and chromatin changes arising from a damage signal at telomeres. *Nature Structural & Molecular Biology, 17*(10), 1218–1225. https://doi.org/10.1038/nsmb.1897.

Peleg, S., Feller, C., Forne, I., Schiller, E., Sévin, D. C., Schauer, T., ... Ladurner, A. G. (2016). Life span extension by targeting a link between metabolism and histone acetylation in *Drosophila*. *EMBO Reports, 17*(3), 455–469. https://doi.org/10.15252/embr.201541132.

Peng, X., & Du, J. (2025). Histone and non-histone lactylation: Molecular mechanisms, biological functions, diseases, and therapeutic targets. *Molecular Biomedicine, 6*(1), 38. https://doi.org/10.1186/s43556-025-00275-6.

Persico, G., Casciaro, F., Amatori, S., Rusin, M., Cantatore, F., Perna, A., ... Giorgio, M. (2022). Histone H3 lysine 4 and 27 trimethylation landscape of human Alzheimer's disease. *Cells, 11*(4), 734. https://doi.org/10.3390/cells11040734.

Quinlan, A. R., & Hall, I. M. (2010). BEDTools: A flexible suite of utilities for comparing genomic features. *Bioinformatics (Oxford, England), 26*(6), 841–842. https://doi.org/10.1093/bioinformatics/btq033.

Rea, S., Xouri, G., & Akhtar, A. (2007). Males absent on the first (MOF): From flies to humans. *Oncogene, 26*(37), 5385–5394. https://doi.org/10.1038/sj.onc.1210607.

Scaffidi, P., & Misteli, T. (2006). Lamin A-dependent nuclear defects in human aging. *Science (New York, N. Y.), 312*(5776), 1059–1063. https://doi.org/10.1126/science.1127168.

Scaffidi, P., & Misteli, T. (2008). Lamin A-dependent misregulation of adult stem cells associated with accelerated ageing. *Nature Cell Biology, 10*(4), 452–459. https://doi.org/10.1038/ncb1708.

Schmidl, C., Rendeiro, A. F., Sheffield, N. C., & Bock, C. (2015). ChIPmentation: Fast, robust, low-input ChIP-seq for histones and transcription factors. *Nature Methods, 12*(10), 963–965. https://doi.org/10.1038/nmeth.3542.

Shumaker, D. K., Dechat, T., Kohlmaier, A., Adam, S. A., Bozovsky, M. R., Erdos, M. R., ... Goldman, R. D. (2006). Mutant nuclear lamin A leads to progressive alterations of epigenetic control in premature aging. *Proceedings of the National Academy of Sciences, 103*(23), 8703–8708. https://doi.org/10.1073/pnas.0602569103.

Sidler, C., Kovalchuk, O., & Kovalchuk, I. (2017). Epigenetic regulation of cellular senescence and aging. *Frontiers in Genetics, 8*, 138. https://doi.org/10.3389/fgene.2017.00138.

Signal, B., Phipps, A. J., Giles, K. A., Huskins, S. N., Mercer, T. R., Robinson, M. D., ... Taberlay, P. C. (2024). Ageing-related changes to H3K4me3, H3K27ac, and H3K27me3 in purified mouse neurons. *Cells, 13*(16), 1393. https://doi.org/10.3390/cells13161393.

Solomon, M. J., & Varshavsky, A. (1985). Formaldehyde-mediated DNA-protein crosslinking: A probe for in vivo chromatin structures. *Proceedings of the National Academy of Sciences of the United States of America, 82*(19), 6470–6474. https://doi.org/10.1073/pnas.82.19.6470.

Strahl, B. D., & Allis, C. D. (2000). The language of covalent histone modifications. *Nature, 403*(6765), 41–45. https://doi.org/10.1038/47412.

Tsurumi, A., & Li, W. (2012). Global heterochromatin loss: A unifying theory of aging? *Epigenetics: Official Journal of the DNA Methylation Society, 7*(7), 680–688. https://doi.org/10.4161/epi.20540.

Villeponteau, B. (1997). The heterochromatin loss model of aging. *Experimental Gerontology, 32*(4–5), 383–394. https://doi.org/10.1016/S0531-5565(96)00155-6.

Wang, Y., Yuan, Q., & Xie, L. (2018). Histone modifications in aging: The underlying mechanisms and implications. *Current Stem Cell Research & Therapy, 13*(2), https://doi.org/10.2174/1574888×12666170817141921.

Wang, Z., Zang, C., Rosenfeld, J. A., Schones, D. E., Barski, A., Cuddapah, S., ... Zhao, K. (2008). Combinatorial patterns of histone acetylations and methylations in the human genome. *Nature Genetics, 40*(7), 897–903. https://doi.org/10.1038/ng.154.

Wiśniewski, J. R., Zougman, A., & Mann, M. (2008). N ε -Formylation of lysine is a widespread post-translational modification of nuclear proteins occurring at residues involved in regulation of chromatin function. *Nucleic Acids Research, 36*(2), 570–577. https://doi.org/10.1093/nar/gkm1057.

Xie, J., Ju, J., Zhou, P., Chen, H., Wang, S., Wang, K., ... Wang, K. (2024). The mechanisms, regulations, and functions of histone lysine crotonylation. *Cell Death Discovery, 10*(1), 66. https://doi.org/10.1038/s41420-024-01830-w.

Xie, Z., Dai, J., Dai, L., Tan, M., Cheng, Z., Wu, Y., ... Zhao, Y. (2012). Lysine succinylation and lysine malonylation in histones. *Molecular & Cellular Proteomics, 11*(5), 100–107. https://doi.org/10.1074/mcp.M111.015875.

Zhang, W., Li, J., Suzuki, K., Qu, J., Wang, P., Zhou, J., ... Belmonte, J. C. I. (2015). A Werner syndrome stem cell model unveils heterochromatin alterations as a driver of human aging. *Science (New York, N. Y.), 348*(6239), 1160–1163. https://doi.org/10.1126/science.aaa1356.

Zhang, X., Bolt, M., Guertin, M. J., Chen, W., Zhang, S., Cherrington, B. D., ... Coonrod, S. A. (2012). Peptidylarginine deiminase 2-catalyzed histone H3 arginine 26 citrullination facilitates estrogen receptor α target gene activation. *Proceedings of the National Academy of Sciences, 109*(33), 13331–13336. https://doi.org/10.1073/pnas.1203280109.

Zhang, Y., Liu, T., Meyer, C. A., Eeckhoute, J., Johnson, D. S., Bernstein, B. E., ... Liu, X. S. (2008). Model-based analysis of ChIP-Seq (MACS). *Genome Biology, 9*(9), R137. https://doi.org/10.1186/gb-2008-9-9-r137.

Zhou, W., Zhu, P., Wang, J., Pascual, G., Ohgi, K. A., Lozach, J., ... Rosenfeld, M. G. (2008). Histone H2A monoubiquitination represses transcription by inhibiting RNA polymerase II transcriptional elongation. *Molecular Cell, 29*(1), 69–80. https://doi.org/10.1016/j.molcel.2007.11.002.

Zippo, A., Serafini, R., Rocchigiani, M., Pennacchini, S., Krepelova, A., & Oliviero, S. (2009). Histone crosstalk between H3S10ph and H4K16ac generates a histone code that mediates transcription elongation. *Cell, 138*(6), 1122–1136. https://doi.org/10.1016/j.cell.2009.07.031.

CHAPTER FIFTEEN

Histone Decatylase 6 and Tau-induced microglial chemotaxis by trans-well migration and wound-scratch assay

Subashchandrabose Chinnathambi* and **Smita Eknath Desale**
Department of Neurochemistry, National Institute of Mental Health and Neuro Sciences (NIMHANS), Institute of National Importance, Bangalore, Karnataka, India
*Corresponding author. e-mail address: subashneuro@nimhans.ac.in

Contents

1. Introduction 386
 1.1 Rationale 388
2. Materials 388
 2.1 Stock and working buffer solutions 388
 2.2 Cell culture 390
 2.3 Tau species preparation 390
 2.4 2-D migration by wound-scratch assay 390
 2.5 3-D migration by Trans-well assay 390
 2.6 Instruments 391
 2.7 Softwares 392
3. Methods 392
 3.1 Preparation of Tau protein (Fig. 1) 392
 3.2 Recombinant HDAC6 ZnF UBP preparation (Fig. 2) 394
 3.3 Mammalian cell culture and maintenance 397
 3.4 Preparation of Tau aggregates for migration assays 398
 3.5 Wound-scratch assay- 2D migration assay 398
 3.6 Transwell-migration assay- 3D migration assay 400
4. Notes 402
 4.1 2-D migration by wound-scratch assay 402
 4.2 3-D migration by Trans-well assay 403
5. Concluding remarks 404
Acknowledgements 404
References 404

Abstract

Development of Alzheimer's is promoted through the accumulation of Tau and Amyloid Beta (Aβ) proteins at various neuronal as well as glial junctions. Tau, a

Methods in Enzymology, Volume 733
ISSN 0076-6879, https://doi.org/10.1016/bs.mie.2026.05.033

microtubule-associated protein, is localized at the axonal region of neurons under physiological conditions. The main function of Tau protein is to stabilize microtubules, mediated by the electrostatic interaction of their surface with the repeat domains of Tau. The post-translational modifications (PTMs) are necessary for the physiological functioning of protein, but upon abnormal phosphorylation of Tau, Neuro-fibrillary Tangles of protein are generated disrupting normal brain functionality. Cell migration is a physiological functioning of the cell necessary for various signalling from development, cellular communication, immune function etc. cellular microenvironment affect the behaviour of the cell, which can be detected by its migration propensity. The in vitro assays assist in understanding the adhesion, migration and invasion strategies of migratory cells in response to extracellular stimuli. The migration property can be linked to phenotypic changes of microglia as anti-inflammatory phenotype of microglia display increase migration and invasion, hence understanding migration propensity over Tau and HDAC6 exposure is necessary.

1. Introduction

Tau protein is an intrinsically disordered protein without any secondary structure with a net positive charge (Bershadsky et al., 1995; Braak et al., 2006; Sonawane et al., 2021). It demonstrates a strong affinity for anionic substances such as cofactors, fatty acids, and phospholipids that can bind to the protein's repeat domain and trigger pathological protein aggregation *Via* charge compensation mechanisms. Extracellular Tau species possess the ability to cause cell death and toxicity through imbalanced calcium metabolism in cultured hippocampal neurons (Capitani & Baldari, 2021; Das et al., 2020a). The bioactive Tau seed can be taken up by microglia, the brain's obligate phagocytes, which then degrade it to neutralize it (Das & Chinnathambi, 2021). The incapability of microglia to clear extracellular pathological Tau in later stages of AD fabricates microglia-mediated Tau pathology by releasing toxic Tau seed into the media (Chidambaram et al., 2022; Das & Chinnathambi, 2019; Day et al., 2015). Microglia are the primary immune cells of the brain with ramified morphology during the resting state for surveillance and gain amoeboid shape over the signal of chemoattractant from surrounding. The chemoattraction phenomenon defines the formation of cellular protrusions and migration toward the signal source (de Oliveira & Mantovani, 1988; Derivery et al., 2009; Desale & Chinnathambi, 2020, 2021a). The changes in shape during migration and phagocytosis are mediated by the actin cytoskeleton. Here, we are interested in studying the ability of HDAC6 to cause pathological aggregation of Tau, which was detected by various biochemical and biophysical techniques.

Wound-scratch assay is a technique use to study migration of cell in a cell monolayer, where distance migrated by cell into the wound is calculated to study the velocity and proficiency during wound closure (Balmik et al., 2020; Desale & Chinnathambi, 2021b; Desale et al., 2021; English et al., 2013).

In Parkinson's disease, several studies have reported elevated expression of HDAC6 at sites of α-synuclein accumulation, particularly in the substantia nigra pars compacta, locus coeruleus, and cortical regions. Recent findings have demonstrated a direct interaction between HDAC6 and α-synuclein, with HDAC6 playing an important role in the autophagic degradation of α-synuclein aggregates. In addition, some studies suggest that HDAC6 facilitates the formation of cytoplasmic α-synuclein inclusions, which are considered less toxic than soluble oligomeric species. In Huntington's disease, when chaperone levels decline, HDAC6 is proposed to enhance chaperone activity via HSF (heat shock factor) to support proper huntingtin protein folding. The same review cites a study showing that inhibiting the HDAC6 deacetylase activity increases α-tubulin acetylation. This reduces striatal neuron vulnerability in HD by improving the autophagic flux and mitochondrial transport. However, more research is needed to clarify HDAC6's role in HD.

HDAC6 plays a controversial role in AD—promoting neurodegeneration according to some studies, while offering protection according to others. For instance, on one hand, HDAC6-mediated deacetylation of α-tubulin reduces microtubule stability, thereby disrupting the cytoskeleton and intracellular transport. On the other hand, deacetylation of tubulin and cortactin promotes actin assembly, fostering axonal growth and migration. Of all Tauopathies, the role of HDAC6 has been most extensively studied in AD, though its precise function and mechanism of action remain unclear.

In this study, we assessed migration of microglia in response to extracellular Tau species and HDAC6, we also measured the effect of microtubule inhibitors Taxol and Colchicine on microglial migration (Galletta & Cooper, 2009; Leng & Edison, 2021; Oliveira et al., 1996; Sonawane et al., 2019a). Trans-well migration assay is used to study cell invasion through porous membrane towards chemoattract gradient at the bottom. The chamber is provided with a well, where the cells are seeded at the top, and treatment media containing specific compounds is placed at the bottom of the well which is separated by a porous membrane of 8 μm pore size (Das et al., 2020b; Gorantla et al., 2020; Sonawane & Chinnathambi, 2018).

1.1 Rationale

Migration of cells through the membrane towards chemical gradient is directly proportional to invasion capacity. Here we analyze the function of podosome formation in microglia to induce invasiveness and the effect of microtubule blockers on 3D migration of microglia. Depending on the directional cues, microglia activation during phagocytosis results in polarization and migration in a specific direction (Sonawane et al., 2019b). Exposure to HDAC6 has been shown to enhance the phagocytic nature of microglia, which may serve as a restorative approach to reduce the spread of Tau.

2. Materials

2.1 Stock and working buffer solutions

10X PBS:

Add 14.2 g Na_2HPO_4, 80 g NaCl, 2 g KCl and 2 g KH_2PO_4 to 800 mL water. Adjust the pH to 7.4 and makeup to 1 L. Filter and store in clean bottle. For preparing 1 L of 1X PBS, take 100 mL 10X PBS and add 700 mL water. Adjust the pH to 7.4 and makeup to 1 L, filter and use.

50X TAE:

Add 24.2 g Tris, 5.71 mL glacial acetic acid and 10 mL 0.5 M EDTA to 800 mL water and adjust the pH 8.0 and makeup the volume to 1 L.

Phosphate buffer:

Prepare 100 mL each of 50 mM Na_2HPO_4 and 50 mM NaH_2PO_4. Titrate Na_2HPO_4 with NaH_2PO_4 till the pH reaches 6.8.

40 % Acrylamide:

Weigh 1 N,N′-Methylenebisacrylamide and dissolve in 40 mL water. Now add 37.5 g acrylamide, makeup the final volume to 96.25 mL with water. Filter and store in the dark at 4 °C.

Lamelli buffer (4Xgel loading dye):

0.375 M Tris pH 6.8, 12 % SDS, 60 % glycerol, 0.6 M DTT and 0.06 % bromophenol blue.

Staining solution:

Add 1 g of Coomassie Brilliant Blue R-250, 450 mL methanol, 90 mL acetic acid and makeup to 1 L with water.

Destaining solution:

Intensive destaining solution: Add 500 mL methanol, 100 mL acetic acid and makeup to 1 L with water.

Normal destaining solution: Add 50 mL methanol, 75 mL acetic acid and makeup to 1 L with water.

SDS-PAGE running buffer:

Weigh 18.8 g glycine and 3.02 g Tris, add to 800 mL water and dissolve completely. Add 10 mL SDS ad makeup to 1 L with water.

Transfer buffer:

Add 5.8 g Tris, 2.9 g Glycine, 200 mL methanol and makeup the volume to 1 L.

1X PBST:

Add 100 mL of 10X PBS to 700 mL water and 1 mL Tween-20. Adjust the pH to 7.4 and makeup the volume to 1 L.

Blocking buffer:

Add 5 g milk powder in 100 mL PBST.

Stripping buffer:

For 100 mL of stripping buffer, add 0.28 g of glycine and 1 g SDS to 70 mL water. Adjust the pH to 2.8 and makeup the volume to 100 mL.

Ponceau S stain:

Add 5 g of Ponceau S and 50 mL glacial acetic acid, makeup the volume to 1 L.

Mounting solution:

50 % glycerol prepared in 1X PBS was used to mount the coverslips for immunofluorescence.

Thioflavin S:

Stock solution of 12.5 mM ThS was prepared in 50 % ethanol and stored at −30° in 25 μL aliquots. Working stock of 8 μM was prepared by in 50 mM ammonium acetate, pH 7.0. The working stock was stored at 4 °C.

ANS:

Stock solution was prepared at a concentration of 10 mM in water. It was filtered and stored at −30 °C in aliquots of 25 μL. Working stock was prepared in 50 mM ammonium acetate, pH 7.0 by diluting the stock to 40 μM and was stored at 4 °C.

Heparin:

The heparin used was of 17,500 Daltons. For preparing the 1 mM stock solution, 17.5 mg of heparin was weighed and dissolved in water and made up to 1 mL. Heparin was filtered and stored in 25 μL aliquots.

Protease inhibitor cocktail:

Dissolve a tablet of protease inhibitor cocktail tablet in 2 mL water and filter.

Uranyl acetate:

Dissolve 2 gm uranyl acetate in 100 mL water. Filter and make aliquots of 50 μL. Store in cool and dark place.

2.2 Cell culture

1. N9 microglia cells.
2. RPMI 1640 media.
3. 1x Pen-strep solution.
4. 1x Anti- anti solution
5. 10 % FBS.
6. 1x PBS.
7. 1x Trypsin-EDTA solution.
8. Tau Species.
9. ALA.
10. Trypan blue
11. Sterile centrifuge tubes, sterile multiwell dishes, and sterile coverslips (12 mm and 18 mm).
12. Transwell plates.

2.3 Tau species preparation

Tau monomer (100 μM), Heparin 17,500 Da (25 μM), NaCl (25 mM), Sodium azide (0.01 %), DTT (1 mM), Protease inhibitor cocktail (1X).

2.4 2-D migration by wound-scratch assay

N9 cells, 6-well plate, 200 μL pipette tip, phase contrast microscopy.

2.5 3-D migration by Trans-well assay

Cell culture inserts of 8 μm pore size in a 24-well plate format, 4 % paraformaldehyde solution, 0.2 % crystal violet solution, 1X PBS, Autoclaved MiliQ water, cotton buds.

2.6 Instruments

1. Laminar air flow with HEPA filter, CO_2 incubator at 5 % CO_2 level.
2. Centrifuge and refrigerator.
3. AKTA pure chromatography system.
4. Optima MAX-XP ultracentrifuge (Beckman Coulter).
5. Avanti high-speed centrifuge.
6. Tecan plate reader Infinite 200 Pro.

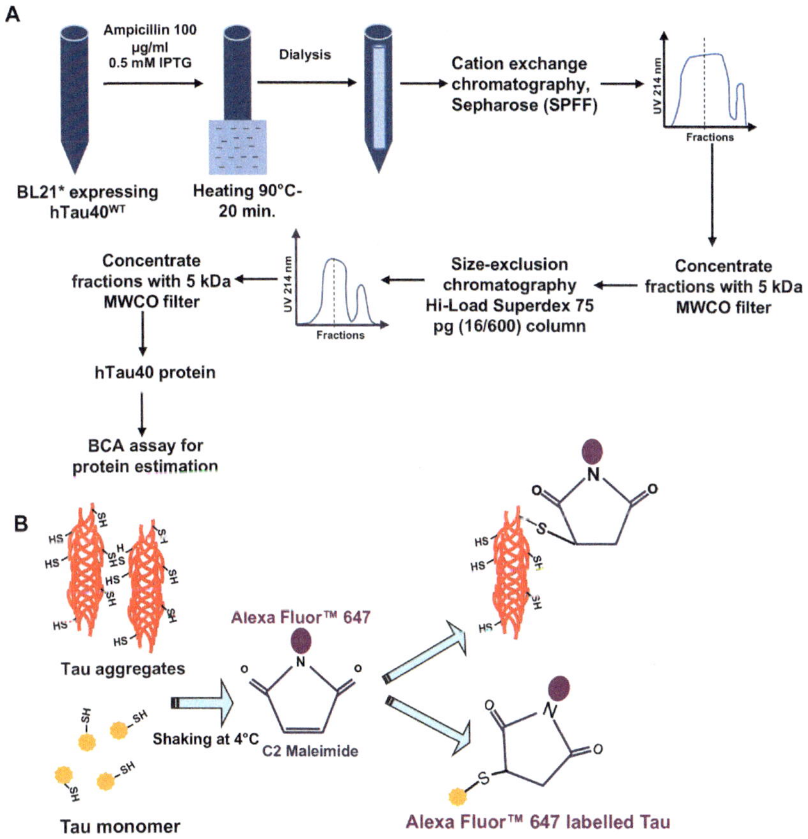

Fig. 1 Preparation of Tau species. (A) The work flow denotes preparation and purification of hTau40 from BL21* with the help of 0.5 mM IPTG and a two-step chromatography procedure *Via* cation-exchange chromatography and size-exclusion chromatography. The purified protein was concentrated and used for further assays. (B) The purified protein was allowed to label with Alexa Fluor 647 conjugated with C2 Maleimide at 4 °C on shaking condition. The labelled Tau species were used for internalization assays.

7. SDS-gel electrophoresis unit (Bio-Rad).
8. Tecnai T20 transmission electron microscope 200 KeV (TEM).
9. FEI high-resolution transmission electron microscope 300 KeV (HR-TEM).
10. Zeiss Axio observer 7.0 with Apotome 2.0 microscope.

2.7 Softwares

1. ZEN 2.3 software.
2. ImageJ software.
3. SigmaPlot 10. Fig. 1

3. Methods

3.1 Preparation of Tau protein (Fig. 1)

Buffers used for Tau preparation

Extraction/lysis buffer
50 mM MES pH 6.8
1 mM EGTA
2 mM $MgCl_2$
5 mM DTT
1 mM PMSF
Protease Inhibitor Cocktail

Dialysis buffer
50 mM MES pH 6.8
50 mM NaCl
1 mM EGTA
1 mM $MgCl_2$
2 mM DTT
0.1 mM PMSF

Cation-exchange chromatography

Wash buffer (Buffer A)	Elution buffer (Buffer B)
50 mM MES pH 6.8	50 mM MES pH 6.8
50 mM NaCl	1 M NaCl
1 mM EGTA	1 mM EGTA
1 mM $MgCl_2$	1 mM $MgCl_2$
2 mM DTT	2 mM DTT
0.1 mM PMSF	0.1 mM PMSF

Size-exclusion chromatography

Gel-filtration buffer
1X PBS
2 mM DTT
Protease Inhibitor Cocktail

1. Tau protein can be recombinantly expressed in high quantities in E coli strain BL21 (D E3).
2. As expression vector, the pNG2 vector is used, a derivative of the PET 3 vector.
3. From culture volume of 10 litre, a yield of 10–100 g protein, depending on the isoform with the purity of>95 % after purification can be expected.
4. For the expression in a 10-litre culture scale, a bio-fermenter can be used.
5. After harvesting the bacterial cultures by centrifugation, the cell pellet is resuspended in ice cold bacterial cell pellet resuspension buffer.
6. Take an aliquot and analyse for tau protein by SDS page.
7. The resuspended cells can either be directly extracted at dialysis before HPLC or frozen at −20 °C, and stored for further processing.
8. Disrupt the cells with a French pressure cell twice. Take an aliquot and analyse for tau protein by SDS page.
9. Add NaCl to final concentration of 500 mM and boil for 20 min. Proteins are denatured, except tau protein which stays in solution and maintains its physiological function.

10. Pellet the denatured proteins and cell debris by centrifugation at 1,27,000 g for 40 min at 4 °C. Take an aliquot of the supernatant and the pellet analysis for tau protein by SDS page.
11. Fill the supernatant into dialysis tubing and dialyze overnight against two changes of cation exchange, chromatography buffer at 4 °C under constant stirring.
12. Clear the dialysate by centrifugation at 1,27,000 g for 40 min at 4 °C. Take an aliquot and analyse for tau protein by SDS page.
13. Apply the clear supernatant onto a cation exchange chromatography column.
14. Wash out unspecific protein with 3−5 vol of buffer A until UV absorption reaches a stable value.
15. Elute the tau protein with 60 % final concentration of buffer B over 6 volumes and fractionate.
16. Take the aliquot of eluted fraction containing protein by UV absorption and analyse it by SDS page.
17. Pool samples containing tau protein and concentrate by Ultra filtration devices to a final volume of 0.5–1 mL, containing 5−10 mg tau protein. Take aliquot and analyse it by SDS page.
18. Apply the tau protein concentrate onto gel filtration column with buffer on low flow rate.
19. After elution of protein, the flow rate can be increased.
20. Take aliquot and analyse by SDS page.
21. Pool suitable tau protein fractions with desired purity regarding amount of degradation and tau dimers.
22. Determine protein concentration and store aliquoted at −80 °C.
23. Cheque an aliquot of the pool by SDS-PAGE for its final purity and as reference (Fig. 2).

3.2 Recombinant HDAC6 ZnF UBP preparation (Fig. 2)

1. HDAC6 ZnF UBP cloned in pET28a-LIC was transformed and expressed in BL21 Codon plus RIL cells and purified in two steps by Ni-NTA affinity chromatography and size-exclusion chromatography.
2. Primary inoculum in Terrific broth (TB), supplemented with 50 μg/mL Kanamycin was prepared from the glycerol stock containing transformed BL21 Codon plus RIL cells by incubating overnight at 37 °C.
3. Secondary culture in TB containing 50 μg/mL Kanamycin inoculated from primary in 1:100 ratio of inoculum and kept in incubation at 37 °C till OD_{600} reaches 0.6.

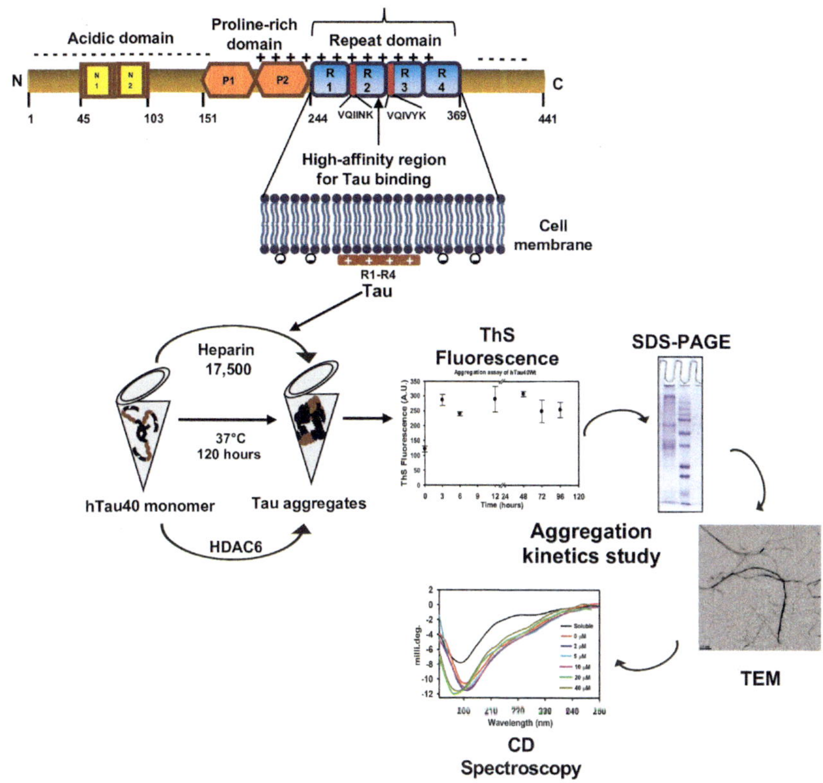

Fig. 2 Aggregation propensity of Tau. Tau is a 441 amino acid protein consisting of N-terminal domain, Proline-rich domain, repeat domain, and C-terminal insert. Among these positively charged repeat domain holds an ability to interact with anionic fatty acids and other cofactors. The workflow denotes the experimental outline to determine ability of HDAC to cause pathological aggregation of Tau. The aggregation kinetics were measured by ThS analysis, SDS-PAGE, TEM and CD spectroscopy.

4. Culture kept at 16 °C for one hour prior to induction by 0.2 mM IPTG.
5. Kept for incubation at 16 °C for 16 h followed by harvesting of cells. Re-suspend the cell pellets in 20–100 mL of lysis buffer depending on the pellet size.
6. Cell pellets were homogenously re-suspended by vortexing and lyzed by subjecting to high pressure (15,000 KPSI) (using Constant cell disruption system).
7. The cell lysate is collected and subjected to centrifugation at 20,000 rpm for 30 min.
8. The supernatant is collected and mixed with Roche Ni-NTA beads in the ratio 1:30 and kept for protein binding for 1 h at 4 °C.

9. Affinity chromatography was performed using 50 mM Tris-Cl pH 8.0 with 20 mM Imidazole for wash and 1000 mM imidazole for elution.
10. The eluted fractions were dialyzed in 50 mM Tris-Cl pH 8.0, 100 mM NaCl, 2 mM TCEP/DTT, 5 % glycerol to remove imidazole.
11. Gel-filtration Chromatography was performed using Superdex 75 Hi-load 16/600 column in 50 mM Tris-Cl pH 8.0, 100 mM NaCl, 2 mM TCEP/DTT which yielded HDAC6 ZnF UBP at retention volume ~110 mL.
12. The fractions containing HDAC6 ZnF UBP were pooled concentrated and stored at −80 °C. The protein concentration was determined by using Bradford assay.

Buffers used for HDAC6 ZnF UBP preparation

Extraction/lysis buffer
50 mM Tris-Cl pH 8.0
500 mM NaCl
1 mM PMSF
20 mM Imidazole
Protease Inhibitor Cocktail

Ni-NTA affinity chromatography

Wash buffer	Elution buffer
50 mM Tris-Cl pH 8.0	50 mM Tris-Cl pH 8.0
500 mM NaCl	500 mM NaCl
20 mM Imidazole	1 M Imidazole
	5 % Glycerol

Dialysis buffer
50 mM Tris-Cl pH 8.0
100 mM NaCl
2 mM TCEP/DTT
5 % Glycerol

Size-exclusion chromatography

Size-exclusion buffer
50 mM Tris-Cl pH 8.0
100 mM NaCl
1 mM TCEP/DTT

3.3 Mammalian cell culture and maintenance

For the cell biology-based studies, N9 microglial cell line: CVCL_0452 was used. N9 is a retrovirus-transformed mouse glial cell line which is widely used for neurobiology and neuroimmunology studies.

Reagents:

Components	Concentration
Cell culture media-	
RPMI 1640 media	
Fetal bovine serum (FBS)	10 %
Pen-strep antibiotic cocktail (Stock 100X)	1X
Anti-anti cocktail (Stock 100X)	1X
For cell passaging-	
PBS solution (Cell biology grade)	1X
Trypsin-EDTA (0.25 %) 10X	1X

Protocol:

Cell growth:

1. The N9 microglial cells were maintained in RPMI 1640 media supplemented with 10 % FBS, 1X pen-strep and an anti-anti cocktail mixture.
2. The cells were incubated at 37 °C in a CO_2 incubator with 95 % humidity.
3. The cells were passaged after reaching 80−90 % confluency in the culture dish.

Cell passage:

1. After reaching to the desired confluency, cells were passaged by removing the growth media and washing with 1X PBS.

2. The trypsin EDTA solution was added to the culture dish and incubated at 37 °C for 5 min which allows the cells to detach from the culture dish.
3. The reaction of trypsin was stopped by adding the growth media containing 10 % FBS.
4. The cell suspension was mixed homogeneously and collected in 15 mL centrifuge tube and centrifuged at 800 rpm for 5 min to pellet down the cells.
5. The supernatant was discarded and the cell pellet was resuspended in fresh growth media.
6. The cells were split to a fresh culture dish at a ratio of 1:3.
7. The culture dishes were incubated at 37 °C in 5 % CO_2 incubator with 95 % humidity.

3.4 Preparation of Tau aggregates for migration assays

1. Tau monomer was used for higher-order Tau aggregates preparation according to well-established protocol in the lab (Fig. 2).
2. Tau aliquot was removed from −80 °C, thawed and subjected to ultracentrifugation at 60,000 rpm for 60 min at 4 °C to verify only monomeric soluble fractions present for in vitro aggregation of Tau.
3. The desired volume of Tau was then diluted in BES buffer pH 7.4 and incubated with DTT for 10 min to ensure monomeric Tau incubation aggregation assay (Fig. 2).
4. After incubation other reagents along with heparin (4:1-Tau:heparin ratio) are added according to mentioned concentrations.
5. The reaction was mixed and incubated at 37 °C for 72 h to attain mature Tau fibril formation.
6. Tau aggregates formation was confirmed with SDS-PAGE, TEM analysis, ThS fluorescence and CD analysis **(*see* Note 1)** (Fig. 2).
7. Protein estimation was carried out with Bicinchoninic Acid (BCA) assay for further use of aggregates for fluorescent labelling during cell-biology assays (Fig. 2).

3.5 Wound-scratch assay- 2D migration assay

Wound scratch assay is a widely used simple technique to determine the speed and polarity of migration in a time-dependent manner by detecting the relative amount of wound closure in a cell monolayer (Fig. 3).

1. Culture N9 microglia cells in RPMI media supplemented with 10 % FBS, 1 % pen-strep solution and 1 % Anti-anti solution, till the confluency reaches to 90 % **(*see* Note 2 and 3)**.

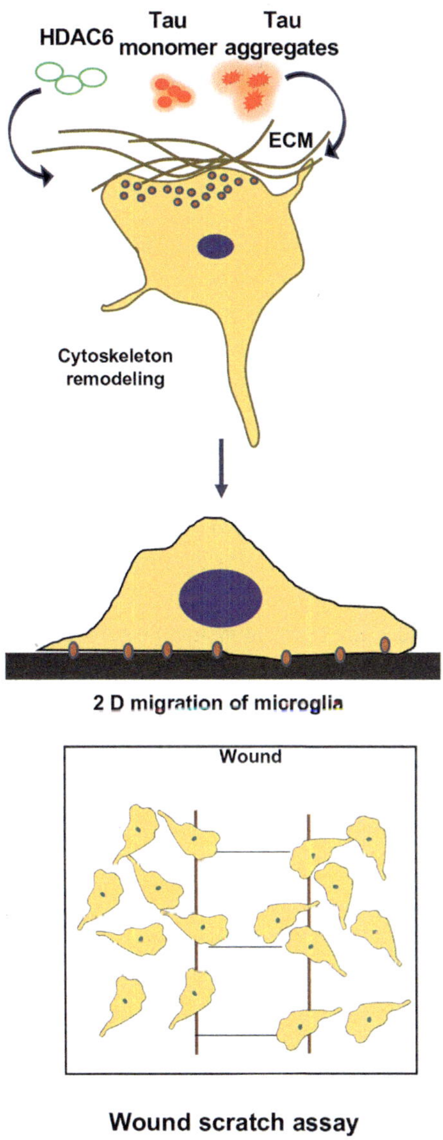

Fig. 3 Microglial chemotaxis-induced by HDAC6. Cell migration is a physiological functioning of the cell necessary for various signalling from development, cellular communication, immune function etc. cellular microenvironment affect the behaviour of the cell, which can be detected by its migration propensity. The in vitro assays assist in understanding the adhesion, migration and invasion strategies of migratory cells in response to extracellular stimuli. Wound-scratch assay is a technique use to study migration of cell in a cell monolayer, where distance migrated by cell into the wound is calculated to study the velocity and proficiency during wound closure.

2. Then, passage cells in the presence of trypsin-EDTA solution and seed on multiwell plate (6 well plate) at a density of 5,00,000 cells/well for wound-scratch assay.
3. Before seeding the cells in the well plate, horizontal lines are given on well with the help of glass marker or laser to mark the same area while imaging the experiment.
4. After overnight incubation, the wound has been created on cell monolayer with 200 μl pipette tip.
5. The media was discard after creation of wound from well plate.
6. Subsequent 2 gentle washes of 1X PBS was given to the cell after wound to remove all dislodged cells occurred during wound creation.
7. Then PBS was removed gently and replaced with 0.5 % serum-deprived RPMI for 24 h, treatment compounds are incorporated in the serum-deprived media with the desired concentration if any (*see* **Note 4**).
8. Imaging of wound closure was carried out at 0, 6, 12, 24 h under phase contrast at 20 × magnification. To ensure that the imaging corresponds to the same position at each time point, imaging should be done at the same location where the vertical wound intersects the horizontal marker line (*see* **Note 6 and 6**).
9. Distanced travelled by cells was counted with the help of captured images and percentage wound-closure was calculated for 6, 12 and 24 h with respect to 0-hour distance of wound (*see* **Note 7 and 8**).

3.6 Transwell-migration assay- 3D migration assay

Tran-well migration assay is also called the Boyden chamber assay, which is a widely used selective migration assay technique that allows the invasion of migratory cells through the porous membrane by following the chemical gradient. In this assay, the migratory cells were placed in an upper chamber containing a porous membrane (5 μm pore size) where the chemoattractant was placed in the lower chamber. The main advantage of this test is selectivity while the disadvantage of this study was prolonged exposure of chemoattractant can equilibrate in both the chamber (Fig. 4).

1. Culture N9 microglia cells in RPMI media supplemented with 10 % FBS, 1 % pen-strep solution and 1 % Anti-anti solution, till the confluency reaches to 90 % (*see* **Note 9**).
2. Transwell inserts are placed in 24 well plate in a sterile condition.
3. Then, passage cells in the presence of trypsin-EDTA solution and seed on the upper well insert of transwell assay with 50,000 cells in RPMI supplemented with 0.5 % serum media.

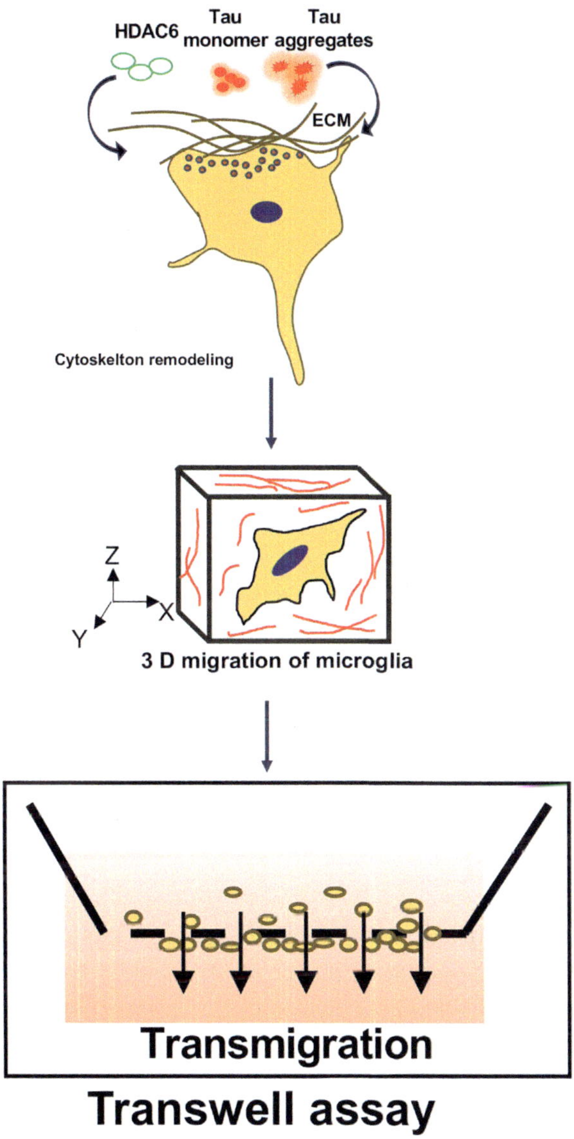

Fig. 4 3-D microglial migration modulated by HDAC6. Trans-well migration assay is use to study the invasion of cell through porous membrane towards chemoattract gradient at the bottom. The chamber is provided with the well, where the cells are seeded at the top and treatment media containing specific compound is placed at the bottom of well which separated by porous membrane of 8 μm pore size. Migration of cell through the membrane towards chemical gradient is directly proportional to the capacity of invasion. Here we analyse the function of podosome formation in microglia to induce invasiveness and the effect of microtubule blockers on 3D migration of microglia.

4. The bottom panel of the transwell insert is supplemented with desired treatment of chemoattractant in 0.5 % serum-starved RPMI media.
5. The chamber was incubated for 24 h at 37 °C and allow the cells to settle and migrate for 24 h from upper panel to lower panel towards the chemoattractant.
6. After 24 h, the insert was removed, and upper side was cleaned with cotton swab completely to remove any no migrated cells (*see* **Note 10 and 11**).
7. Latter the entire insert was dipped into 400 μl of 4 % paraformaldehyde solution for 20 min for fixing the cells migrated at lower panel of the insert (*see* **Note 12**).
8. Washing of inserts was done with 1X PBS twice in the same 24 well plate to remove the traces of PFA solution.
9. The staining of fixed cells was done with 0.2 % Crystal violet or any other primary stain for 15 min.
10. After staining the cleaning of upper well of insert was completely cleaned to remove any traces of stain, which would add on the background while imaging (*see* **Note 13 and 14**).
11. Before imaging with bright-field 20x under a microscope, the lower surface of the insert was allowed to air dry for 2−4 h.
12. Number of cells per field on the lower side of the filter was counted for estimation of transmigration in the presence of chemoattractant (*see* **Note 15**).

4. Notes

4.1 2-D migration by wound-scratch assay

1. Tau species are SDS stable filaments which could be visualized in SDS-gel electrophoresis as a smear of higher order proteins with longer filaments deposited at the wells. ALA sample was freshly prepared before the experiment to avoid oxidation. During solubilization at 50 °C the microcentrifuge tube is covered with parafilm to avoid evaporation of ethanol.
2. Wound-scratch assay is a technique use to study migration of cell in a cell monolayer, where distance migrated by cell into the wound is calculated to study the velocity and proficiency during wound closure.
3. In this study, we assessed migration of microglia in response to extracellular Tau species and ALA, we also measured the effect of inhibitors on microglial migration.

4. For the assay, (5,00,000 cells/well) N9 cells were seeded in a 6-well plate and maintained in RPMI media for 24 h till the confluency reached to 80 %.
5. Scratch was created with a sterile 200 μL pipette tip, and the well was given a thorough wash with 1X PBS to remove any unbound cell, followed by treatment with Tau species and ALA in 0.5 % FBS basal RPMI media.
6. Cells were incubated further for 24 h to study the migration of N9 cells into the wound. A number of cells migrated into the wound were observed at 0, 6, 12, and 24 h under phase-contrast microscopy.
7. The % wound closure was calculated for different areas of culture, and the average was calculated to quantify the migration. The graph is plotted as a number of cells travelled into the wound at different time intervals 0, 6, 12, and 24 h or percentage wound closure with respect to 0th hour reading of the same wound.
8. % wound closure = (Mean wound closure for treatment/ mean wound closure of cell control) *100 Mean wound closure = (length of the wound at t = 6, 12, 24 h) – (length of the wound at t = 0th hour).

4.2 3-D migration by Trans-well assay

1. Trans-well migration assay is used to study cell invasion through porous membrane towards chemoattract gradient at the bottom.
2. The chamber is provided with a well, where the cells are seeded at the top, and treatment media containing specific compounds is placed at the bottom of the well which is separated by a porous membrane of 8 μm pore size.
3. Migration of cells through the membrane towards chemical gradient is directly proportional to invasion capacity. Here we analyze the function of podosome formation in microglia to induce invasiveness and the effect of inhibitors and extracellular Tau on 3D migration of microglia.
4. The bottom panel of the trans well insert is supplemented with Tau species and ALA in 0.5 % serum-starved RPMI media. The upper well insert is provided with 5×10^4 of N9 cells in RPMI supplemented with 0.5 % serum media.
5. The chamber was incubated for 24 h at 37 °C before fixing and staining the insert.
6. The insert was removed, and the upper side was cleaned with a cotton swab followed by fixation with 4 % paraformaldehyde (20 min) and staining with 0.2 % crystal violet (15 min).

7. The number of cells (fields per filter) on the lower side of the filter was counted after drying under phase contrast at 20X magnification using phase contrast microscope Zeiss.

5. Concluding remarks

Cell migration assay is an important determinant of the rate of migration and execution of migration to a selective microenvironment by a particular type of cell to function accurately. Cell migration plays an important role in various cellular processes- embryonic development, neuronal cell migration and nervous system development, tissue homeostasis and immune cell migration and tissue invasion; also, in the pathological state- such as cancer cell metastasis, Wiskott-Aldrich syndrome and osteoporosis, etc. Microglia as prime-most CNS immune cells surveil the tissue environment and also migrate at the site of injury by detecting chemical gradients. We have previously reported that the treatment of neuro2a cells with HDAC6 ZnF UBP demonstrates enhancement in the formation of neurites and migratory structures such as the lamellipodia, filopodia and podosome-like structures, potentially regulating actin dynamics in the neuro2a and microglial cells. In the present study, we have explored the possible signalling involved in the effect of the HDAC6 domain on the actin dynamics and established a further understanding of how this helps to facilitate the navigation and migration of the neuronal cell systems.

Acknowledgements

We are grateful to Chinnathambi's lab members for their scientific discussions, helpful suggestions and critical reading of the manuscript. We also greatly acknowledge internal support from the Department of Neurochemistry, National Institute of Mental Health and Neuro Sciences (NIMHANS), Institute of National Importance, Bangalore.

References

Balmik, A. A., Das, R., Dangi, A., Gorantla, N. V., Marelli, U. K., & Chinnathambi, S. (2020). Melatonin interacts with repeat domain of Tau to mediate disaggregation of paired helical filaments. *Biochimica et Biophysica Acta (BBA) - General Subjects, 1864*(3), 129467.

Bershadsky, A. D., Gluck, U., Denisenko, O. N., Sklyarova, T. V., Spector, I., & Ben-Ze'ev, A. (1995). The state of actin assembly regulates actin and vinculin expression by a feedback loop. *Journal of Cell Science, 108*, 1183–1193.

Braak, H., Alafuzoff, I., Arzberger, T., Kretzschmar, H., & Del Tredici, K. (2006). Staging of Alzheimer disease-associated neurofibrillary pathology using paraffin sections and immunocytochemistry. *Acta Neuro, 112*, 389–404.

Capitani, N., & Baldari, C. T. (2021). F-Actin dynamics in the regulation of endosomal recycling and immune synapse assembly. *Frontiers in Cell and Developmental Biology, 9*, 670882.

Chidambaram, H., Das, R., & Chinnathambi, S. (2022). G-Protein coupled purinergic P2Y12 receptor interacts and internalizes TauRD-mediated by membrane-associated actin cytoskeleton remodelling in microglia. *European Journal of Cell Biology*, 151201.

Das, R., Balmik, A. A., & Chinnathambi, S. (2020a). Phagocytosis of full-length Tau oligomers by actin-remodeling of activated microglia. *Journal of Neuroinflammation, 17*(1), 1–15.

Das, R., Balmik, A. A., & Chinnathambi, S. (2020b). Effect of melatonin on Tau aggregation and Tau-mediated cell surface morphology. *International Journal of Biological Macromolecules, 152*, 30–39.

Das, R., & Chinnathambi, S. (2019). Microglial priming of antigen presentation and adaptive stimulation in Alzheimer's disease. *Cellular and Molecular Life Sciences: CMLS, 76*(19), 3681–3694.

Das, R., & Chinnathambi, S. (2021). Microglial remodeling of actin network by Tau oligomers, via G protein-coupled purinergic receptor, P2Y12R-driven chemotaxis. *Traffic (Copenhagen, Denmark), 22*(5), 153–170.

Day, C. A., Baetz, N. W., Copeland, C. A., Kraft, L. J., Han, B., Tiwari, A., ... Davidson, M. W. (2015). Microtubule motors power plasma membrane tubulation in clathrin-independent endocytosis. *Traffic (Copenhagen, Denmark), 16*, 572–590.

de Oliveira, C. A., & Mantovani, B. (1988). Latrunculin A is a potent inhibitor of phagocytosis by macrophages. *Life Sciences, 43*, 1825–1830.

Derivery, E., Sousa, C., Gautier, J. J., Lombard, B., Loew, D., & Gautreau, A. (2009). The Arp2/3 activator WASH controls the fission of endosomes through a large multiprotein complex. *Developmental Cell, 17*, 712–723.

Desale, S. E., & Chinnathambi, S. (2020). Role of dietary fatty acids in microglial polarization in Alzheimer's disease. *Jour of Neuro, 17*, 1–14.

Desale, S. E., & Chinnathambi, S. (2021a). α-Linolenic acid induces clearance of Tau seeds via Actin-remodeling in Microglia. *Molecular Biology, 2*, 1–14.

Desale, S. E., & Chinnathambi, S. (2021b). α-Linolenic acid modulates phagocytosis and endosomal pathways of extracellular Tau in microglia. *Cell Adhesion and Migration, 15*, 84–100.

Desale, S. E., Dubey, T., & Chinnathambi, S. (2021). α-Linolenic acid inhibits Tau aggregation and modulates Tau conformation. *International Journal of Biological Macromolecules, 166*, 687–693.

English, J. A., Harauma, A., Föcking, M., Wynne, K., Scaife, C., Cagney, G., ... Cotter, D. R. (2013). Omega-3 fatty acid deficiency disrupts endocytosis, neuritogenesis, and mitochondrial protein pathways in the mouse hippocampus. *Frontiers in Genetics, 4*, 208.

Galletta, B. J., & Cooper, J. A. (2009). Actin and endocytosis: Mechanisms and phylogeny. *Current Opinion in Cell Biology, 21*, 20–27.

Gorantla, N. V., Das, R., Chidambaram, H., Dubey, T., Mulani, F. A., Thulasiram, H. V., & Chinnathambi, S. (2020). Basic limonoid modulates chaperone-mediated proteostasis and dissolve Tau fibrils. *Scientific Reports, 10*(1), 4023.

Leng, F., & Edison, P. (2021). Neuroinflammation and microglial activation in Alzheimer disease: Where do we go from here? *Nature Reviews Neuroscience, 17*, 157–172.

Oliveira, C. A., Kashman, Y., & Mantovani, B. (1996). Effects of latrunculin A on immunological phagocytosis and macrophage spreading-associated changes in the F-actin/G-actin content of the cells. *Chemico-Biological Interactions, 100*, 141–153.

Sonawane, S. K., Ahmad, A., & Chinnathambi, S. (2019a). Protein-capped metal nanoparticles inhibit tau aggregation in Alzheimer's disease. *ACS Omega, 4*(7), 12833–12840.

Sonawane, S. K., Balmik, A. A., Boral, D., Ramasamy, S., & Chinnathambi, S. (2019b). Baicalein suppresses repeat Tau fibrillization by sequestering oligomers. *Archives of Biochemistry and Biophysics, 675*, 108119.

Sonawane, S. K., & Chinnathambi, S. (2018). Prion-like propagation of post-translationally modified tau in Alzheimer's disease: A hypothesis. *Journal of Molecular Neuroscience, 65*, 480–490.

Sonawane, S. K., Dubey, T., Balmik, A. A., Das, R., & Chinnathambi, S. (2021). Alzheimer's disease pathology: A Tau perspective. In T. Govindaraju (Ed.). *Alzheimer's disease: Recent findings in pathophysiology, diagnostic and therapeutic modalities*. The Royal Society Of Chemistry.

CHAPTER SIXTEEN

Experimental techniques to evaluate IA-MSCs–mediated alleviation of vasogenic edema following ischemic stroke

Aishika Datta, Nikita Malik, Gautam Karmarkar, Anirban Barik, Deepaneeta Sarmah, Vikas Khajekar, Akshada Dubey, Ushmita Mukherjee, Anita Kumari, Swagata Bhattacharya, Bijoyani Ghosh, and Pallab Bhattacharya*

Department of Pharmacology and Toxicology, National Institute of Pharmaceutical Education and Research (NIPER), Ahmedabad, Gandhinagar, Gujarat, India

*Corresponding author. e-mail address: pallab.bhu@gmail.com; pallab.bhattacharya@niperahm.res.in

Contents

1. Introduction 408
2. Rationale 410
3. IA-MSCs in combination with SIRT-1 Activator and NFkB Inhibitor: Implication in post stroke edema 411
   3.1 Induction of middle cerebral artery occlusion model 411
   3.2 Administration of intra-arterial mesenchymal stem cells (IA-MSCs), SIRT 1 activator, inhibitor and NFkB Inhibitor after ischemic stroke 414
   3.3 Histological assessment 416
   3.4 Behavioural assessment 417
   3.5 Cerebral infarct size and percentage edema measurement 419
   3.6 Western blotting 419
   3.7 RT-PCR 423
   3.8 Immunofluorescence analysis 424
   3.9 Oxidative stress measurement 425
   3.10 Mitochondrial functional assessment in post-stroke vasogenic edema 426
4. Conclusion 429
Acknowledgement 429
References 429

Abstract

Ischemic stroke induced cerebral edema is a devastating complication which leads to excessive swelling, increased intracranial pressure (ICP) and rapid neurological deterioration. The mortality rate of post-stroke cerebral edema is approximately 80 %. However, therapeutic options for post-stroke cerebral edema are sparce, emphasizing

Methods in Enzymology, Volume 733
ISSN 0076-6879, https://doi.org/10.1016/bs.mie.2026.06.021

an urgent need to explore new therapeutic strategies for post-stroke edema. In past few decades stem cell therapy has been a promising therapeutic strategy in decreasing infarct volume and promoting functional recovery by various mechanisms following stroke. Previous studies from our lab demonstrated that IA-MSCs administration were beneficial in mitigating post-stroke edema by regulating PKCδ/ MMP9/ AQP4 axis and promoting blood-brain-barrier (BBB) integrity. It has also been reported that IA-MSCs administration mitigate post-stroke edema by maintaining BBB integrity, mitochondria function and regulating AQP4 expression by modulation of the SIRT-1/PKCδ/NF-κB axis. In this chapter, we illustrated a multi-level methodological approach to assess therapeutic outcomes of IA-MSCs in alleviating post-stroke cerebral edema by utilizing behavioural analyses, histological staining, biochemical assays, mitochondrial functional assays and molecular investigations by RT-PCR, western blotting and immunofluorescence staining.

1. Introduction

Stroke is the leading cause of disability and mortality, with ischemic stroke accounting for 83 % of all cases (Ji et al., 2021). Brain edema occurs following interruption of cerebral blood flow, which contributes to higher mortality and disability post-stroke (Ji et al., 2021). Severity of cerebral edema is associated with worsening of long-term functional outcomes following stroke (Ji et al., 2021). Within 24–48 h following ischemic insult, water and ions are abnormally transported into injured neurons, resulting in cytotoxic edema (Ji et al., 2021; Stokum et al., 2015). While after 72–96 h of ischemic injury, enhanced degradation and permeability of the blood-brain barrier leads to formation of vasogenic edema (Ji et al., 2021; Michinaga & Koyama, 2015).

Over 90% of stroke patients fail to receive tissue plasminogen activator (tPA), the conventional treatment for stroke, due to its narrow therapeutic window and risk of hemorrhagic transformation (Sarvari et al., 2020). Stroke associated BBB disruption and hyperpermeability has been reported to reduce the therapeutic efficacy of tPA. In contrast, tPA administration further exacerbates BBB disruption by enhancing MMP-9 expression and phosphorylation of connexin 43, a gap junction protein (Sarvari et al., 2020). Hyperosmolar therapy and glucocorticoids are primarily used to manage cytotoxic edema and vasogenic edema respectively (Datta et al., 2025). It has been reported that mannitol leads to rebound rises in intracranial pressure (ICP) and steroids are not useful in treatment of cytotoxic edema an d worsens outcomes in patients with brain ischemia (Rabinstein, 2006).

Current therapeutic interventions for post-stroke cerebral edema are limited, so there is an urgent need to explore novel therapeutic agents (Ji et al., 2021).

Stem cell therapy has been established as a potential therapeutic strategy to support the endogenous regenerative potential of brain and promote functional recovery following stroke (Asgari Taei et al., 2022). In various pre-clinical studies, it has been demonstrated that administration of MSCs can reduced infarct volume and improve functional outcomes in experimental stroke model via numerous mechanisms such as alleviating cerebral edema (Tang et al., 2014), suppressing inflammation, enhancing angiogenesis, mitochondrial health, reducing apoptosis and enhancing neurogenesis Asgari Taei et al., 2022).

Mitochondria regulate cellular energy metabolism and various cellular functions, including reactive oxygen species (ROS) production, calcium balance, cell proliferation, programmed cell death, cell survival and autophagy (Sadeghian et al., 2025). Due to high energy consumption, brain is susceptible to mitochondrial dysfunction. In ischemic stroke, disrupted mitochondrial activity leads to elevated oxidative stress and reduced ATP synthesis which enhances neuronal injury and neurodegeneration (Sadeghian et al., 2025). Further, ischemic stroke mediated mitochondria dysfunction in brain endothelial cells leads to energy failure, activation of matrix metalloproteinases (MMPs), inflammatory responses and enhanced oxidative stress that collectively degrades the tight junction proteins. This cascade results into disrupted BBB integrity and poor neurological outcomes (Sadeghian et al., 2025). Additionally, excessive mitochondrial ROS production is also hallmark for dysfunction of TCA, resulting in the activation of several pro-inflammatory cytokines and transcription of cellular detrimental genes, which further leads to activation of MMPs and BBB disruption (Datta et al., 2025; Sarmah et al., 2019). PKCδ triggers the mitochondrial dysfunction by supressing pyruvate dehydrogenase complex (PDH complex) causing TCA cycle failure and elevated ROS generation (Datta et al., 2025; Kim & Hammerling, 2020). Sirtuin 1 (SIRT1) is a NAD dependent deacetylase enzyme plays a crucial role in increasing mitochondrial integrity and alleviating oxidative stress. It acts as a cellular energy sensor and redox state and are controlled by cellular metabolic state (Sarmah et al., 2019).

Furthermore, elevated mitochondrial ROS generation induces the activation of pro-inflammatory cytokines through inflammasome-dependent and independent mechanisms (Marchi et al., 2023; Datta et al., 2025).

In inflammasome-independent pathway, elevated mitochondrial ROS leads to activation of the transcriptional regulator nuclear factor kappa B (NF-κB), which translocates to the nucleus and promotes the transcription of numerous genes such as aquaporin-4 (AQP4) and MMPs. This mechanism contributes to BBB degradation and decreased expression of tight junction proteins such as claudin-5 (Ito et al., 2006; Yan et al., 2004; Datta et al., 2025).

Aquaporin 4 (AQP4) is water channel protein that facilitates the exchange of water across cellular compartments. It is highly expressed at interface of BBB and cerebrospinal fluid (CSF). After ischemic stroke, AQP4 expression is significantly altered due to ischemia–reperfusion (I/R) injury. Although, increased and decreased expressions of AQP4 have been reported post-stroke; however, its upregulation is widely recognized as a key factor leading to cerebral edema formation following ischemic stroke (Datta et al., 2022).

Emerging evidence indicated that SIRT1 may regulate AQP4 expression in various studies. In subarachnoid hemorrhage and ischemic stroke models, SIRT1 activation has been associated with attenuation of edema formation (Datta et al., 2025). Previous studies from our lab have reported that intra-arterial administration of mesenchymal stem cells (IA-MSCs) can activate SIRT1 and regulate NF-κB signalling. We have also observed that IA-MSCs modulated BDNF/TrkB-mediated endoplasmic reticulum stress, calcineurin-dependent mechanisms, and acid-sensing ion channel regulation to provide neuroprotection post-stroke (Datta et al., 2025; Kaur et al., 2021; Sarmah et al., 2022; Vats et al., 2021). Additionally, IA-MSCs also mitigated post-stroke edema by regulating protein kinase C delta (PKCδ)-mediated modulation of AQP4 (Datta et al., 2025).

In this chapter, we provide comprehensive protocols for precise delineation of molecular events, demonstrating the therapeutic potential of IA-MSCs in modulating mitochondrial function, alleviating oxidative stress, reducing NF-κB activation and regulating AQP4 expression, thereby preserving BBB integrity and decreasing cerebral edema.

2. Rationale

Following ischemic stroke mitochondria dysfunction occurs, which leads to generate excessive ROS, which is hallmark for TCA dysfunction, leads to activation of numerous pro-inflammatory cytokines, causing activation of MMPs and BBB disruption. SIRT-1 plays a crucial role in maintaining mitochondrial health and regulating the AQP4 expression. Hence, present study explored how modulation of SIRT-1/PKCδ/NFκB loop by IA-MSCs administration leads to alleviate cerebral edema following stroke. A combination of different technique was employed such as assessment of biochemical, molecular, behavioural, histological, and mitochondrial functional methods to validate different components of the proposed mechanism. The middle cerebral artery occlusion (MCAo) model was used in ovariectomized Sprague–Dawley rats because it closely mimics human focal ischemic stroke and reproducible. Female rats were Ovariectomized to eliminate the neuroprotective effect of estrogen, thereby ensuring, accurate investigation of stem cell effect on post-stroke edema. MSCs was administered through intra-arterial route as it provides targeted delivery of MSCs into ischemic brain region, enhancing therapeutic efficacy in comparison to systemic administration. Pharmacological modulators were used for investigating the role of SIRT-1/PKCδ/NFκB loop in post-stroke edema. To evaluate neurological and motor functional recovery post stroke behavioural assessment was performed. Histological assessment was performed to assess cell viability using haematoxylin and eosin staining. Further, edema volume was measured to quantify brain swelling.

Biochemical assay was performed to assess oxidative stress and antioxidant defence systems. RT-PCR analysis, western blotting and Immunofluorescence studies was performed to quantify mRNA and protein expression respectively. Further performing mitochondrial assay to determine whether IA-MSCs administration preserve mitochondrial bioenergetics and decreasing oxidative damage. These methods provided direct evidence corelating mitochondrial protection with edema reduction and BBB disruption. These techniques enabled assessment of neuroprotective efficacy of IA-MSCs therapy through modulating SIRT-1/PKCδ/NFκB loop in post-stroke edema (Fig. 1).

3. IA-MSCs in combination with SIRT-1 Activator and NFkB Inhibitor: Implication in post stroke edema

3.1 Induction of middle cerebral artery occlusion model

3.1.1 Animals

1. The study was performed using ovariectomized female Sprague Dawley (SD) rats weighing between 220–250 g.

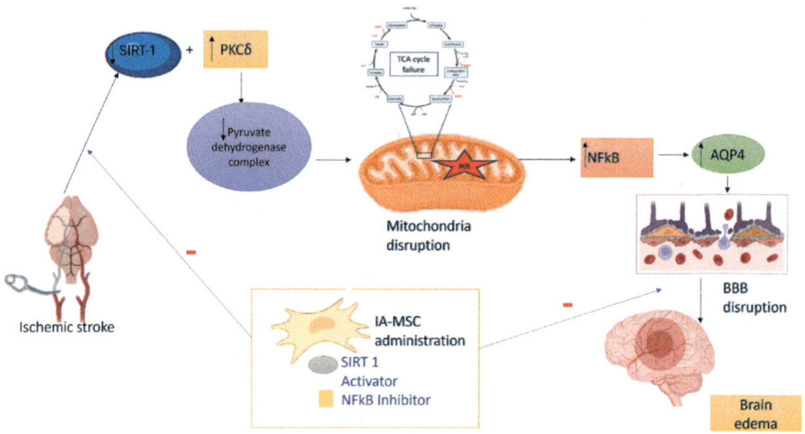

Fig. 1 Role of IA-MSCs in modulating post-stroke edema.

2. The animals were procured from NIPER-A breeding facility and CCSEA Approved commercial breeding sources. All experimental procedures were approved by the Institutional Animal Ethical Committee (IAEC) of NIPER-Ahmedabad, in accordance with Committee for Control and Supervision of Experiments on Animals (CCSEA) guidelines.

3.1.2 Reagents

0.9 % normal saline, povidone iodine and buprenorphine.

3.1.3 Equipment and surgical

Small Animal Anesthetizer (Harvard Apparatus, USA), Blood gas analyser, Radiometer, PE-50 catheter (BD Biosciences), Glucometer (One stop), Laser Doppler flowmetry (LDF) (PowerLab AD Instruments), Near infrared spectroscopy (moorVMS-NIRS monitor), Bead sterilizer, Surgical instruments including scissors, forceps, muscle clamps, arterial clamps, micro-scissors etc., Silicone-coated nylon filament (Doccol Corp.), sutures, suturing needles.

3.1.4 Procedure

1. The present study uses ovariectomized female SD rats. Prior to 14–20 days of ovariectomy, vaginal smear of animals was examined to assess the stages of their estrous cycle.
2. 5–7 days prior to MCAo, ovariectomy was performed and proper post-operative care was given to prevent infection, reduce pain and accelerate healing.

3. Rats were anesthetized using a combination of isoflurane/oxygen combination (Small Animal Anesthetizer, Harvard Apparatus, USA, 4 % at induction and 1.5 % as maintenance dose) and the incision site was cleaned off of hair and other debris.
4. To monitor blood gas parameters (using blood gas analyser, Radiometer), the right femoral artery was catheterized using PE-50 catheter (BD Biosciences).
5. Under an operating microscope, an incision was made on the neck, the surrounding tissues (glandular and tracheal muscles) and the vagus nerve were separated to isolate the left common carotid artery (CCA) using forceps.
6. The left external carotid artery (ECA) was isolated and superior thyroid artery (STA) was cauterized.
7. A suture (5−0) was tied at the distal end of ECA from the CCA bifurcation and the CCA and internal carotid artery (ICA) were clamped.
8. A 5–0 suture was tied loosely around the segment of ECA.
9. A 3–0 silicone-coated intraluminal nylon filament was inserted into the distal segment of ECA by making a small incision on the artery and was further advanced into the ECA lumen to reach the bifurcation.
10. The filament was then turned into the ICA and was advanced into the distal ICA until resistance is felt in hitting the bifurcation of MCA.
11. This was further confirmed by a significant drop of regional cerebral blood flow (CBF) by the LDF signal, indicating successful occlusion of the MCA.
12. The filament was withdrawn following a 90-minute occlusion period.
13. Blood pressure, blood gas parameters, blood glucose, cranial and rectal temperatures were maintained during the entire procedure and while reperfusion (Fig. 2).

3.1.5 Notes

1. A commercially available glucometer (One stop) was used to measure blood glucose before surgery.
2. To measure the CBF, Laser Doppler flowmetry (LDF) (PowerLab ADInstruments) was performed during the surgery.
3. The near infrared spectroscopy (moorVMS-NIRS monitor) was used to monitor the cerebral oxygen saturation of the region (rcSO2)

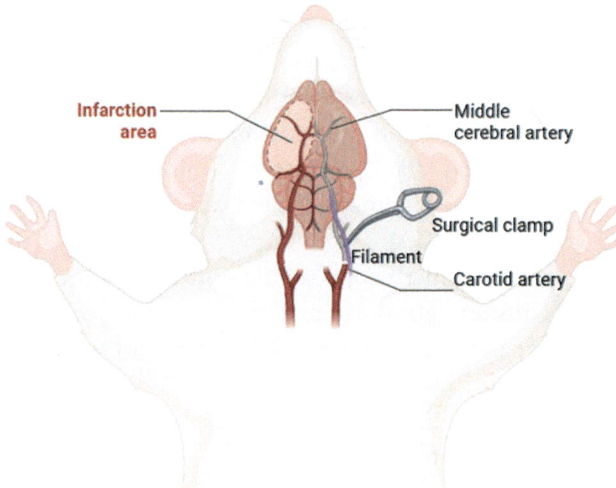

Fig. 2 MCAO model induction in rat followed by 90 occlusions.

4. Following proper suturing, povidone iodine was applied topically on the wound to accelerated healing, and they were allowed to recover at room temperature nearby a heating lamp.
5. At the time of recovery, pain reliever (Buprenorphine) was administered. Temperature was regulated and hydration was maintained.
6. In another set of animals, all other surgical procedures were performed except filament insertion and designated as Sham animals.
7. Following 24 hrs of surgery, rats with non-significant neurodeficit score, small or no infarct size, non-significant CBF and brain oxygenation drop along with inconsistent physiological parameters were excluded.

3.2 Administration of intra-arterial mesenchymal stem cells (IA-MSCs), SIRT 1 activator, inhibitor and NFkB Inhibitor after ischemic stroke

3.2.1 Reagents

Phosphate buffered saline (PBS), Rat Bone-marrow MSCs (Merck Millipore), α-MEM media, Fetal bovine serum (FBS), HEPES, penicillin/streptomycin, Trypsin, Trypan blue, DMSO, Resveratrol, EX-527, Parthenolide.

3.2.2 Equipment

Laminar hood, Incubator, Water bath, Refrigerator, Centrifuge, Microscope, Cell counter.

3.2.3 Procedure

1. MSCs were used between passages 3–5 for all experiments.
2. MSCs were cultured in 10% FBS containing α-MEM media and penicillin/streptomycin under aseptic conditions in the central cell facility of NIPER-A.
3. Cells were maintained in a humidified atmosphere with 5% CO_2 at 37 °C.
4. Media for cells were changed every alternate day.
5. Following trypsinization, cells were centrifuged and pelleted down.
6. Trypan blue assay was performed to quantify live cells.
7. 1^*10^5 cells were suspended in PBS and transported in ice for infusion.
8. MSCs were administered intraarterially at 6 h post-reperfusion via left ICA using a 1 ml hypodermic syringe with 30 G needle connected to a PE-10 catheter.
9. Coagulation and formation of air bubble were avoided by filling the catheter with heparinized physiological saline prior to MSCs infusion.
10. Further, the SIRT-1 activator, Resveratrol was administered at a dose of 15 mg/kg via intraperitoneal (IP) route prophylactically for 7 days prior to MCAo.
11. The SIRT-1 inhibitor, EX-527 was administered via IP route at a dose of 5 mg/kg in the last 20 min during MCAo.
12. The NFkB inhibitor, Parthenolide was administered via IP route at the end of the occlusion at a dose of 500 μg/Kg.

3.2.4 Notes

1. Following surgery, post-operative care was taken under cautious supervision using saline infusion, analgesic administration (buprenorphine, 0.05 mg/kg dose) as per approved schedules and optimum supply of normal pellet diet with sufficient water.
2. The outcome evaluation of all animals allocated in different groups (via randomization) were executed by the investigators in a blinded manner.
3. Animals were excluded from the study due to mortality, nonsignificant neurodeficit scores, arterial perforation following surgery, smaller infarct size and non-significant cerebral blood flow as well as brain oxygenation levels.
4. All animals were euthanized as per approved protocol at 24 h post-MCAo following behavioural assessment.
5. Once euthanized, the brains were harvested and stored in 80 °C refrigerator until further use.

3.3 Histological assessment

To evaluate the effects of post-stroke IAMSCs administration on neuronal integrity associated with edema, histological analyses were performed using hematoxylin and eosin staining on cryostat brain sections.

3.3.1 Equipment

Cryostat (Thermo scientific), embedding molds, specimen discs, brush and forceps, superfrost slides, coplin jars, coverslips, −80 °C freezer (Thermo scientific), and light microscope (Leica DM 750).

3.3.2 Reagents

Xylene, ethanol (70 %, 80 %, 90 %, and 100 %), distilled water, Harris hematoxylin and eosin solution (Abcam ab245880), acid alcohol (1 % HCl in 70% ethanol), bluing agent, and DPX mounting medium.

3.3.3 Procedure

1. Rat brains were perfused 24 h post-stroke using ice-cold PBS and fixed using 4 % PFA.
2. Further, processed the brains using gradient sucrose concentration, 15 %, and 30 %.
3. Ten μm thick brain sections were cut using cryostat and placed on poly-L-lysine-coated slides.
4. The slides were treated with xylene (2 × 5 min) followed by rehydration through a descending ethanol series (100 %, 95 %, 80 %, and 70 %) subsequently rinsed in distilled water.
5. Sections were stained using hematoxylin for 5–8 min and washed under running tap water.
6. Following that, section was kept for few seconds in 1 % acid alcohol for differentiation, followed by immediate rinsing and bluing in tap water for 1–2 min
7. Counterstained the sections with eosin for 1–2 min, then dehydrate through ascending grades of ethanol (70 %, 80 %, 90 %, and 100 %).
8. Cleared the sections using xylene (2 × 5 min), and mount with DPX.
9. Sections were allowed to dry for ample time and imaged using light microscope.

3.3.4 Notes

H&E staining is used to assess gross tissue architecture, infarct regions, and edema-associated morphological alterations. Neuronal injury is a characteristic feature including eosinophilic cytoplasm, nuclear pyknosis, and cytoplasmic shrinkage. Whereas in edematous regions vacuolation and loss of structural integrity can be seen. This staining provides a comprehensive overview of IA-MSCs mediated attenuation of tissue damage and edema.

3.4 Behavioural assessment

3.4.1 Neurodeficit score (NDS)

It is the 12-point neurological scale use to assess the brain damage after stroke, it evaluates proprioception, postural reflex, visual and tactile placement. Where, higher score indicates severe neurological damage and lower score shows least neurological damage.

3.4.1.1 Procedure

1. NDS scoring was assessed 24 h post- reperfusion.
2. Postural reflex: the rat was placed in open field for 2–3 min and the spontaneous activity was observed.
3. A postural reflex test: rats were gently held by tail 1 m above floor and forelimb extension towards the floor was observed.
4. Circling behaviour: rats were placed on plane platform surface and spontaneous circling was observed
5. Forelimb placing test: this test was used to assess sensorimotor integration.
6. Visual placing: The animals was slowly lowered from each side towards the cage and forelimb extension was assessed.
7. Tactile placing: The animals were slowly lowered towards the cage corners lightly contacting with dorsal and lateral surface of rat's forepaw.
8. Proprioceptive placing: rats forepaws were placed against the table edge to stimulate the limb muscles.

   (Notes: lower values indicate less neurodeficit).

3.4.2 Rotarod test

This test was used to assess the balance and motor coordination post stroke.

3.4.2.1 Equipment

Rotarod apparatus (Rotamax, Columbus apparatus).

3.4.2.2 Reagent

70 % IPA.

3.4.2.3 Procedure

1. Animals were kept in behavioral room for 30–60 min for habituation before starting the test.
2. Assessment was performed after 24 h post-stroke.
3. Rotarod apparatus was cleaned with 70 % IPA to remove olfactory cues.
4. Speed was set at 5 RPM and placed the animal on rotating rod with cut off time of 180 s and time to fall from rod was recorded.
5. The gap between 2 consecutives cycles was 5 min and 3 reading taken per animal.
6. Repeated the same protocol for 10 and 20 RPM.

Notes: During selection period 3 cycles of each animal should perform at each RPM (5, 10, 20). The animals which retain on rods till cutoff time should be included used for study and otherwise excluded (Fig. 3).

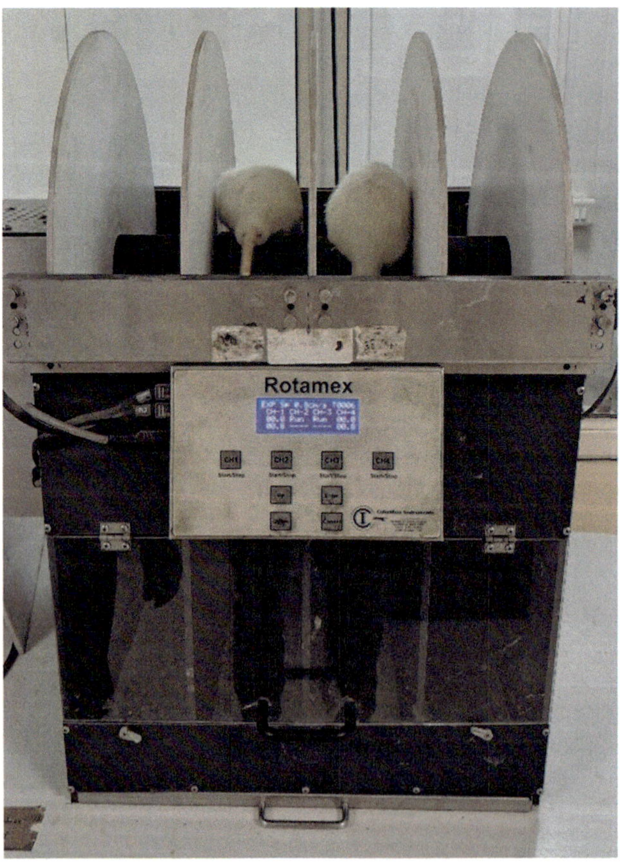

Fig. 3 Rotarod apparatus assessing motor coordination post-stroke.

3.4.3 Grip strength

This test used to evaluate the forelimb neuromuscular strength and motor deficit post stroke.

3.4.3.1 Apparatus

Electronic grip strength meter.

3.4.3.2 Reagents

70 % IPA.

3.4.3.3 Procedure

1. Animals were acclimatised for 30–60 min in behavioural room before starting the trials.
2. Turned on the grip strength meter and calibrate it to zero.
3. Test was performed after 24 h of reperfusion.
4. Rats were gently held from the base of the tail and allow only forepaws to grasp the metal grid.
5. Readings were taken of ipsilateral, contralateral and both forepaws.
6. First ensured the grip was firm, then rat was pulled until it releases from the bar.
7. Peak reading appeared on the meter was recorded.

Notes: Repeat the test for 3 times at an interval of 30 min and take baseline measurement one hour before surgery (Fig. 4).

3.5 Cerebral infarct size and percentage edema measurement

Cerebral infarct size was measured in 2 mm thick coronal brain sections of rats cut using a brain matrix following euthanasia. The sections were incubated in 0.5 % TTC solution in 0.1 M PBS at 37 °C for 30 min and then fixed in 4 % paraformaldehyde (4 % PFA) solution for 30 min at room temperature. The white portion of the brain indicates the infarct region while the red coloured part is the viable region. The images were taken and analysed using NIH ImageJ software. The edema volume was measured using the same software as per previously reported study.

3.6 Western blotting

The Western blot analysis is a quantitative molecular method for assessing changes in protein expression in the brain cortex tissue in response to ischemic stroke and MSCs treatment. This approach allows researcher to monitor changes in crucial signaling pathways.

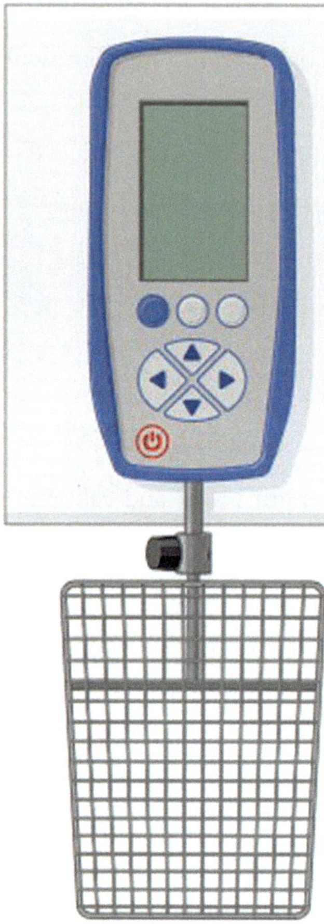

Fig. 4 Grip strength apparatus.

3.6.1 Equipment and additional material

Sonicator, Homogenizer, Centrifuge, Spectrophotometer, Electrophoresis Apparatus (Electrophoresis Tank, Power Supply), Gel Doc System, Glass Plates, Comb, PVDF Membrane, Stacks, Tubes, Micro pestle.

3.6.2 Reagents

Lysate preparation:

- RIPA buffer (Tris-HCl, NaCl, 0.5 % Sodium Deoxycholate and SDS)
- PMSF
- BCA reagents (A and B)
- Laemmli buffer (SDS, β-mercaptoethanol, Glycerol and bromophenol)

Gel Electrophoresis (SDS-PAGE):
- Acrylamide
- Bis-acrylamide
- APS
- TEMED
- Stacking buffers
- Separating buffers
- Running buffer (Tris, Glycine and SDS)

Electrotransfer:
- Transfer buffer (Tris, Glycine and Methanol)

Detection steps:
- Blocking solution (4 % Skim milk or BSA in TBST)
- Washing solution: TBST (Tris-Buffered Saline with tween 20).
- Antibodies (Target-specific primary antibody and HRP-conjugated secondary antibody)
- ECL (Enhanced Chemiluminescence)

3.6.3 Procedure

1. Preparation of tissue lysate
   - **a)** Brain tissue (75–80 mg) obtained from the penumbra region was triturated in 1 ml of RIPA buffer.
   - **b)** The triturated tissue was homogenized for 10 s in each cycle at a speed of 11000 RPM for 3 cycles.
   - **c)** The homogenized brain tissue was sonicated for 3 cycles (8 s "on" and 8 s "off") under cold conditions.
   - **d)** The samples were centrifuged at 12,000 rpm for 20 min at 4°C to pellet any cellular debris.
   - **e)** Supernatants were collected as protein extracts.
   - **f)** Protein concentration in samples was determined by BCA assay.
   - **g)** Loading samples were prepared by adding equal amount of protein in each sample.
   - **h)** The sample was heated in 95–100°C for 5 min.

     Note: Lysate can be stored at −20 °C or −80°

2. **Gel Preparation**
   - **a)** Glass plates were cleaned and positioned on the casting stand.
   - **b)** Solution of the separating gel (with high acrylamide concentration, pH of 8.8) was prepared and poured to allow protein separation according to their size.

c) Water was poured on top to level the surface and bubbles were removed; gel was left to set for 15–30 min.
d) Water was drained, and solution of the stacking gel (with low acrylamide concentration, pH of 6.8) was poured to concentrate the protein bands.
e) The comb was placed avoiding bubbles.
f) Stacking gel was left to set completely.

3. **Gel Electrophoresis (SDS-PAGE)**
   a) Gel was placed in the electrophoresis chamber filled with running buffer.
   b) Molecular weight markers and protein samples were loaded.
   c) Electrophoresis was conducted by applying voltage starting from 60 V and then increased to 80–100 V, until the dye reached the bottom.
4. **Protein transfer**
   a) PVDF membrane was first activated in methanol for 5 mins, while stacks or filter papers of transfer sandwich were pre-wet in transfer buffer.
   b) Transfer sandwich was placed in the electro transfer system without introducing any air bubbles.
   c) Transfer of proteins from gel to membrane was performed.
5. **Blocking and Antibody Incubation**
   a) Blocking of the membrane was done using blocking buffer for 1–1.5 h at room temperature while shaking.
   b) Incubation of primary antibodies was done overnight at 4 °C in TBST (blocking buffer).
   c) The membrane was washed thrice using TBST for 5 min each time.
   d) Incubation of secondary antibodies was done for 1 h at room temperature.
   e) The membrane was washed thrice using TBST.
6. **Detection and Analysis**
   a) The ECL substrate was incubated on the membrane for 1–2 min.
   b) The images were obtained using the Gel Doc system.
   c) The data analysis was performed by detecting the bands based on molecular weight and protein concentration compared with the loading control (for instance, β-actin).

3.6.4 Notes

1. Normalization: The use of a loading control is essential for normalizing the amount of proteins in each lane for even sample loading.

2. Phosphorylated Proteins: In case of phosphorylation studies, always use phosphatase inhibitors and block the membranes with BSA because the casein content in milk gives high backgrounds.
3. Transfer Efficiency: This can be checked after blotting with the help of Ponceau S, a reversible stain that stains protein bands on the membrane.
4. Troubleshooting "No Signal": This may result from incorrect transfer direction, incompatibility between primary and secondary antibodies, or absence of the target protein.
5. Troubleshooting "Multiple Bands": Multiple bands appear when the protein is cleaved, forms multimers, or undergoes proteolysis in the absence of protease inhibitors.

3.7 RT-PCR

3.7.1 Reagents

TRIzol reagent (Invitrogen), cDNA synthesis kit (Bio-Rad iScript cDNA synthesis kit), Primer, iQ SYBR Green Supermix (Bio-Rad).

3.7.2 Equipment

Nanodrop spectrophotometer 2000c, Real-Time PCR system (CFX Real-Time PCR, Bio-Rad).

3.7.3 Procedure

1. RNA was extracted from about 20 mg of cortical brain tissue that had been homogenized using 1 ml of TRIzol reagent.
2. 300 ul of chloroform was added per ml of trizol and incubated at room temperature for 10 mins.
3. The sample was then centrifuged at 12,000 *g* for 15 min at 4 °C to separate the phases.
4. Three separate layers developed following centrifugation: RNA-containing upper aqueous phase, DNA-containing interphase, and proteins-containing lower organic phase
5. A fresh MCT with 600 ul of cooled IPA was carefully filled with the clear aqueous phase (500 μL). To enable RNA precipitation, the samples were incubated for ten minutes at room temperature.
6. The mixture was centrifuged at 12,000 *g* for 10 min at 4 °C, resulting in a visible RNA pellet.
7. After discarding the supernatant, 500 uL of 75 % chilled ethanol was used to wash the RNA pellet.
8. The samples were centrifuged at 12,000 *g* for 10 min at 4 °C

9. The ethanol was carefully removed, and the RNA pellet was air-dried
10. The pellet was dissolved in 30 μL of RNase-free water by gentle pipetting and incubation at 65 °C for 5 min.
11. Nanodrop spectrophotometer was used for measuring absorbance ratios (260/280 and 260/230) to assess the quantity and purity of RNA
12. cDNA was synthesized from isolated RNA using a commercial cDNA synthesis kit
13. Gene-specific primers were used for amplification of AQP4, NFκB, SIRT-1, and PKCδ.
14. 18S rRNA was used as an internal control to normalize the levels of gene expression.
15. The CFX Real-Time PCR system with iQ SYBR Green Supermix was used to conduct real-time PCR. (Biorad)
16. The cDNA samples were diluted in a 1:10 ratio and the experiment was performed in triplicate to avoid error.

3.7.4 Notes

1. Only samples with acceptable purity ratios (260/280 and 260/230) were used for downstream analysis.
2. Triplicate measurements were used to minimize technical variability.

3.8 Immunofluorescence analysis

3.8.1 Reagents

Acetone (−20 °C), Phosphate Buffered Saline (PBS), PBST (1 % solution), Bovine Serum Albumin (BSA), TBST solution, Glycine (3 M), Primary antibodies, Alexa Fluor 488, Alexa Fluor 647, DAPI nuclear stain.

3.8.2 Equipment

Cryostat for brain sectioning, Confocal laser scanning microscope, Standard fluorescence microscopy accessories.

3.8.3 Procedure

1. Cryostat-cut brain sections were mounted on the charged slides.
2. Sections were fixed/permeabilized using acetone at −20 °C for 10 min.
3. Slides were washed three times with PBS to remove excess solvent.
4. Blocking was performed using 4 % BSA-TBST solution with 3 M glycine for 30 min followed by blocking without glycine for 90 min
5. Sections were incubated overnight at 4 °C with primary antibodies (1:200 dilution) targeting AQP4, PKCδ, SIRT-1, NFκB, and Claudin-5.
6. The next day, sections were washed with 1 % PBST solution.

7. Secondary antibody incubation was performed in the dark for 1 h at room temperature using Alexa Fluor 647 or Alexa Fluor 488 conjugated antibodies (1:500 dilution).
8. Sections were then counterstained with DAPI for 5 min to label nuclei.
9. Immunofluorescence signals were visualized and captured using a confocal microscope.

3.8.4 Notes

1. All fluorescence-related steps were carried out in the dark to prevent photobleaching.
2. Proper washing between steps was maintained to reduce background signal.

3.9 Oxidative stress measurement

3.9.1 Reagents

RIPA buffer, Extraction buffer, Protease inhibitors, Bicinchoninic acid, Griess reagents, Sodium dodecyl sulfate, Phosphate buffer, Glutathione, Thio barbituric acid, DTNB, Pierce™ BCA protein assay kit (Thermo scientific, 23225).

3.9.2 Equipment

Sonicator, Centrifuge, UV spectrophotometer, Homogeniser.

3.9.3 Procedure

1. Cortical brain regions from peri-infarct zone of 75 mg were weighed, homogenized in 1 ml of radioimmunoprecipitation assay (RIPA) buffer, and SDS free extraction buffer with the appropriate protease inhibitors and sonicated in ice (8 s on and 8 s off) for 3 cycles.
2. The lysed tissue samples were centrifuged and the supernatant was collected.
3. The protein content for each sample was calculated by Bicinchoninic acid (BCA) assay.
4. Levels of malondialdehyde (MDA), nitrite, reduced glutathione (GSH), catalase, and SOD were assessed.
5. MDA levels were assessed by using the thiobarbituric acid (TBA) assay. 20 % v/v acetic acid, 8.1 % v/v sodium dodecyl sulfate, 0.8 % w/v thiobarbituric acid, and 150 μl of water were added into required amount of the sample. The mixture was incubated at 95 °C for 1 h and the absorbance was checked at 532 nm. The concentration of MDA was represented as μM/mg of protein.

6. To check nitrite levels, Griess method was used. 1:1 Griess working reagent (1 % sulfanilamide and 0.1 % naphthyl ethylenediamine in 5 % orthophosphoric acid) was added in the sample. The absorbance was measured at 540 nm and expressed in μM/mg of protein
7. GSH estimation was performed by DTNB method. A solution of 1.5 ml of DTNB reagent in 0.1 M phosphate buffer (pH 7.4) was added to the sample and incubated at room temperature for 10 min. The absorbance was checked at 412 nm expressed in μM/mg of protein
8. Catalase kinetic assay was performed in a UV spectrophotometer at 240 nm. First-order rate constant (K) values of samples reacting with hydrogen peroxide were calculated to determine the catalase activity expressed in U/ml/mg.
9. SOD assay was performed as per the manufacturers' information by commercially available SOD kit (Sigma Aldrich 19160-1KT-F). Samples with the assay buffers and working solution were incubated for 20 min at 37°C. Absorbance was then assessed at 450 nm (Datta et al., 2025; Datta et al., 2022; Datta et al., 2026; Sarmah, et al., 2024).

3.10 Mitochondrial functional assessment in post-stroke vasogenic edema

Mitochondrial dysfunction is a central driver of post-stroke vasogenic edema through excessive reactive oxygen species (ROS) generation, impaired oxidative phosphorylation, and disruption of endothelial and astrocytic homeostasis. The following protocols demonstrate standardized approaches to evaluate mitochondrial respiration, mitochondrial ROS production, and mitochondrial membrane potential in peri-infarct brain regions (Datta et al., 2025; Datta et al., 2025; Datta et al., 2024; Datta et al., 2026; Sarmah et al., 2024).

3.10.1 Equipment

High-resolution respirometer (e.g., Oroboros Oxygraph-2k), Fluorescence microscope with appropriate filter settings, Refrigerated centrifuge with variable speed (up to 10,000 x g), Dounce homogenizer (glass), Temperature-controlled incubation system (37 °C), Microcentrifuge tubes (pre-chilled), Pipettes with low-retention tips, Ice bucket and cold room access.

3.10.2 Reagents

For mitochondria isolation:

Mitochondrial extraction buffer (125 mM sucrose, 250 mM mannitol, 10 mM HEPES, 10 mM EGTA, 0.01 % BSA, 1 μL/ml protease inhibitors), Digitonin (for selective membrane permeabilization).

For mitochondrial respiration:

Pyruvate (5 mM), malate (2 mM), glutamate (10 mM), ADP (1 mM), Succinate (10 mM), Oligomycin (ATP synthase inhibitor; 2 μg/ml), FCCP titration (uncoupler; 0.5 μM), Rotenone (complex I inhibitor; 0.5 μM), Antimycin A (complex III inhibitor; 2.5 μM).

For ROS measurement:

Phosphate-buffered saline (PBS), Dimethyl sulfoxide (DMSO) (for dye preparation), MitoSOX Red mitochondrial superoxide indicator (Invitrogen; M36008).

For membrane potential:

Phosphate-buffered saline (PBS), Dimethyl sulfoxide (DMSO) (for dye preparation), JC-1 dye (Invitrogen; T3168).

3.10.3 Procedure

1. Mitochondrial Isolation
   - **a)** Rapidly dissect the ipsilateral cortical region (peri-infarct) following euthanasia.
   - **b)** Mince the tissue using sterilized scissor.
   - **c)** Homogenize tissue in ice-cold 3 ml mitochondrial extraction buffer using a Dounce homogenizer (10 strokes through arm A and 10 strokes through arm B).
   - **d)** Centrifuge twice at 700 × g for 10 min at 4°C to remove nuclei and debris.
   - **e)** Transfer supernatant and centrifuge at 10,000 × g for 10 min at 4 °C.
   - **f)** Resuspend mitochondrial pellet in 1 ml mitochondrial extraction buffer.
   - **g)** Treat with low concentration digitonin (0.02 %) to remove synaptosomal contamination.
   - **h)** Centrifuge at 10,000 × g for 10 min at 4 °C
   - **i)** Collect mitochondrial pellet and resuspend in 100 μl mitochondrial extraction buffer and use immediately for downstream assays.
   - **j)** Protein content was assessed using BCA assay

2. Mitochondrial Respiration Analysis
   a) Calibrate the respirometer according to manufacturer instructions.
   b) Add freshly isolated mitochondria (300 μg protein) into the respiration chamber containing assay buffer.
   c) Sequentially add substrates and inhibitors as per SUIT protocol:
      - Pyruvate + malate + glutamate → basal respiration
      - ADP → oxidative phosphorylation (OXPHOS state)
      - Succinate → combined complex I + II respiration
      - Oligomycin → leak respiration
      - FCCP titration→ maximal uncoupled respiration
      - Rotenone + Antimycin A → non-mitochondrial respiration
   d) Record oxygen consumption and oxygen concentration continuously in the DatLab software.
   e) Calculate respiratory control ratio (RCR) as RCR= ADP stimulated respiration/respiration after oligomycin addition.
3. Mitochondrial ROS Measurement (MitoSOX Assay)
   a) Prepare a stock solution of 5 mM by dissolving the vial's content in 13 μl of anhydrous DMSO.
   b) Prepare a solution of 1 ml of PBS containing 1 μl of MitoSOX stock solution.
   c) Incubate 50 μg of freshly isolated mitochondria in the above solution at 37 °C for 10 min in the dark.
   d) Wash gently with PBS to remove excess dye.
   e) Mount samples on slides.
   f) Capture fluorescence images at excitation/emission 510/580 nm.
   g) Quantify fluorescence intensity relative to control groups.
4. Mitochondrial Membrane Potential (JC-1 Assay)
   a) Prepare stock solution of 2 mM in DMSO.
   b) Incubate 50 μg of mitochondrial sample with 1 μM JC-1 dye (in PBS) at 37 °C for 30 min
   c) Wash with PBS to remove excess dye.
   d) Capture fluorescence images at:
      - Green (monomer): 485/540 nm
      - Red (aggregate): 535/590 nm
   e) Calculate red/green fluorescence ratio as an indicator of membrane potential.

3.10.4 Notes

1. Perform all steps rapidly and on ice to preserve mitochondrial integrity.
2. Avoid freeze–thaw cycles, which significantly impair mitochondrial function.
3. Ensure equal mitochondrial protein loading across samples for accurate comparison.
4. The dyes are highly sensitive to light and oxidation; prepare fresh working solutions and minimize light exposure.
5. JC-1 interpretation: A decrease in red/green ratio indicates mitochondrial depolarization, commonly observed in post-stroke conditions associated with vasogenic edema.
6. Ensure proper calibration and oxygen saturation of the respirometer before each run.
7. Troubleshooting: High variability in respiration may arise from incomplete mitochondrial isolation or contamination with non-mitochondrial fractions

4. Conclusion

This chapter describe about the comprehensive and integrative methodology procedures for assessing the efficacy of IA-MSCs in modulating post-stroke edema. This chapter describe the reproducible model for development of ischemic stroke in-vivo, procedure for administration of IA-MSCs and assessing motor functional outcome through behavioural assessment. In addition, this chapter also provides detailed information regarding the methods used for molecular analysis, oxidative stress assessment and mitochondrial function assay assessment. Collectively this chapter provided methodological insight for the evaluation of therapeutic potential of IA-MSCs in alleviating post-stroke edema by modulating SIRT-1/PKCδ/NF-κB axis.

Acknowledgement

We acknowledge BioRender.com for image creation. Current affiliation of Ms. Swagata Bhattacharya is Department of Pharmaceutical Engineering & Technology, Indian Institute of Technology (BHU) Varanasi.

References

Asgari Taei, A., Khodabakhsh, P., Nasoohi, S., Farahmandfar, M., & Dargahi, L. (2022). Paracrine effects of mesenchymal stem cells in ischemic stroke: Opportunities and challenges. *Molecular Neurobiology, 59*(10), 6281–6306.

Datta, A., Ghosh, B., Barik, A., Karmarkar, G., Sarmah, D., Borah, A., ... Bhattacharya, P. (2025). Stem cell therapy modulates molecular cues of vasogenic edema following ischemic stroke: Role of sirtuin-1 in regulating aquaporin-4 expression. *Stem Cell Reviews and Reports, 21*(3), 797–815.

Datta, A., Rahane, D., Bhurle, G., Akundi, S., Mukherjee, U., Dubey, A., ... Bhattacharya, P. (2025). Stroke in pregnancy brings epigenetic changes in correlation with affected mitochondrial dynamics and inflammasome mediated apoptosis in rodents. *Journal of the American Heart Association, 14*(9), e039411. https://doi.org/10.1161/JAHA.124.039411.

Datta, A., Sarmah, D., Ghosh, B., Rana, N., Bhattacharya, P., & Borah, A. (2024). High-resolution respirometry for mitochondrial function in rodent brain. *Methods in Molecular Biology (Clifton, N. J.), 2761*, 49–55. https://doi.org/10.1007/978-1-0716-3662-6_4.

Datta, A., Sarmah, D., Kaur, H., Chaudhary, A., Mounica, K. L., Kalia, K., ... Bhattacharya, P. (2022). Post-stroke impairment of the blood–brain barrier and perifocal vasogenic edema is alleviated by endovascular mesenchymal stem cell administration: modulation of the PKCδ/MMP9/AQP4-mediated pathway. *Molecular Neurobiology, 59*(5), 2758–2775.

Datta, A., Sehgal, K., Sarmah, D., Dhiman, S., Shah, B., Dhakne, P., ... Bhattacharya, P. (2026). Stroke in persistent chronic kidney disease condition alters innate-immunity to escalate mitochondrial dysfunction and aging. *npj Aging*. https://doi.org/10.1038/s41514-026-00361-1.

Ito, H., Yamamoto, N., Arima, H., Hirate, H., Morishima, T., Umenishi, F., ... Sobue, K. (2006). Interleukin-1β induces the expression of aquaporin-4 through a nuclear factor-κB pathway in rat astrocytes. *Journal of Neurochemistry, 99*(1), 107–118.

Ji, C., Yu, X., Xu, W., Lenahan, C., Tu, S., & Shao, A. (2021). The role of glymphatic system in the cerebral edema formation after ischemic stroke. *Experimental Neurology, 340*, 113685.

Kaur, H., Sarmah, D., Veeresh, P., Datta, A., Kalia, K., Borah, A., ... Bhattacharya, P. (2021). Endovascular stem cell therapy post stroke rescues neurons from endoplasmic reticulum stress-induced apoptosis by modulating brain-derived neurotrophic factor/tropomyosin receptor kinase B signaling. *ACS chemical neuroscience, 12*(19), 3745–3759.

Kim, Y.-K., & Hammerling, U. (2020). The mitochondrial PKCδ/retinol signal complex exerts real-time control on energy homeostasis. *Biochimica et Biophysica Acta (BBA)-Molecular and Cell Biology of Lipids, 1865*(11), 158614.

Marchi, S., Guilbaud, E., Tait, S. W., Yamazaki, T., & Galluzzi, L. (2023). Mitochondrial control of inflammation. *Nature Reviews. Immunology, 23*(3), 159–173.

Michinaga, S., & Koyama, Y. (2015). Pathogenesis of brain edema and investigation into anti-edema drugs. *International Journal of Molecular Sciences, 16*(5), 9949–9975.

Rabinstein, A. A. (2006). Treatment of cerebral edema. *The Neurologist, 12*(2), 59–73.

Sadeghian, N., Panahpour, H., Alipour, M. R., Farajdokht, F., & Shadman, J. (2025). Investigating the therapeutic potential of nasal administration of mitochondria on blood-brain barrier integrity and vasogenic brain Edema in a rat ischemic stroke model. *Experimental Neurology*, 115609.

Sarmah, D., Datta, A., Kaur, H., Kalia, K., Borah, A., Rodriguez, A. M., ... Bhattacharya, P. (2022). Sirtuin-1-mediated NF-κB pathway modulation to mitigate inflammasome signaling and cellular apoptosis is one of the neuroprotective effects of intra-arterial mesenchymal stem cell therapy following ischemic stroke. *Stem Cell Reviews and Reports, 18*(2), 821–838.

Sarmah, D., Datta, A., Rana, N., Suthar, P., Gupta, V., Kaur, H., ... Bhattacharya, P. (2024). SIRT-1/RHOT-1/PGC-1α loop modulates mitochondrial biogenesis and transfer to offer resilience following endovascular stem cell therapy in ischemic stroke. *Free Radical Biology and Medicine, 225*. https://doi.org/10.1016/j.freeradbiomed.2024.09.022.

Sarmah, D., Datta, A., Rana, N., Suthar, P., Gupta, V., Kaur, H., ... Yavagal, D. R. (2024). SIRT-1/RHOT-1/PGC-1α loop modulates mitochondrial biogenesis and transfer to offer resilience following endovascular stem cell therapy in ischemic stroke. *Free Radical Biology and Medicine, 225*, 255–274.

Sarmah, D., Kaur, H., Saraf, J., Vats, K., Pravalika, K., Wanve, M., ... Wang, X. (2019). Mitochondrial dysfunction in stroke: Implications of stem cell therapy. *Translational Stroke Research, 10*(2), 121–136.

Sarvari, S., Moakedi, F., Hone, E., Simpkins, J. W., & Ren, X. (2020). Mechanisms in blood-brain barrier opening and metabolism-challenged cerebrovascular ischemia with emphasis on ischemic stroke. *Metabolic Brain Disease, 35*(6), 851–868.

Stokum, J. A., Kurland, D. B., Gerzanich, V., & Simard, J. M. (2015). Mechanisms of astrocyte-mediated cerebral edema. *Neurochemical Research, 40*(2), 317–328.

Tang, G., Liu, Y., Zhang, Z., Lu, Y., Wang, Y., Huang, J., ... Wang, Y. (2014). Mesenchymal stem cells maintain blood-brain barrier integrity by inhibiting aquaporin-4 upregulation after cerebral ischemia. *Stem Cells, 32*(12), 3150–3162.

Vats, K., Sarmah, D., Datta, A., Saraf, J., Kaur, H., Pravalika, K., ... Dave, K. R. (2021). Intra-arterial stem cell therapy diminishes inflammasome activation after ischemic stroke: A possible role of acid sensing ion channel 1a. *Journal of Molecular Neuroscience, 71*(2), 419–426.

Yan, C., Wang, H., Aggarwal, B. B., & Boyd, D. D. (2004). A novel homologous recombination system to study 92 kDa type IV collagenase transcription demonstrates that the NFκB motif drives the transition from a repressed to an activated state of gene expression. *The FASEB Journal, 18*(3), 540–541.

9780443473326